Certified Wireless Design Professional (CWDP)

Study Guide

(CWDP-304)

Certified Wireless Network Professionals

Errata, when available, for this study guide can be found at: www.cwnp.com/errata/

First printing: September 2021, version 2.0

ISBN: 978-1-7372166-3-6

In addition to the authors of this book, listed in the About the Authors section of the Introduction, CWNP would like to say a special thanks to all those involved in the development of materials for CWDP-304 from the Job Task Analysis (JTA) through to materials review and feedback. These individuals include Robert Bartz, Ian Beyer, Tom Carpenter, Jonathan Davis, Landon Foster, Manon Lessard, Peter Mackenzie, Troy Martin, Scott McNeil, Phil Morgan, Ferney Munoz, Jim Palmer, and Djamel Ramoul. If we have left out your name it is only because so many helped and not because you were not appreciated. Many thanks to all of you.

Table of Contents

Introduction

The Certified Wireless Design Professional (CWDP) is an individual who can gather requirements, design, deploy, validate, and optimize a WLAN. The CWDP-304 exam tests the candidate's knowledge to verify his or her ability to perform the duties of a CWDP.

The Certified Wireless Design Professional (CWDP) has the knowledge and skill set required to manage the entire WLAN design life cycle: defining, designing, deploying, and diagnosing. Tasks within these stages include gathering necessary information and requirements, creating a design, implementing the network, and validating and optimizing to ensure objectives are met. A CWDP can contribute to any stage of the life cycle and is able to take responsibility for any or all the stages within this process.

The CWDP-304 exam consists of 60 multiple choice, single correct answer questions and is delivered through Pearson VUE. The candidate can register for the exam at the Pearson VUE website (https://home.pearsonvue.com/). The candidate will have 90 minutes to take the exam and must achieve a score of 70% or greater to earn the CWDP certification. If the candidate desires to become a Certified Wireless Network Trainer (CWNT) the passing score must be 80% or greater. A CWNT is authorized to teach official CWNP courses for certifications in which they hold the CWNT credential.

Book Features

The CWDP Certified Wireless Design Professional Study and Reference Guide includes the following features:

- End of chapter Facts to Remember. These lists of important facts help you retain the information learned in the chapter.
- End of chapter review quizzes. These quizzes help you test the knowledge you've acquired from the chapter. Each chapter contains 10 quiz items.
- Notes with special indicators. The notes throughout the book fall into one of three categories as outlined in Table i.1.
- Extensive footnotes throughout.
- CWNP universal glossary. A glossary of terms provided at the end of the book that helps you as a reference while reading.
- Complete coverage of the CWDP-304 objectives. Every objective is covered in the book and each chapter lists the objectives covered within.

Icon	Description
	**Note:** A general note related to the current topic.
	**Defined Note:** A note providing a concise definition of a term or concept.
	**Exam Note:** A note providing tips for exam preparation.

Table i.1: Book Note Icons

About the Authors

Tom Carpenter is the CTO of CWNP and has 30 years of experience in the Information Technology industry. He has written 26 previous books and developed more than 50 eLearning programs in the past fifteen years. He is a CWNE and holds several other industry certifications as well. As the CTO of CWNP, Tom is responsible for setting the direction for certifications and managing product development projects through their lifecycles. He can be reached at tom@cwnp.com and is heard on a monthly webinar series presented by CWNP and archived at YouTube on the CWNPTV channel. Tom lives in Ohio with his wife Julie and loves books and all things tech.

Brett Hill is CWNE #147. After earning an Electronics Engineering Degree, Brett began his technology career in the early 90's repairing business computer systems, pulling cable and polishing fiber. Seeing the opportunity to learn various wireless technologies, he became a tower climber and wireless technician where he was introduced to IEEE 802.11. Twenty-five years later, Brett Hill is a Sr. Wireless Engineer for World Wide Technology, Inc. and is appointed as a Subject Matter Expert in wireless designs to the largest public power utility in the United States. Now, with a recent degree in information Technology and a passion for public speaking, he volunteers his time teaching wireless at Universities, serving on the CWNP Board of Advisors, and assisting in the development of training material for wireless certifications.

Jason D. Hintersteiner, CWNE #171 is an expert in the small-to-medium business (SMB) Wi-Fi market, possessing a unique perspective on the industry by having worked in engineering and marketing roles for both wireless internet service providers and access point equipment vendors. For more than a decade, Mr. Hintersteiner has been a principal network architect and analyst for several hundred wired, Wi-Fi, and point-to-multipoint wireless networks, spanning multiple verticals including hospitality, student housing, assisted living, residential apartments, wireless surveillance, religious non-profit, education, warehouses, factories, commercial offices, and retail. Mr. Hintersteiner has also taught dozens of vendor-specific training courses and has served as an expert witness in legal cases involving network forensic analysis and patent litigation. Mr. Hintersteiner holds both a Bachelor of Science and a Master of Science from the Massachusetts Institute of Technology, and a Master of Business Administration from the University of Connecticut. He has also served on the CWNE Board of Advisors.

Manon "Mae" Lessard is a Canadian network administrator for a large university. With 16 years' experience in networking, 14 of which have been mostly spent exploring her passion for all things wireless, she is an active member of the wireless community. She is involved in various organizations such as the Wireless Lan Association and the Slack Wi-Fi Pros community, holds several industry certifications and is CWNE #275. You can reach her on Twitter (@Mae149) or through her blog at missmaeswifi.com. She has also served on the CWNE Board of Advisors.

The authors would like to dedicate this book to future CWDPs: you and your passion for Wi-Fi is what makes this all worthwhile.

CWDP-304 Objectives

The CWDP-304 exam tests your knowledge against four knowledge domains as documented in Table i.2. The CWDP candidate should understand these domains before taking the exam. The CWDP-304 objectives follow.

Knowledge Domain	Percentage
Define Specifications for the WLAN	25%
Design the WLAN	40%
Deploy the WLAN	10%
Validate and Optimize the WLAN	25%

Table i.2: CWDP-304 Exam Knowledge Domains with Percentage of Questions in Each Domain

1.0 Define Specifications for the WLAN – 25%

1.1 Collect and use business requirements

- 1.1.1 Business use cases and justification
- 1.1.2 User requirements
- 1.1.3 Regulatory compliance
- 1.1.4 Industry compliance
- 1.1.5 Budget
- 1.1.6 Aesthetics
- 1.1.7 Architectural constraints
- 1.1.8 Mounting restrictions
- 1.1.9 Access restrictions
- 1.1.10 Time constraints
- 1.1.11 Building codes and safety codes

1.2 Collect and define technical requirements

- 1.2.1 Vendor selection
- 1.2.2 Location services such as RTLS
- 1.2.3 Latency requirements
- 1.2.4 Signal strength requirements
- 1.2.5 Capacity requirements

1.2.6 Security requirements

- BYOD and guest access
- Roaming
- Monitoring
- Authentication and encryption

1.2.7 Applications and their specific requirements

1.2.8 WLAN upgrade requirements, when applicable

1.2.9 Bridge link requirements, when applicable

1.2.10 Voice over WLAN (VoWLAN), when applicable

1.2.11 Client devices including most important and least capable device

1.2.12 Requirement areas

1.3 Collect project documentation

1.3.1 Validated floor plans

1.3.2 Network infrastructure

- Network diagrams
- AP locations
- Existing network services including DNS, DHCP, NTP, and authentication servers
- Switch capabilities and capacity

1.3.3 Cabling infrastructure

- Cabling maps and plans
- Wiring closet locations

1.3.4 Power availability and PoE capabilities

1.3.5 Existing wireless systems

1.3.6 Previous design/survey documentation

1.4 Define requirement areas including essential metrics for each requirement

1.4.1 Client device types and capabilities

1.4.2 Applications and their requirements

1.4.3 User and device density

1.4.4 SSIDs

1.4.5 Security settings

1.4.6 Understand common vertical markets

1.5 Gather information on environmental factors

1.5.1 Building materials
1.5.2 Attenuation values
1.5.3 Ceiling heights
1.5.4 Site annotations (photos, notes, plans)
1.5.5 Wireless environment scan
- Packet captures
- Spectrum captures
- Wi-Fi scanners

2.0 Design the WLAN – 40%

2.1 Define WLAN architectures and select the appropriate architecture for a design

2.1.1 Controller-based (physical and virtual) architectures
2.1.2 Distributed (cloud-based and local WNMS)
2.1.3 Standalone/Autonomous APs
2.1.4 Dynamic vs. static channel assignment
2.1.5 Dynamic radio management
2.1.6 Software defined radio
2.1.7 RF profiles
2.1.8 Select and/or recommend the appropriate equipment for the design and selected architecture (APs, antennas, controllers, managed services)

2.2 Produce a design to meet requirements

2.2.1 Select and use appropriate design tools
- Design and survey software and hardware
- Spectrum analysis software and hardware
- Access points and antennas
- Portable power source
- Tripods
- Measuring tools
- Cameras
- Personal Protective Equipment (PPE)

2.2.2 Select and use appropriate design methodologies
- WLAN predictive design (new builds/site or area not accessible)

- Validated RF modeling
- AP-on-a-Stick (APoS) measurements
- Bridge and mesh planning

2.2.3 Understand and use the common features of wireless design software

- Import and scale floor plans
- Model attenuation of the site (including calibration)
- Select and place APs and antennas
- Adjust AP and antenna settings
- Define requirement areas and parameters
- Define channel and power settings

2.2.4 Select and use common vendor features and make configuration recommendations

- Band steering
- Automatic/static channel selection
- Load balancing
- RF/AP templates

2.2.5 Design for different client and application types

- VoIP handsets
- Laptops
- Handheld scanners
- Smartphones and tablets
- IoT and smart devices
- Location tracking systems
- Voice and video systems

2.2.6 Ensure end-to-end QoS is properly implemented

- Wi-Fi Multimedia (WMM)
- Wired and wireless QoS mappings
- QoS markings, classifications, and queues

2.2.7 Define and recommend security solutions

- Monitoring (detection and prevention)
- Authentication servers
- EAP methods
- Authentication types
- Encryption types

2.2.8 Design and secure roaming

- Secure BSS transition (roaming)
- Vendor roaming solutions
- Client support issues

2.3 Create, distribute, and communicate design documentation

- 2.3.1 Bill of Materials (BoM)
- 2.3.2 Design report
- 2.3.3 Physical installation guide

3.0 Deploy the WLAN – 10%

3.1 Ensure proper understanding and implementation of the design

- 3.1.1 Implementation meeting (explain design decisions to implementers and ensure understanding of design deployment)
- 3.1.2 Distribute required documents

3.2 Recommend or perform essential deployment tasks

- 3.2.1 Understand and perform installation procedures for different WLAN architectures (cloud-based, controller-based, WNMS, autonomous)
- 3.2.2 Infrastructure configuration supporting the WLAN (DHCP, DNS, NTP, switches, and routers)
- 3.2.3 Channel assignment, automatic radio management, and transmit power configuration
- 3.2.4 Installation procedures for cloud-based APs, controller-based APs, WNMS APs, and autonomous APs

3.3 Perform an installation audit for quality assurance

- 3.3.1 Verify proper AP and antenna location and orientation
- 3.3.2 Verify aesthetic requirements are met
- 3.3.3 Verify physical security of the installation

4.0 Validate and Optimize the WLAN – 25%

4.1 Confirm the WLAN system is operational

- 4.1.1 AP status
- 4.1.2 Verify PoE provisioning of power requirements are met

4.2 Perform an RF validation survey

- 4.2.1 Ensure coverage requirements
- 4.2.2 Evaluate impacts of contention and interference

4.3 Perform client performance testing

- 4.3.1 Connectivity testing
- 4.3.2 Application testing
- 4.3.3 Roaming testing
- 4.3.4 Capacity testing
- 4.3.5 Security testing

4.4 Recommend appropriate physical adjustments

- 4.4.1 AP
- 4.4.2 Antennas and connectors

4.5 Recommend appropriate RF adjustments

- 4.5.1 RF channel assignment
- 4.5.2 RF channel bandwidth
- 4.5.3 RF coverage (transmit power, radio count, antennas)
- 4.5.4 RF interference issues

4.6 Recommend remediation for application issues

- 4.6.1 Connectivity issues
- 4.6.2 Application issues
- 4.6.3 Roaming issues
- 4.6.4 Capacity issues
- 4.6.5 Security issues

4.7 Implement knowledge transfer and hand-off

- 4.7.1 System training
- 4.7.2 Solution documentation and assets (validation documentation, digital or physical assets, guides, floorplans, configuration documents)

4.7.3 Final meeting (Q&A and hand-off)

Each chapter of the CWDP Study and Reference Guide lists the objectives covered in that chapter on the chapter title page.

Extended Table of Contents

Chapter 1: Introduction to WLAN Design

Objectives Covered

1.1 Collect business requirements and constraints (introductory coverage)
1.2 Collect and define technical requirements (introductory coverage)
1.3 Collect project documentation (introductory coverage)

Proper Wireless LAN (WLAN) design is achieved through an awareness of the technology and experience in its operation and behavior in varied environments. Good WLAN designers are first good WLAN technicians. They know the how the technology works and what can be implemented. Discovering the details of signal propagation, modulation, and reception (to name a few) is essential to learning the behavior of wireless signals in an environment and designing a network that accommodates this behavior. This knowledge is imperative to becoming a successful Certified Wireless Design Professional (CWDP). *The CWDP Certified Wireless Design Professional Study and Reference Guide* will help you build the essential skills needed for success. Additionally, the intermediate knowledge you learned in your CWNA studies will be priceless along this journey.

Learning the proper skillset to become a CWDP doesn't happen overnight, and it certainly cannot all be obtained from theoretical knowledge alone. To reach the level of a CWDP, one should have a foundation of WLAN knowledge, found in *The CWNA Certified Wireless Network Administrator Study and Reference Guide*, coupled with experience in the design and troubleshooting of WLANs. The more issues you become familiar with, the better you will become at anticipating those issues in an environment for which you are tasked to design a WLAN. Whether it is a 2.4 GHz, 5 GHz, 6 GHz, multi-band, or mesh WLAN, the basic principles of designing a WLAN will always apply.

This chapter introduces several important concepts that will prove to be relevant in other areas throughout the book. You will start off with the significance of a good design and move on to understanding some issues that may occur with a bad design or no design at all. Next, you will walk through the different phases of the design process and gain an understanding of why this process is important to a successful WLAN design. This chapter provides an overview of WLAN design and a foundation for the remainder of the book.

1.1 The Significance of a Good Design

The idea of just mounting wireless access points (APs) and transmitting a blanket of coverage at maximum power shows up in far too many deployments. *"The radio will transmit 200 milliwatts on maximum power, so 200 milliwatts is what we are going to use."* The concept of using more power to get through more walls or push the signal farther is more common than it should be. Some truly believe that turning up the radio transmit power solves a coverage issue, but it often produces more issues that cannot be resolved until the power level is turned back down, leaving the original coverage issue unresolved. A good design is essential for a performant WLAN, and there is no substitute for it.

The design process begins with the CWDP having the proper education and experience to know what questions to ask the customer, such as these:

- What is the business use case?
- Can you describe the environment in which the design will function?
- How many devices will connect and from where?
- What types of devices will be used?
- Is there an existing WLAN or is this a new installation?
- What special capabilities are required?
- What applications will be used?

Of course, these questions do not represent an exhaustive list, but gathering information is just as important as the design itself. Possessing accurate information about the design location will make the eventual design more accurate, as it provides important details that allow the CWDP to make proper decisions[1].

Dealing with a Bad Design

When wireless networks are designed poorly, many issues arise and persist throughout the life of the WLAN. Problems may include:

- **Insufficient Coverage:** This phrase can have many implications in wireless networks, and it reflects on the person or persons who designed the WLAN. When a survey is performed, improper surveys often result in insufficient coverage. The use of surveying tools is often important for accurate designs, but you first must know and understand the behavior of the wireless frequencies you are surveying; and second, you must know and understand how to use and

[1] Asking the right questions is the key to eliciting requirements, the primary topic of chapter 2. If you do not ask the right questions, you will not get the needed information to design an effective WLAN for the client or your organization.

interpret the data received from the surveying tools; and third, you must understand how to properly use the surveying tools. Failing in any of these can lead to an improper survey and insufficient coverage.

- Given that coverage should not be define as "the signal is there" but as "the signal is there with sufficient SNR for the required data rate", performing spectrum analysis or using site survey tools that report the achievable data rate is important to proper coverage design. After meeting with the customer to gather essential information and requirements, the designer should analyze the target coverage areas to discover possible interference sources. The information gathered will be used to select appropriate channels in the coverage area. When automatic channel selection is used, during the validation process, the designer or installation technician should validate that an acceptable channel has been selected in coverage areas with potential interference.
- Insufficient coverage can and will exist unless the designer is educated and experienced enough to plan for issues that could arise, and part of that planning is knowing how to provide sufficient coverage. This ability starts with the knowledge found in the CWNA materials and extends to the more advanced information covered throughout this book.
- For purposes of this book and the CWDP exam, sufficient coverage is defined as the availability of a wireless connection that provides the required data rate and stability demanded by the clients. Figure 1.1 depicts insufficient coverage, assuming that coverage is desired in the areas without signal.

▪ **Insufficient capacity:** Another problem with poorly designed WLANs is the lack of needed capacity. This issue arises when a WLAN is designed well based solely on signal coverage of the facility, but no consideration is made for the number of client devices within the requirement areas or for the applications being used on those devices. It is much easier to design for coverage alone, but capacity is an essential factor to consider. It is also a common mistake to analyze square footage and users while leaving out device counts and applications. A user does not equal a device and many users will employ several devices. Additionally, many applications require much more bandwidth than others. A good comparison is an instant messaging client contrasted with a streaming video conference. The instant messaging client is likely to require far less data throughput than the streaming video conference. Overlooking key use cases of the WLAN can lead to a poor design with insufficient capacity planning, and the WLAN will not perform to the customer's expectations. Proper requirements analysis and planning is essential to meeting these expectations.

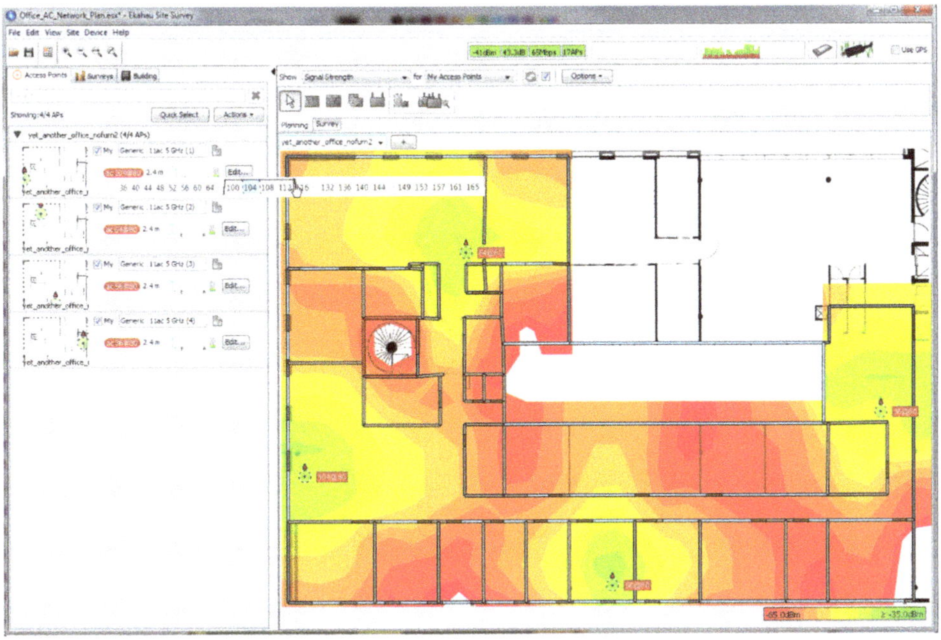

Figure 1.1: Floorplan Showing Areas without Coverage

- **Lack of scalability:** When a WLAN is designed for client devices and applications in use at the time of the design, and no consideration is made for future technology changes, device upgrades, or added employees, lack of scalability is the common result. Scalability is the ability to expand. Elasticity is the ability to expand and contract. In WLANs, elasticity is rarely a requirement, but scalability is a requirement in most deployments. Failure to plan for growth and change can result in projects that need early refresh and in unplanned network expenses as growth occurs. One should design just for what is in use now and for technologies, applications, devices, and users that will result from growth. This planning will allow for scalability of the WLAN without excessive costs in the future.

- **Lack of required functionality:** Functional capabilities of a WLAN include roaming, metrics measurement, support for multicast and streaming, and other functional capabilities. Lack of required functionality is often the result of budget constraints and a fear of going over budget with equipment purchases, but sometimes it is simply a lack of planning by the designer. Great coverage is in the design, capacity and future growth are built into the new design as well, but needed functions are not available. It is essential to both gather these functional requirements during the initial meetings with the customer and to know how to fulfill them. Learning the science behind radio frequency is not enough to become a professional WLAN designer. The WLAN must be a solution for multiple needs that reach beyond the physical layer of RF. It is the job of a designer to provide a solution with all the necessary capabilities.
- **Using the wrong equipment:** Designs that use the wrong equipment may result in several problems. One example is using an AP with internal antennas in an environment with high temperatures. APs with internal antennas will not sustain functionality as well under higher temperatures, and improper equipment choice in this scenario can lead to WLAN failure. Remember that it is a best practice to reference the product specifications before selecting devices for use in a high-temperature area or in areas with other environmental considerations such as dust and humidity. Another common mistake is selecting the wrong antenna for a design. Sometimes a customer has a few large boxes full of APs and plenty of antennas to go on them, so they want to save cost by utilizing them in their new WLAN. Having equipment on hand does not mean it is the right equipment for the task. Be sure your equipment choice fits the environment.

As you can see, the primary task the prevents bad design is good requirements engineering. For this reason, chapters 2 and 3 provide in-depth coverage of the requirements engineering process based on standards that have evolved over decades and provide sound guidance for the process.

Installing from "Cookie-Cutter" Designs

"It worked for the last location, so it will be fine here," says the new designer. Using the same design for every location that has similar characteristics can get you into trouble in several ways. Just because both locations are 2,000 square meters of office space does not mean that the walls, floors, or neighboring networks are the same. No two environments are exactly alike and using a "cookie-cutter" design[2], because one site looks like another, is not a wise practice. Of course, exceptions to this advice exist, such as the deployment of WLANs in buildings constructed with nearly identical floorplans and identical materials, such as store and restaurant chains, but, even then, the designer should be careful to verify true duplication as existing buildings are often retrofitted for such chain installations. A good WLAN designer will never assume anything. Instead, the designer will use proper engineering techniques to assess the environment and then select the correct equipment for each design based on the educated assessment.

A common pitfall for wireless design professionals is giving into non-professional suggestions, such as "*one AP in every other hospital room*" or "*one AP in every classroom.*" It is easy to go along with a suggestion like this from a customer and avoid lengthy assessments of the environment, but when these plans are followed, the need for a design professional is seen as diminished or unnecessary because the RF design is already planned out in the suggestion. Many risks are associated with avoiding the use of a design professional and going with a cookie-cutter design. The risks include these:

- **Missed interferers**: When a design is pre-planned, with a "cookie-cutter" design as in the previous examples, it is easy to set aside analyzing the wireless spectrum to locate potential interferers, because equipment selection and placement usually come after the spectrum analysis.
- **Increase in labor and hardware costs:** "Cookie-cutter" designs will often call for more APs than are required, and the costs of the installation are significantly

---

[2] The phrase "cookie cutter" design is borrowed from the fact that a cookie cutter results in the same cookie shape every time or at least it does when someone other than me uses one. In WLAN design, you do not want to simply retrofit existing designs into new facilities. Even if the floorplan is similar, the furniture may be different, the wall materials may be different, the clients and applications are often different. All of these things demand a proper design for the specific target environment.

increased. In addition to the added APs, extra cable pulls, and possibly a need for more switch installations, which might require more fiber installations also increase the expense. These added components will also require more labor. A good WLAN design will balance between the requirements and cost while, of course, considering future scalability. Over-engineering is really just a term used for the lack of proper engineering.

- **Poor performance:** Often, a "cookie-cutter" design will result in poor performance. The performance degradation is due to the lack of proper engineering and, possibly, an attempt to ensure "blanket coverage" throughout the facility. Many things can happen once a cookie-cutter designed network is put into service. The WLAN controller (WLC) may change power levels to the point that transmit power in some of the APs may be too high for optimal device connectivity or power levels on some of the APs could be turned down by the WLAN controller, resulting in coverage holes. "Cookie-cutter" designs typically provide very inconsistent overlapping coverage between APs, resulting in poor device roaming between coverage cells. With so many issues, a lot of work will have to be performed in the WLC configurations or other management interfaces to patch the design, but it still may not function as it should. The overall RF coverage will most likely be less than optimal, resulting in lower data rates throughout the entire facility. Nothing replaces the time spent performing good design practices.

It is often argued that even if "cookie-cutter" designs can result in over-engineering, they may also save money on the front-end in consulting fees for surveys and designs. In some scenarios, this trade-off may be true and beneficial, but that is certainly not the most common reality. This is the position of CWNP.

Problems with No Design

It is not uncommon practice for network administrators to receive a request for Wi-Fi access from a department manager and then simply order an AP, pull a cable, and provide Wi-Fi access as requested. This practice gives no consideration to existing coverage in other areas or how the new AP's RF may propagate into and interfere with other coverage areas. These types of WLAN expansions evolve as requests come in, and the issues caused by such installations evolve as well. As more of these ad-hoc installations take place in an environment, the probability of severe problems increases.

Be sure to test; if new areas within a facility need WLAN access, the existing coverage areas should be verified and documented so that the new expansion of the WLAN will become a properly functioning part of the existing WLAN and coverage will overlap properly with the existing APs. Anytime a WLAN is extended considerations must be made for the impact it will have on the existing network. Increased Co-Channel Interference (CCI) may result, and existing channel plans may no longer suffice. It will not require full-scale design procedures, but it does require a professional who understands the design concepts.

"Shotgun" Implementations

In a "shotgun" implementation, APs are simply placed where you think people will need wireless access without any thought for how the APs will impact each other. "Shotgun" implementations can result in over-engineering (too many APs) and poor roaming support, among other problems.

These implementations are notorious for inconsistency and for either providing too much coverage from too many APs, or not providing enough coverage with dependable data rates. This is also called a "best-guess effort" because there is no science, training, or skill involved in this methodology. Many of these individual coverage cells will expand far beyond the necessary distance for adequate data rates resulting in overloading APs with too many client devices connected at low data rates. Other areas will have too much coverage from too many APs which is an unneeded expense. It can also result in problems from CCI and/or inadequate coverage where it is needed the most.

While these "shotgun" implementations often result in poor performs and extended budgets, they also leave no reasonable solution for tuning and modifying the WLAN to correct the performance issues apart from simply turning off some AP radios when overengineering is in play. More importantly, growth and expansion of the WLAN is problematic without surfacing more performance issues.

1.2 The Design Process

The basic design process used in WLAN design is explained in this section. The process can be explained in four phases with tasks performed in each phase. However, with modern WLANs increasing in complexity, it is important to remember that some recursion may be required in many projects. For example, you may pass through the Define phase, where business requirements and stakeholder requirements are initially developed, only to encounter unexpected issues in the Design phase requiring a revisit of the business and stakeholder requirements. In fact, it is also possible that some requirements engineering will still occur in the Deploy phase of the design process. As you will see in chapters 2 and 3, where requirements engineering is explained in detail, this recursion is a normal and expected part of systems engineering, including WLAN systems. Therefore, you should not think of the four phases of the design process as purely sequential. The design process phases, the recursive nature of requirements engineering, and the flow of a WLAN implementation project are illustrated in Figure 1.2.

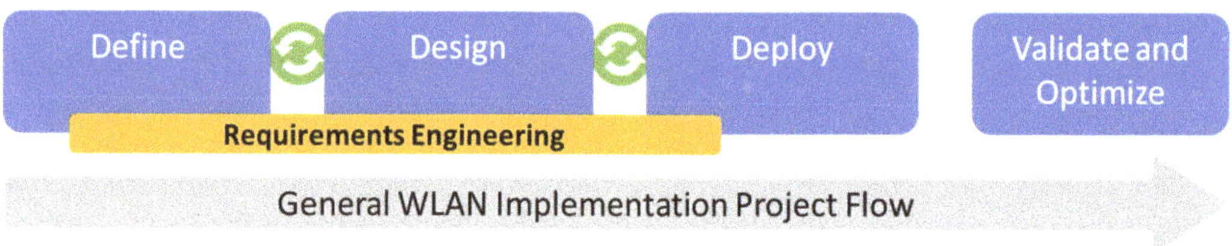

Figure 1.2: The WLAN Design Process

The design process, then, can be described in four phases: Define, Design, Deploy, and Validate and Optimize. In fact, the knowledge domains of the CWDP exam follow these four phases: Define Specifications for the WLAN, Design the WLAN, Deploy the WLAN, and Validate and Optimize the WLAN. The remaining chapters of the book explore the various tasks performed within this high-level design process in detail and the following sections provide a foundational overview of them.

Define

The Define phase of the WLAN design process consists of several important tasks that are critical to creating a correct design. Regardless of project size, the definition of the project and requirements for it should not be taken lightly. Here are some important tasks to implement when defining a WLAN:

- **Information gathering** is the typical first step in the define phase of the design process. If a project charter or request for proposal (RFP) exists, this is the first piece of information used and that on which is built all future information and documentation. As a WLAN designer with expertise in wireless design, it is your responsibility to communicate with the business stakeholders concerning technical capabilities of a WLAN. If the project charter or RFP makes requests that are impossible or impractical, this should be communicated to the stakeholders with options for alternate solutions. Starting on a proper, realistic foundation is key to project success.
- **Requirements engineering** is essential to the definition of a WLAN solution. This stage of defining a WLAN can consists of multiple tasks. The customer, whether internal or external, submits a request for a WLAN to be designed and installed and the right information must be gathered to determine the type of installation and the equipment needed for the installation. All business requirements will be documented in the define phase. Given that site analysis may be required to fully document technical requirements, some such requirements will not be documented until the design phase begins; however, before the WLAN designer begins defining AP locations, channels, output power levels, and other wireless specifications, all requirements should be elicited and documented.

  For example, if a customer has a large warehouse of over 750,000 square feet, and they want to have wireless access to their company's corporate network while in the warehouse, it may sound simple, but many questions need answers before designing the WLAN to fulfill this request. What applications will be used on the WLAN? What types of client devices will run these applications? Will VoIP clients be utilized? Is locationing required now or, possibly, in the future? Does the RF need to be contained, as much as possible, within the warehouse for security reasons, or is it acceptable if the RF covers the areas external to the warehouse? How high are the ceilings? How high will the APs be mounted? What is the highest temperature reached where the APs will be mounted? What types of shelving are in the warehouse and how tall are these shelves? Is there a network switch in or near the warehouse with enough PoE ports available and enough power budget?

These questions represent some of the questions that might be asked for such a warehouse design and several more may surface before the wireless designer has enough information to feel comfortable to start the design. This type of information helps the designer to determine the equipment needed to achieve the customer's expectations. Many of these questions can be answered in the define phase; however, requirements engineering typically continues into the design phase as the expertise of the designer is required to answer many of the questions and they will be answered during surveys and analysis of the target location of the WLAN.

- **Documenting the requirements** is just as important as generating the requirements. Requirements engineering includes documentation and the process is not complete or accurate without the documentation. Chapters 2 and 3 provide more information on this important aspect of the define phase.

A designer must be skilled in several areas to be successful in performing the tasks required in the define phase. These areas include:

- **Communication Skills:** While beyond the scope of this book and the CWDP certification (as it is a much-needed area of study, application, and certification on its own), communication skills are significantly more important than any other skillset for the WLAN designer, assuming the professional level knowledge of WLAN operations is possessed. The WLAN designer must be able to communicate the capabilities of vendor solutions to the customer and understand the customer's needs. Without effective communication skills, the designer will face great challenges throughout the project. For example, if the designer fails to gather the proper information up front, multiple meetings with the customer may be needed during the design process, leaving the customer feeling a lack of confidence in the outcome of the design.
- **WLAN Knowledge:** The WLAN designer must have in-depth knowledge of WLAN technologies. This knowledge will serve the designer well as business needs and proposed solutions are explained. It will help the designer

communicate the capabilities of current technologies as they apply to the customer's needs and provide alternative solutions when unforeseen technical issues arise in the middle of a design.

- **Wired LAN Knowledge:** It is not enough to be skilled in WLANs. The WLAN designer must also have a strong understanding of wired LAN technologies as well. Knowledge of wired LANs will be of great benefit as you plan for network connections, roaming solutions, security configurations, and other required capabilities of the WLAN. Required knowledge includes switching and routing, network services, IP addressing, and network segmentation through VLANs, firewalls, and access control lists (ACLs), and more.
- **Cabling Infrastructure Knowledge:** It is greatly beneficial to know fiber optic and cabling standards so that the WLAN design does not fail when APs are mounted, and cable runs are too far from the nearest available switch location. This often becomes a problem for new designers and overlooking the need for a new switch installation can become expensive as well as time-consuming. A WLAN designer should have complete awareness of potential cabling issues when designing new WLANs so that new switch locations can be added to the design documentation up front instead of during the installation.
- **Customer feedback and agreement** – This area is important as a follow-up after all the requirements engineering and information gathering is complete. Documenting your findings and creating a presentation to support your ideas and suggestions for the customer's WLAN design is an important stage before beginning the actual design. You should explain your proposed design based on the information gathered from the customer and receive feedback from them to ensure that you have accurately defined the need and that both parties agree.

After the need is clearly defined, documented, and approved, the Design phase can begin.

Design

The Design phase of the WLAN design process is where the technical expertise of the designer is used to translate the business and stakeholder requirements into pure technical requirements and to generate the WLAN designs. Many tasks will be performed in the Design phase and may include:

- Performing a spectrum analysis of the facility
- Performing a site survey for AP placement
- Making proper equipment selections to support the customer's needs
- Creating design documents using wireless network design software
- Defining equipment configurations
- Designing or recommending the WLAN's supporting infrastructure
- Creating a Bill of Materials (BoM)

The purpose of the spectrum analysis is to determine if there are interferers in the area, existing or rogue wireless APs present, or any other issues that may cause limitations on band and channel selection. Proper use of spectrum analysis tools can save a lot of time during the design phase and can also save a lot of money in person-hours and cable moves caused by re-designing around issues found after the installation.

The site survey is performed by measuring the RF signal propagation (based on necessary power levels and channel settings) of an AP throughout the facility. Its purpose is to determine correct placement, cell size, and overlapping coverage of each AP needed to meet the requirements of the new WLAN, which were determined during the Define phase. During the site survey, proper AP model selections are made based on the environment being surveyed. If needed, additional network switches and locations will be selected along with backup power, WLCs, and security mechanisms. Recognizing the need for other network support items is important to getting a design done right the first time without repeat visits and surprise add-on costs.

The Design phase is broad and requires significant knowledge and skill in multiple areas of networking. There are times when the RF coverage is the only portion of the network a WLAN designer will be required to provide, but it is certainly not common, and more is expected from wireless professionals every day. Best practices for a WLAN designer would be to build on the knowledge of every aspect of networking technologies. In doing so, you will avoid being caught by surprise when asked to design and implement a new WLAN along with all its supporting infrastructure.

It is important to understand the significance of an accurate site survey. Equipment and software cannot make up for bad RF surveys, nor should they be expected to do so. Take the time to learn the proper techniques required to provide accuracy in this stage to avoid problems with the WLAN once it is in operation. If the customer's WLAN requirements, determined in the Define stage, specify an expected minimum data rate, a bad survey can and will hamper the RF performance, causing areas of coverage not to meet those expectations.

Deploy

Documentation containing a bill of materials (BoM) is usually provided with a complete design package, and once the design is completed and approval is granted, the equipment is ordered for installation. Deployment of a large WLAN is usually a complex stage in the process that can involve multiple contractors for various parts of the project. This means that different people or groups may be performing tasks in every part of the implementation, making it easy to run into problems if a good plan is not followed. Here is where a WLAN designer should anticipate the potential for issues and plan accordingly.

In a WLAN design plan, a ***Bill of Materials (BoM)*** is a listing of the hardware and possibly software required to deploy the designed WLAN. This will include everything required, like antennas, cables, mounting kits, poles, lightning arrestors, and anything else needed to deploy a proper WLAN.

This plan should follow steps like the following:

- **Install network cabling:** This stage almost always takes longer than all others because of the amount of work involved in getting a cable from point A to point B. There could be holes to drill, permits acquired, conduit to run, and cable trays to mount. Contractors or in-house telecom staff commonly perform this work, so it can usually be carried out in parallel to other work.

- **Configure Infrastructure:** Before installing the APs, it is always a good idea to provision the wired LAN to support the WLAN. This means configuring the infrastructure to support the new or upgraded WLAN equipment and client devices. Items such as DNS, DHCP, switch ports, port security, QoS, and routing must be configured to support the newly scaled network and the extra demands that the WLAN will place on the infrastructure. Have this information in a written plan so that it can be shared with a network administrator and any contractors taking part in this portion of the deployment. This will provide a way for two-party verification of the infrastructure's configurations.
- **Provide PoE:** Most infrastructure switches manufactured now provide Power Over Ethernet (PoE). You should always check the technical specifications of both the switch and the APs being deployed to make sure the switch is providing appropriate power. Depending on the model of your hardware, the switch may only provide enough power for twenty-four ports on a forty-eight-port switch. This depends on the power required from the APs. Many APs require 15 watts of power, while others can require up to 30 watts. If you are deploying new APs on existing network switches, this is something you need to discover and plan for during the Define stage. Power injectors can be utilized to remedy a lack of sufficient PoE from the switch, although having a large number of injectors requires a lot of space and a lot of power outlets. This could prove to be a challenge for both rack space and aesthetics.
- **Test the Infrastructure Configurations:** After configurations are complete, it is always a good idea to provision a single AP and test it before deploying the entire project. This can be carried out by simply configuring an AP and plugging it into a switch port via a patch cable, and then connecting a client device that will be used on the WLAN by the customer. If the client device has no issues connecting, you may want to repeat this step at other switch locations as needed to confirm that switch and network configurations are correct. This step also depends on the size of the deployment and whether

time permits. If the cabling team is not finished with the cable installations, use the extra time for these configuration tests.

- **Configure the APs:** With lightweight APs and WLAN controllers, most APs (depending on the model) can simply be plugged into the network and find a WLC to join. This is not advised and should be avoided. Each AP should be configured, for several reasons:
  - ***A unique name*** should be configured to identify the AP with a matching label placed on the outside. This provides easier tracking for future troubleshooting in large deployments. The label can be left out if required for security purposes, although the design documentation should display the unique name with the location of the AP, so the network administrator can tell where each one is located.
  - ***Controller names and addresses*** are configured in most APs to provide primary, secondary, and tertiary controllers for AP failover. This depends on the deployment type and number of WLCs deployed within the network.
  - ***Access Point Mode*** is another setting required based on the model of AP used and the type of deployment. Typically, the "Local Mode" is the most common configuration, but there are various other deployments that require Bridge Mode, Monitor Mode, Flex Mode, and so on. This setting coincides with the deployment type needed in your design.
- **Deploy APs***:* Knowing that your APs are configured and working properly, you can now send them out for installation. At this point, they should be "Plug-and-Play," and one great plus to this arrangement is that you have already tested the infrastructure's hardware and the APs. If you or a contractor are mounting the APs and they do not come online, then you can narrow down your troubleshooting procedures to the cabling and patch panels.

Many other settings are available for management and control over individual APs, but those just listed are the most common. These settings can be configured by console cable, through a WLC, or with some type of management software such as a cloud manager. This choice is also determined by the type of deployment and vendor used. It is always best practice to provide as much detail as possible to support easy troubleshooting in the future.

Validate and Optimize

The final phase of a WLAN implementation project is the Validation phase. Here is where the WLAN is thoroughly tested to ensure that it meets the customer's requirements that were detailed during the Define phase[3]. The validation or post-installation survey process utilizes an active site survey performed after the installation of the WLAN is complete and all AP radios are turned on, ready for connectivity. At a minimum, the following three items should be verified:

- Required coverage
- Desired capacity
- Necessary capabilities

Required coverage means sufficient WLAN coverage (typically measured by signal strength) throughout all the areas listed in the defined requirements. Capacity ensures that the WLAN provides the desired throughput defined in the requirements. Capabilities include features like fast secure roaming, guest registration, onboarding, security, and more.

This phase can become repetitive if multiple changes are needed to bring the WLAN into compliance with the customer's requirements. You may test an area and find that adjustments need to be made and then go back and test the area again after you make those adjustments only to find that more corrections are needed. This is known to

[3] It's odd that many organizations want to skip over the Define phase and either perform no real requirements engineering or perform them so poorly and quickly to be of little value only to ask for an exhaustive evaluation at the end. One has to ask, "Evaluate what?" If there were no requirements to be met, we cannot evaluate the success in meeting them.

happen several times before a validation or post-installation survey is complete and, based on the size of the WLAN, it could carry on for several days before you are confident the WLAN is at its optimal performance. It is a necessary step and should always be taken seriously, because the result of failing to implement the post-installation survey can be poor WLAN performance.

Many WLAN product vendors provide variations of automated management to help with optimization of the WLAN. This is often called radio resource management or RRM, and as mentioned earlier, this software solution is no replacement for a CWDP's skill set and experience. RRM can control power levels, channelization, data rates, and even modulation types, and it usually comes into play in an attempt to correct issues resulting from bad design or lack of design. It is not advised to depend on this technology to optimize your WLAN. The CWDP should optimize the design through the design process. A WLAN design is only as good as the engineer that sees it through to the end.

1.3 Implementing Effective Project Management

In many cases, the person designing the WLAN will also have to perform some or all the project management tasks. With this as a given, it is a good practice to familiarize yourself with the project management process. CWNP does not expect you to be a Project Management Professional (PMP)[4] certified individual to pass the CWDP exam, but you should know basic concepts related to managing WLAN design and installation projects.

All projects usually start with an idea for a deliverable. The deliverable, in this case, is a WLAN. The idea is pitched to stakeholders as a business use case, and if it is agreed upon, a project charter is drawn up and signed by all necessary parties; you have an official project. The project charter is a document that officially recognizes the project, and this book will not go into a lot of detail covering a project charter, since you only need to be familiar enough to understand what it is for. Figure 1.2 shows a portion of a project charter document.

[4] In addition to the PMP certification, CompTIA offers the Project+ certification, which is a less intensive certification process, but it teaches the principles and concepts important for most project managers. Interestingly, it started out as IT Project+ many years ago and was simply rebranded as Project+ along the way.

WLAN Project Charter

General Project Information

Project Name	2019 WLAN Project
Project Sponsor	John Smith
Project Manager	Dale Carpenter
Email Address	wlanproject@asdhajshd.com
Phone Number	867-5309
Organizational Unit	Information Technology
Process Impacted	All departments
Expected Start Date	February 1, 2019
Expected Completion Date	April 30, 2019
Expected Savings	**$700,000**
Estimated Costs	**$648,500**
Green Belts Assigned	no
Black Belts Assigned	no

IMPORTANT REMINDER
A narrative written charter must be circulated and signed by the project sponsors. You can attach a completed version of this charter to your narrative written charter in an effort to keep it short and concise.

Please make sure you meet with the project team and sponsors before completing this charter. Much of the information required will need to come from a discussion with team members and sponsors.

Describe the Problem or Issue, Goals, Objectives, and Deliverables of this Project

Problem or Issue	The current WLAN is based on 802.11n on the infrastructure and includes many 802.11g clients. Several applications in use are experiencing performance problems.
Purpose of Project	Resolve the WLAN performance problems.
Business Case	By increasing the performance of several mission-critical applications, we will reduce time-to-entry for hundreds of users and optimize use of time. We will also be capable of implementing several newer applications that have been on hold due to the performance of the current WLAN and its clients.
Goals / Metrics	90% of devices on 802.11ac and 10% of devices on 802.11n. No remaining 802.11g devices.
Expected Deliverables	WLAN survey report; WLAN design plan; WLAN implementation; WLAN validation report

Figure 1.2: WLAN Project Charter

According to the Project Management Institute, a ***project charter*** is a document issued by the project initiator or sponsor that formally authorizes the existence of a project and provides a project manager with the authority to apply organizational resources to project activities.

Many companies will require a non-disclosure agreement (NDA) to be signed by you and any other personnel taking part in the project. This document simply signifies that a confidential relationship between parties has been established and all information about the company and its project that is shared or learned during the project must remain confidential between the parties. This is very common in the corporate world to protect information that could damage the corporation if others accessed it outside the company.

Once you have a project, you are in the Initiation phase where the project objectives are solidified, the business goals are clarified, and the scope of work is fully defined. You are now ready for project planning. A lot of this phase of the project lifecycle is parallel to the defining phase of the WLAN design process. A customer provides you with a business goal and reason for the project, and then you, the designer, perform a requirements analysis that gathers information vital to developing the scope of work.

The scope of work or statement of work (SoW) defines the work to be done, and it often represents a high-level scope initially but then develops into a more detailed scope as the project progresses. This is not always the case, but the agile approach to large corporate projects is now very common, as changes could develop during many stages of the project that can affect time and cost. The project manager will develop a work breakdown structure (WBS) that will divide the project into multiple tasks that can be assigned to various individuals along with a task cost and time to complete. An example of work breakdown would look like this at a high level:

1. Site survey
2. Install fiber for new switches
3. Install cabling for new APs
4. Install new switches
5. Install new APs
6. Test WLAN

This is a simple list for demonstration purposes only. The WBS can become much larger than this. In fact, each of the list items can now be broken down into smaller tasks:

1. **Site Survey**

   1.1 Perform spectrum analysis.

1.2 Locate all existing network switches.

1.3 Document existing fiber.

1.4 Document existing Wi-Fi (if any).

2. **Install fiber for new switches**

   2.1 Obtain drilling permits.

   2.2 Hang and mount conduit.

   2.3 Rack-mount fiber fan-out boxes.

   2.4 Pull fiber through conduit.

   2.5 Terminate fiber.

   2.6 Test fiber.

3. **Install cabling for new APs**

   3.1 Locate and identify cabling routes.

   3.2 Install conduit (if needed).

   3.3 Pull and label cable drops.

   3.4 Rack-mount cabling patch panels.

   3.5 Dress and terminate cables.

   3.6 Test cables.

4. **Install new APs**

   4.1 Configure and label new APs.

   4.2 Document mac addresses and serial numbers of new APs.

   4.3 Test new AP configurations.

   4.4 Mount new APs.

5. **Test WLAN**

   5.1 Perform an RF coverage test.

5.2 Test user device connectivity.

5.3 Test data rate requirements.

5.4 Test capacity requirements.

5.5 Test capabilities requirements.

The preceding WBS is only an example and not meant to be used as a real-world WBS, because every project is different and should be managed on an individual basis. The WBS is the "list" that drives your project and acts as a checklist for you along the way. According to *Gawande, Atul, The Checklist Manifesto, 2001*, "Good checklists, on the other hand, are precise. They are efficient, to the point, and easy to use even in the most difficult situations. They do not try to spell out everything – a checklist cannot fly a plane. Instead, they provide reminders of only the most critical and important steps – the ones that even the highly skilled professionals using them could miss. Good checklists are, above all, practical." The same is true for a WBS. It is the work checklist for your WLAN design and deployment project and should be useful rather than simply showing how much you know about WLAN design.

With this WBS example, you can see that the phases of the project can be broken down into smaller individual tasks, and these tasks can be assigned to different individuals to be responsible for. With this information, a project team is formed with competent employees and contractors who possess the proper skill set to complete the work listed. The WBS helps with time, cost, and resource management, as each task can be given a cost and time to complete from the responsible individual assigned to the task.

According to the Project Management Institute, a ***project resource*** can be skilled human resources (specific disciplines either individually or in crews or teams), equipment, services, supplies, commodities, material, budgets, or funds.

To help with resource management, a communications matrix should be developed. This document will detail the roles and responsibilities of each team member, how documentation should be distributed to and from each team member, and how each team member and stakeholder prefers to be contacted. It will also list the contact information of each person.

At this point, we have a project plan, a breakdown into individual tasks of the work to be done, competent and responsible team members assigned to each task, a cost and time frame assigned to each task, and a document providing detailed communication methods for each person involved in the project as well as their roles and responsibilities. With all of this in place, you can effectively manage the project as it progresses while keeping track of your resources.

A ***communications matrix*** is a document containing the communications management plan for the project. If should list all required contact information and the kind of reporting required to and from the various contacts.

1.4 Chapter Summary

In this chapter, you learned the WLAN design process from a high-level viewpoint and covered detailed areas of the design phases: Define, Design, Deploy, and Validate/Optimize. You should also be familiar enough with the steps of project management so that you can effectively manage your project from start to finish within the time frame and budget given.

1.5 Facts to Remember

Be sure to remember the following facts as you prepare for the CWDP certification, and be sure that you can explain the details related to them:

- A bad design results in insufficient coverage, no scalability, insufficient capacity, and lack of required functionality.
- Cookie-cutter designs often result in increased hardware costs and poor performance.
- Failing to perform any design functions often results in excessive co-channel contention, poor performance, and limited capabilities.
- The WLAN design process can be broken down into four basic phases: define, design, deploy, and validate.
- The Define phase includes requirements analysis.
- The Design phase includes site surveys, WLAN design, wired LAN design at times, WLAN hardware selection.
- The Deploy phase involves the configuration of the infrastructure, power provisioning, network connection provisioning, and AP configuration and installation.
- The Validation phase includes validation of the requirements and tuning or adjusting to ensure they are properly met.
- Effective project management requires understanding of the project lifecycle, its stages, and documentation, along with the ability to manage all project resources throughout the life of the project.

1.6 Review Questions

1. If you are configuring access points for a new WLAN, what phase of the design process are you in?
    a. Define
    b. Design
    c. Deploy
    d. Validate
2. If you are trying to locate possible WLAN interferers before starting your site survey, which WLAN tool would you use to find these interferers?
    a. Spectrum analyzers
    b. Aircrack-ng
    c. Wireshark
    d. Protocol analyzer
3. To determine correct placement of access points, cell size, and overlapping coverage of each access point to be deployed within the new WLAN, you may perform a ___________________.
    a. Validation survey
    b. Site survey
    c. Project initiation
    d. Throughput test

4. When you simply hang access points in an area where you believe WLAN coverage is needed, what issues may result from it? Choose all that apply.
    a. Over-engineering
    b. Problems roaming
    c. Higher costs
    d. Incompatible 802.11 functions
5. A post-installation survey is part of which phase of the WLAN design process?
    a. Define
    b. Design
    c. Deploy
    d. Validation
6. If you are required to design a WLAN for a customer in a multi-tenant office complex, what are some things you might find to be a challenge for the design? Choose all that apply.
    a. Thick glass doors
    b. Neighbor WLAN
    c. Co-channel interference
    d. Rogue devices connecting
7. In the Deploy phase, configuring and testing the wired LAN, and configuring the access points before deploying them, are good practices to prevent ________________________.
    a. Unanticipated network outage
    b. AP licensing issues
    c. MAC addresses blocked on switchports
    d. Unnecessary troubleshooting

8. In order to have a properly functioning WLAN, there is no substitute for a ____________________.

    a. WLAN controller

    b. RRM

    c. Spectrum analyzer

    d. Good design

9. A solid WLAN is not designed on customer requirements alone, but through a thorough ____________________ during the ____________________ phase.

    a. Requirements analysis, Design

    b. Spectrum analysis, Define

    c. Coverage analysis, Design

    d. Requirements analysis, Define

10. A Post-Installation Survey should take place during the ____________________ phase and confirms that the WLAN has satisfactory ____________________, ____________________, and ____________________.

    a. Discovery, RF Coverage, Performance, and Capacity

    b. Validation, Capacity, Performance, and Features

    c. Verification, Coverage, Capacity, and Capabilities

    d. Validation, Coverage, Capacity, and Capabilities

1.7 Review Answers

1. **The correct answer is C.** After you have reached the point where you are configuring APs, you are in the "Deploy Phase." The WLAN design has been defined and agreed upon, the design is complete, and materials have been ordered and received. You are not ready to validate the WLAN until all configurations, cable installations, and network hardware has been installed and is ready for testing.

2. **The correct answer is A.** When locating interferers, you need a Spectrum Analyzer. Although survey software like Air Magnet Survey Pro can locate rogue APs, other interferers such as cordless phones, baby monitors, jamming devices, microwave ovens, and security cameras can only be effectively located and identified with a Spectrum Analyzer. Aircrack-ng is a password discovery tool and Wireshark is a tool used for packet or protocol analysis.

3. **The correct answer is B.** The validation survey or post-installation survey is performed after the WLAN has been completely installed, meaning that proper location of APs and their cell-sizes have already been determined. The spectrum analysis should be performed during the requirements analysis and before determining AP locations and cell-sizes. A throughput test can be performed after the installation of the WLAN is complete or parts of it complete. The site survey is where AP placement is determined through an RF measuring/testing process utilizing software that displays the AP's signal strength, the noise floor of the area, and the signal to noise ratio. This procedure is used to determine AP locations, cell-sizes, and proper overlapping coverage between cells.

4. **The correct answers are A, B and C.** Hanging APs without a proper site survey can result in too many APs throughout the entire facility. This is considered over-engineering, and all of the extra equipment and cabling involved in the over-engineering makes the project cost more than necessary. When a proper site survey is not performed, uneven coverage may also result, leaving a lack of overlapping coverage between APs as well as too much overlapping coverage in other areas. This has proven to cause roaming issues when devices move between coverage cells. Overlapping coverage between cells should be designed consistently so that this issue can be avoided. Lower data rates also become a factor in uneven coverage. This

most commonly happens when APs are placed too far apart, and modulation types change and data rates down shift to sustain a reliable connection between the APs and user devices.

5. **The correct answer is D.** The post-installation survey, also known as a validation survey, is performed during the validate phase. Here the WLAN is tested, adjusted, and tested again to ensure customer requirements are met. This phase comes after requirements are defined and the design is completed and deployed.

6. **The correct answers are B and C.** In a multi-tenant office complex, chances are that you will find several existing WLANs. These WLANs nearly always have RF propagating into other surrounding office spaces, making it difficult to design a new WLAN without co-channel interference from neighboring WLANs becoming a challenge. You can design around any issues caused by thick glass doors, and you shouldn't have issues with rogue devices connecting with proper security measures put into place.

7. **The correct answer is D.** There are many issues that become factors to look into when APs do not come up and function properly in a new design and installation. One main factor is various stages of the work being performed by multiple departments and/or contractors. When you can configure network hardware and APs, and then test those configurations, usually by using a patch cable to connect a configured AP to the various switches and connecting a mobile device to the AP, you eliminate a lot of troubleshooting steps once the APs are mounted and connected to the new cabling. You wouldn't experience an unanticipated network outage, because the new APs are not considered part of the functional network until they are installed, tested, and turned over to the customer. Testing a single AP from switch to switch will not prevent licensing issues, because it will only utilize one license. If 100 licenses were purchased and 125 APs are being installed, the licensing will not cause an issue until the installation of at least 101 APs is complete. Testing configurations will reveal any existing switchport security that limits the number of MAC addresses allowed on each port, and new switches typically do not limit MAC addresses by default.

8. **The correct answer is D.** You can have a properly functional WLAN without a WLC by implementing autonomous APs or by using some form of cloud management. Radio resource management is not necessary for a WLAN to function properly if the design process is followed and performed accurately. A good design is the result of following the design process to which there is no substitute. Utilizing proper tools and skills such as the proper use of the spectrum analyzer are included in having a good design.

9. **The correct answer is D.** In many cases the customer knows what they want but not what they need to get what they want. A requirements analysis is needed to discover other important information that will help in defining the "need(s)" for the WLAN design. The Define phase is where the requirements analysis is performed before the design begins. A spectrum analysis can provide some of the information needed for the design as it relates to the rf spectrum. A coverage analysis is not an official name of any part of a WLAN design, although it could be considered part of defining coverage or part of validating coverage after an installation.

10. **The correct answer is D.** The post-installation survey is performed after the installation of all equipment and cabling. This survey validates the design by testing for adequate wireless coverage, based on the information given in the customer requirements and other information discovered during the requirements analysis. The WLAN is also tested to ensure that capacity requirements are met and all performance and capabilities of the WLAN are functioning properly.

Chapter 2: Defining Specifications for the WLAN

Objectives Covered

1.1 Collect business requirements and constraints
1.2 Collect and define technical requirements
1.3 Collect project documentation
1.5 Gather information on environmental factors

It is an everyday occurrence in many organizations to have complaints about the WLAN, such as these:

- "The network is running slow."
- "I have zero Wi-Fi in my office."
- "I can't walk down the hall on my phone without dropping the call every time."
- "It takes me forever to connect to the wireless network."

If you have heard them, you've dealt with unhappy users. They are all the result of requirements analysis not being executed properly before designing the WLAN or simply not designing the WLAN at all or not validating the WLAN after installation. Creating a WLAN that will provide proper capabilities for the users is the primary goal of the Define phase and requirements analysis.

This chapter introduces requirements analysis in general and then addresses specific issues of the process for building a better WLAN.

When looking for the business use case and justification, you must think back toward the end of Chapter 1, where we discussed effective project management. Every project starts with an idea. This could be a great idea to install a WLAN at the office complex, but the corporate leaders are going to ask one very important question about the WLAN when you present the idea for a project, "What are the business use case and justification"? That question is asked to find out if there is a need this WLAN can fulfill or some use that will make work easier, more reliable, faster, or cheaper. Something has to "justify" the time and investment into the WLAN to make it worth doing. So business stakeholders will ask their project manager to give a presentation that explains the details of the business use of the proposed WLAN project and then provide some forecast of numbers that gives justification for approval. Once the project is approved, and the idea becomes a charter, it is time to go to work. We will first look at the business requirements.

2.1: Introduction to Requirements Engineering

Requirements engineering is an interdisciplinary function that mediates between the domains of the acquirer and supplier to establish and maintain the requirements to be met by the system, software, or service of interest[5]. This statement is a technical way of saying that requirements engineering requires skillsets in and out of the technology domain and is used to match the requirements of a project to the capabilities of a system. Requirements engineering is concerned with discovering, eliciting, developing, analyzing, determining verification methods, validating, communicating, documenting, and managing requirements[6]. A requirement is a statement which translates or expresses a need and its associated constraints and conditions[7].

Rather than provide opinions of individuals that vary greatly on effective requirements engineering, CWNP has chosen to adopt the *ISO/IEC/IEEE 29148 Systems and Software Engineering – Life Cycle Processes – Requirements Engineering (2018)* standard as the model for WLAN and IoT wireless network requirements. This standard is based on nine different standards before it and it represents more than 30 years of thinking related to requirements engineering. Hundreds of expert-level engineers were involved in the development and maturation of these standards over the decades resulting in a concise and effective set of processes.

The standard defines three levels of requirements process:

- Organization Environment
- Business Operation
- System Operation

It also considers the external environment and the impact it may have on a solution. Given that we are focused on a specific system (a WLAN), our focus will remain at the business operation and system operation level. For more information on the other levels, see the relevant sections of the standard[8].

In fact, IEEE 15288-2015, on which 29148-2018 largely depends, defines several processes that should be performed to define the requirements for a system, to transform the requirements into an effective product, to permit consistent reproduction of the product

---

[5] IEEE 29148-2018, clause 4.1.19

[6] Ibid

[7] Ibid (clause 4.1.17)

[8] In addition to the relevant sections of 29148-2018, the reader may find information from IEEE 15288-2015 and IEEE 15289-2019 useful. These standards focus on stakeholder requirements definition and the requirements analysis process as well as documentation in requirements engineering.

where necessary, to use the product to provide required services, to sustain the provision of those services, and to dispose of the product when it is retired from service[9]. These technical processes include:

- **Business or Mission Analysis:** define the business problem or opportunity and the environment in which the solution will operate.
- **Stakeholder Needs and Requirements Definition:** define the stakeholder requirements for the system that will provide the capabilities needed within the environment.
- **System Requirements Definition:** convert stakeholder and user view of the desired capabilities into a technical view of the solution that meets the stakeholder and business requirements.
- **Architecture Definition:** create a system architecture, or system architecture alternatives, from which one or more may be selected in compliance with system requirements to express a set of consistent views.
- **Design Definition:** provide detailed data and information about the system and its components that allows for implementation according to the architecture definition and in compliance with requirements.
- **System Analysis:** analyze the designed system to determine if assumptions are correct and if requirements, including quality and validity, are met.
- **Implementation:** deploy each designed system component by transforming requirements, architecture, and design into actions that create the system.
- **Integration:** synthesize the system components into a complete system through interconnections, data communications, and configuration elements.
- **Verification:** provide evidence that the system fulfills its specified requirements and characteristics.
- **Transition:** using an orderly and define process, move the system into operations while considering other systems impacted.
- **Validation:** provide evidence that the system now in use indeed fulfills the specified requirements in the operational state.
- **Operation:** use the system as intended.
- **Maintenance:** sustain the system to ensure continued capability.

---

[9] IEEE 15288-2015 Systems and Software Engineering – Systems Life Cycle Processes, clause 6.3, Technical Management Processes

- **Disposal:** remove or replace the system while considering loss of information, sensitive configuration data, and other considerations during deactivation[10].

Figure 2.1a illustrates how these concepts fit into the Define, Design, Deploy, and Validate/Optimize model CWNP recommends for WLAN design. The grey areas are those where the WLAN designer is least involved, though the designer is more likely to be involved in portions of the Deploy phase than ongoing Operations. Of course, in smaller organizations, it is not at all uncommon for the same person to be the designer, administrator, troubleshooter, security administrator, and disposal manager.

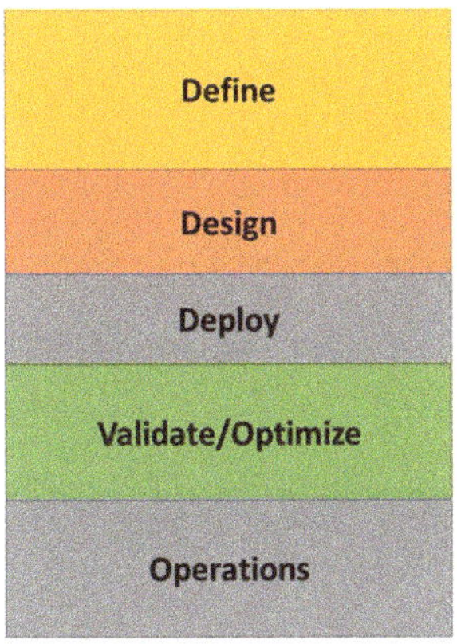

- Business or Mission Analysis
- Stakeholder Needs and Requirements Definition
- System Requirements Definition
- Architecture Definition
- Design Definition
- System Analysis
- Implementation
- Integration
- Verification
- Transition
- Validation
- Operation
- Maintenance
- Disposal

Figure 2.1a: The IEEE 15288-2015 Technical Processes Related to WLAN Design[11]

---

[10] While the Disposal process will not be a major part of this book, it is sometimes the responsibility of the WLAN designer to plan for the removal of the existing system during the deployment.

[11] Any system or process must be applied to a specific case. In the case of WLAN design, particularly for the CWDP certification, the primary focus is on the WLAN itself. While the CWDP may be asked to recommend wired-side solutions, application-based solutions, and other components beyond the WLAN, they are not the primary goal of the certification or this book. The requirements engineering processes defined in 29148-2018 each apply to lesser and greater degrees in WLAN design. For example, while Architecture Definition is considered part of requirements definition and documentation, due to the constraint of designing a WLAN, the number of architectures become limited. Therefore, the task of Architecture Definition may be simpler in some WLAN design projects than others, at times, as simple as selecting from one or two architectures available from the vendor in which the target organization is already locked. The point is to be aware of the processes and use those applicable to a given scenario in the level of depth they demand in that context. Additionally, it is entirely possible that some processes may overlap the four phases we are defining in this book or that they may occur in a different phase on some projects. Figure 2.1a should be taken as a general reference map and not a set of hard rules.

From the business or mission analysis come the business requirements. From the stakeholder identification and needs analysis come the user or stakeholder requirements. With these, the WLAN designer can begin to form system requirements, which are primarily focused on mapping business and used requirements to system features and capabilities.

Before proceeding through the chapter, it would be helpful to define some important terminology related to requirements engineering. These terms are defined within the IEEE 29148-2018 standard and those definitions are used here[12].

- **Requirement:** statement which translates or expresses a need and its associated constraints and conditions.
- **Requirements Elicitation:** process through which the acquirer and the suppliers of a system discover, review, articulate, understand, and document the requirements on the system and the life cycle processes.
- **Requirements Management:** activities that ensure requirements are identified, documented, maintained, communicated, and traced throughout the life cycle of a system, product, or service.
- **Requirements Validation:** confirmation by examination that requirements (individually and as a set) define the right system as intended by the stakeholders.
- **Requirements Verification:** confirmation by examination that requirements (individually and as a set) are well formed.
- **Requirements Traceability Matrix:** table that links requirements to their origin and traces them throughout the project life cycle.
- **Acquirer:** stakeholder that acquires or procures a product or service from a supplier.
- **Supplier:** organization or individual that enters into an agreement with the acquirer for the supply of a product or service.
- **Condition:** measurable qualitative or quantitative attribute that is stipulated for a requirement.
- **Constraint:** externally imposed limitation on system requirements, design, or implementation or on the process used to develop or modify a system.

[12] All terms are defined in clause 4.1 of IEEE 29148-2018.

- **Stakeholder:** individual or organization having a right, share, claim, or interest in a system or in its possession of characteristics that meet their needs and expectations.
- **System-of-Interest:** systems whose life cycle is under consideration in the context of requirements engineering.
- **Trade-Off:** decision-making actions that select from various requirements and alternative solutions on the basis of net benefit to the stakeholders.
- **User:** individual or group that benefits from a system during its utilization.
- **Validation:** confirmation, through the provision of objective evidence, that the requirements for a specific intended use or application have been fulfilled. The right system has been built.
- **Verification:** confirmation, through the provision of objective evidence, that specified requirements have been fulfilled. The system has been built right.
- **Business Requirements Specification:** structured collection of the business requirements.
- **Stakeholder Requirements Specification:** structured collection of the business requirements.
- **System Requirements Specification:** structured collection of the business requirements.

The last three definitions, Business, Stakeholder, and System Requirements Specifications are the outcomes of the requirements engineering process. Additionally, while the 29148-2018 standard makes no direct distinction between requirements engineering and requirements analysis, the distinction is seen with careful reading of the standard. Requirements engineering is the umbrella concept that includes requirements elicitation, requirements analysis, requirements change management, requirements reuse, and requirements quality assessment. Requirements analysis is the process of evaluating requirements to ensure they define the right system (requirements validation) and are well formed (requirements verification). Therefore, requirements definition is not, strictly speaking, requirements analysis. You cannot analyze that which does not exist either logically or literally. Requirements definition is about creating requirements statements from business and user needs. Requirements analysis is about analyzing those requirements to ensure accuracy before one begins designing a system based upon them.

With this brief introduction to standards-based requirements engineering processes, you're ready to begin exploring the processes in greater detail. We will begin with business and user requirements.

2.2: Business and User (Stakeholder) Requirements

Business requirements are the high-level driving force behind a technology project of any can. For WLANs, the technology can help businesses achieve efficiency gains, cost reduction, profit margin increases, and more. The stakeholder and user requirements help the WLAN designer to ensure that the system requirements or technical requirements meet the needs from a technical and capability perspective at a more granular level than the business requirements. We will explore business requirements first and then investigate user and stakeholder requirements.

The Business Requirements

The Business Analysis Process provides the definition of the business problem or opportunity, the solution space or environment in which the solution will operate, and potential solution classes that can solve the problem or take advantage of the opportunity[13]. It is about gaining a clear picture of the business drivers that are behind the desire for a new WLAN or a WLAN upgrade. The business drivers act as the justification or the reason for the project. It results in a business requirements specification in relation to the WLAN design and deployment project.

Business requirements come from two primary sources:

- **Internal business requirements:** These are based on the business objectives and policies of the company. Some of these policies cover things like information security, employee use of resources, and personal device use such as bring your own devices (BYOD)[14].

[13] IEEE 15288-2015, clause 6.4.1.1, Business or Mission Analysis Process, Purpose

[14] These are driven by internal environmental factors. The factors include business objectives, long-term plans, and other environmental factors that may impact the WLAN design.

- **External business requirements:** These are based on meeting and complying with regulations from regulatory bodies, such as federal and local governments and their agencies[15].

Business requirements refers to the concept of eliciting and documenting the requirements of the organization early in the development cycle of a system to guide the design of the future system.

A business requirements specification document should include specific details of what a system is expected to do and what it is going to cost. It should also clarify how the technical objectives (in the infrastructure design) help to reach the business objectives (the company mission). It will not define technical details of the system, which are reserved for the system requirements specification document.

The document should also identify the general characteristics of the solution that will meet customer and business needs, give details on which business needs will be met through this solution, and be informative enough that all stakeholders accept it.

The customer, project sponsor, or other members of the project team will often provide the business requirements. The engineer given the assignment to design this WLAN should ask for a copy of these business requirements so that the design can be built on the requirements and focus on a scalable expansion of the design for future needs. By doing so, the design will accomplish what the customer expects to get out of the WLAN as well as provide for future growth of the wireless infrastructure. When no such requirements are available, if time and budget allow, the WLAN designer should seek a meeting with the appropriate stakeholder(s) to elicit a minimal set of business requirements such as business goal or objective that related to the WLAN project and any known constraints that will impact decisions in the project.

Other things that may be included or accompany the business requirements or as part of them are:

---

[15] These are driven by external environmental factors. Regulatory compliance is discussed in detail in section 2.3 of this chapter.

- **The project charter**: A document that gives official recognition of a project and gives a project manager written approval to begin working on the named project.
- **The project purpose**: The project purpose is the reason an organization or business establishes a project and the primary objective for the project to achieve once completed.
- **The project vision statement**: A document that provides the project owner(s) with a well-written description of the project outcome that will result in positive changes in the business.
- **A Scope of Work (SoW) or Statement of Work**: A written statement listing the work to be performed under a contract to complete a project.

An example of a business requirement would be this:

"The WLAN must provide access within the warehouse for all mobile devices used for inventory management with the same performance levels experienced with the wired terminals currently in use."

Ultimately, the business requirement may be derived from a larger digitalization project or simply from a desire to increase efficiency or some other objective. In this case, the organization wants to move forward in the area of mobility, but not backward in the area of performance (data throughput). Modern applications are far more throughput-intensive, and it is almost always unacceptable to sacrifice data throughput to expand network functionality.

Part of business requirements analysis is the gathering of business use cases. A business use case, from the perspective of a WLAN designer, is a description of a scenario or group of actions that will include the WLAN system. For example, inventory personnel using Wi-Fi barcode scanners to add and remove items from inventory is a business use case. This use case does not immediately reveal the technical requirements, but it tells you where to look to find them.

29148-2018[16] provides a template for the Business Requirements Specification document. Such a document may be provided to you, or you may have to solicit the portions of

[16] IEEE 29148-2018 is only 98 pages including the cover page and end pages. It's an easy one hour read and many who have done so have completely redesigned their requirements engineering processes. It is well worth your time to go beyond the portions covered here and explore the entire

greatest importance to your project, if it is not provided. Figure 2.1b shows the outline of the document. Each section is defined in detail within the standard and you are encouraged to read the standard to learn more.

1. Introduction
- 1.1 Business purpose
- 1.2 Business scope
- 1.3 Business overview
- 1.4 Definitions
- 1.5 Major stakeholders

2. References

3. Business management requirements
- 3.1 Business environment
- 3.2 Mission, goals, and objectives
- 3.3 Business model
- 3.4 Information environment

4. Business operational requirements
- 4.1 Business processes
- 4.2 Business operational policies and rules
- 4.3 Business operational constraints
- 4.4 Business operational modes
- 4.5 Business operational quality
- 4.6 Business structure

5. Preliminary operational concept of proposed system
- 5.1 Preliminary operational concept
- 5.2 Preliminary operational scenarios

6. Other preliminary life-cycle concepts
- 6.1 Preliminary acquisition concept
- 6.2 Preliminary deployment concept
- 6.3 Preliminary support concept
- 6.4 Preliminary retirement concept

7 Project Constraints

8. Appendix
- 8.1 Acronyms and abbreviations

Figure 2.1b: Business Requirements Specification (BRS) Template

---

specification and additionally consider IEEE 15288-2015 and IEEE 15289-2019 for further clarification on processes and documentation.

This, or a similar, document should be provided to you or created and agreed upon by the acquirer. This acquirer is the decision maker, the one with authority to greenlight the project.

The User or Stakeholder Requirements

The stakeholder or user requirements are the capabilities that help users achieve business or organizational requirements and assist stakeholders in achieving their goals. User requirements should include the goals or tasks that users must perform when using the WLAN[17]. Stakeholder requirements should ensure that managers, department leaders, outside firms, and others with a stake in the project have their needs considered in the system as well.

All users are stakeholders but not all stakeholders are users[18]. However, organizations will often group the stakeholder and user requirements together because users are a set of the group stakeholders.

The first step in this process is to identify and communicate with the stakeholders. The acquirer may be able to provide an excellent starting list of managers, outside individuals, and users that can fill the stakeholder role for your requirements elicitation.

---

[17] These requirements can be elicited from a representative group of users rather than requiring that all users be individually interviewed or otherwise involved in requirements definition. It is often useful to have a mixture of long-term users and short-term users involved. Long-term users bring expertise to the process. Short-term users bring a fresh perspective to the project. Additionally, long-term users are sometimes known to overlook many tasks they perform as they are on autopilot during their work. Short-term users are less likely to fall prone to this.

[18] The book *Project Management, Planning and Control, Albert Lester, 2017* defines a stakeholder as a person or organization who has a vested interest in the project. This interest can be positive or negative. *A Guide to the Project Management Body of Knowledge (PMBOK Guide), Seventh Edition, 2021* defines it as an individual, group, or organization that may affect, be affected by, or perceive itself to be affected by a decision, activity, or outcome of a project, program, or portfolio. IEEE 29148-2018 defines it as an individual or organization having a right, share, claim or interest in a system or in its possession of characteristics that meet their needs and expectations. It goes on to say that stakeholders include, but are not limited to, end users, end user organizations, supporters, developers, producers, trainers, maintainers, disposers, acquirers, customers, operators, supplier organizations, accreditors, and regulatory bodies. So, all users are stakeholders but not all stakeholders are users.

From these stakeholders, you can identify characteristics of the system, contexts of use, constraints, and the stakeholder and, specifically, user needs.

Here are some examples of user requirements:

- Printing from mobile devices anywhere on site
- Transferring large files quickly from anywhere in the facility across the WLAN
- The ability to utilize VoIP phones throughout the facility
- Access to all the same resources on mobile devices as they do on desktop PCs

Here are some examples of stakeholder requirements:

- Privacy standards must be met according to company policy
- Security standards must be met according to company policy
- Outside vendors' vehicles should be able to connect to the system when on location

You will be more likely to meet the requirements of the users if you gather specifics rather than vague user requirements. Be sure to ask enough questions during the Define stage to ensure that you have a grasp of what is required for user performance on the WLAN. If the answer isn't detailed enough for complete understanding, ask for clarification. You may need to ask several questions before getting a clear view of what is needed to fulfill the requirement. Here is a real-world example:

- Stakeholder: "We want to be able to use VoIP technology on the Wi-Fi here."
- You: "What exactly do you have in mind when you say, 'VoIP Technology'?"
- Stakeholder: "You know, we want to talk using the Wi-Fi instead of having to use desktop phones or run up our cell phone bill."
- You: "So are you planning on using wireless handsets, or do you plan on utilizing Wi-Fi calling on your cell phones?" After asking this question, you may get sidetracked into other questions if the customer wasn't aware that one or the other options existed.
- Stakeholder: "We want to use wireless handsets with phone numbers and extensions just like our desktop phones have."
- When you get this answer, it is common to make suggestions about products at this point, and you will have to explain the technology and differences in each option.
- You: "Do you have a brand or specific phone in mind?"

- Stakeholder: "Yes. We want to use Brand X for our phones because we heard this product works with our existing phone structure."
- You: "Great. Brand X has two models. Have you picked out one or the other?"
- Stakeholder: "What's the difference? Model A was $200 cheaper and all we saw was that the color was different."
- You: "Model A has slower data rates and less resiliency to interference while only supporting the older versions of Wi-Fi and model B has higher speeds and more resiliency to interference while supporting newer versions of Wi-Fi. You are more likely to get longer use out of model B and I can explain the technical details to you, if you desire."

At this point you are going to sound as if you are a salesperson for the phone itself, but it will be wise to explain why the customer should invest in model B instead of model A. After you explain the benefits of the better phone, the customer may still pick Model A because of the price tag, and now you will be facing a design requirement for that phone.

Other questions will arise as you go through the requirements elicitation process, and some of the requirements may entail a need for other requirements, so be sure to research the equipment and user devices chosen by the customer before finalizing anything. You may also run into a conflict with your design when, for example, the customer tells you they already have 250 VoIP handsets and are just waiting on the WLAN to be installed so they can use them. One requirement may require you to install an 802.11ax WLAN to fulfill the requirements of the customer, and then you find out the 250 phones they already have will only function on 802.11abg networks. This will require you to be skillful in your design process and network selections and setup.

During this stakeholder requirements definition process, you should also define critical performance needs. At this stage, you will not define how you will meet them, but will define what must be met.

In addition to the elicitation of requirements from the business and stakeholders, at some point throughout these first two process of requirements engineering, you will use your expertise to identify *enabling systems*. These systems are those that support the development of a new system during its development but do not necessarily contribute directly to its ongoing operations. The IT stakeholders are most likely to be the source of enabling system identification. For example, they may require that you provide design plans in a particular format (PDF, network diagramming software, etc.) or even that you use a specific WLAN design tool (iBwave, AirMagnet, ekahau, TamoSoft, etc.). Such a requirement should be identified, if it exists. If the acquirer is expecting the design documents to be delivered in a specific format, including design tool file types, and you are unaware, it can create serious problems or delays as you transition to deployment.

1. Introduction

- 1.1 Stakeholder purpose
- 1.2 Stakeholder scope
- 1.3 Overview
- 1.4 Definitions
- 1.5 Stakeholders

2. References

3. Business management requirements

- 3.1 Business environment
- 3.2 Mission, goals, and objective
- 3.3 Business model
- 3.4 Information environment

4. System operational requirements

- 4.1 System processes
- 4.2 System operational policies and rules
- 4.3 System operational constraints
- 4.4 System operational modes and states

5. User requirements

6. Detailed Life-cycle concepts of proposed system

- 6.1 Operational concept
- 6.2 Operational scenarios
- 6.3 Acquisition concept
- 6.4 Deployment concept
- 6.5 Support concept
- 6.6 Retirement concept

7 Project Constraints

8. Appendix

- 8.1 Acronyms and abbreviations

Figure 2.1c: Stakeholder Requirements Specification (StRS) Document Outline

Finally, a WLAN implemented for one or a set of purposes often reveals additional use cases; therefore, in addition to the current needs, you must plan for future growth, so having a conversation with different departments in the company may reveal new projects that are planned. These projects could demand more throughput from the WLAN than is required for your current design in progress. Figure 2.1c shows the recommended outline for a Stakeholder Requirements Specification (StRS) document based on the 29148-2018 standard.

2.3: Regulatory Compliance

Regulatory compliance is a restraint or restriction imposed on an organization by governing bodies and parent companies. It is an external environmental factor as it is imposed on the organization by external regulations or policies. These restrictions enforce specific laws and regulations that have some form of authority over certain areas of business operation. In the case of a company needing a WLAN, in the United States there are the FCC regulations and the FAA regulations (if applicable) that you must comply with. With the FCC there are areas of the spectrum(s) not available for use. There are also power output restrictions on wireless transmitters based on the channel and band being used. The EIRP (Effective Isotropic Radiated Power) of a wireless transmitter is limited by the FCC. You must understand how to convert between mW and dB so that you can calculate the correct EIRP based on the transmit power of your radio, the dB loss in the antenna cable you are using, a lightning arrestor (if applicable), and the antenna.

```
EIRP = transmit strength + antenna gain - cable loss
```

The FCC defines power limitations for WLANs in FCC Part 15.247.

The FAA (Federal Aviation Administration) places restrictions on frequencies used in and around an airport. The design of any wireless bridge that must propagate a signal near an airport must comply with FAA restrictions placed on Wi-Fi spectrums. Therefore, WLANs have DCA (Dynamic Channel Assignment) options to prevent frequencies within the WLAN from interfering with aviation radar frequencies and DFS (Dynamic Frequency Selection) to detect radar events on a channel and move from that channel.

You will find more on FAA policies and regulations through the following link: cwnp.link/faapolicy

Other regulating bodies around the world have varying policies and restrictions. Depending on the country you are working in, you will encounter similar regulations and policies to which you will have to adhere to perform the work.

Be sure to contact the appropriate governing body for clarity about policies and regulations in the event you are not sure.

2.4: Industry Compliance

In the wireless industry, there are several areas where equipment and supporting infrastructure should comply with standards set to ensure product interoperability with other vendor products. Industry compliance is also an external environmental factor, though they are often enforced through pressure rather than law. For example, if you do not use certified products, you have no assurance of interoperability. If you do not comply with industry best practices, litigation cases against the organization can be stronger.

The most commonly known certification body related to Wi-Fi hardware is the Wi-Fi Alliance. When a wireless product possesses the seal of approval from the Wi-Fi Alliance, you know it has undergone a series of independent tests in multiple configurations to ensure that it has interoperability with other Wi-Fi-certified products and has been given the right to display the logo shown in Figure 2.1d.

Figure 2.1d: Wi-Fi Alliance Wi-Fi Certified Logo

Another compliance area involving a WLAN is the supporting infrastructure. PoE (Power over Ethernet) is a well-known necessity in the world of wireless products. Many wireless APs today require 30 watts of power to function properly. Because of this, it is important to ensure that all supporting category cabling complies with industry standards. BICSI (Building Industry Consulting Service International) plays a big role in the development of current standards for structure cabling, such as the ANSI/TIA – 568 standards. These standards address several important installation procedures for infrastructure cabling and distance. The maximum allowed distance for any category cabling in 90 meters of horizontal cabling, with a total of 100 meters after patch panels and patch cords have been installed.

One very important reason for WLANs to adhere to these standards is the PoE factor. Longer distances of wiring impose more resistance, causing a voltage drop. The longer the cable run, the better chance there is of having insufficient power to run the AP plugged in at the other end. You should always adhere to the 100-meter total distance standard to provide sufficient PoE from the switch to the AP.

Another example industry standard is Payment Card Industry Data Security Standard (PCI-DSS)[19] for payment card processing. In some countries, it is mandatory. In others, it is not. However, even where it is not mandatory by law, pressure to comply with it from payment card companies is significant and most retail outlets choose to comply with it or simply use systems that are already in compliance. Figure 2.1e shows the high-level overview of the PCI-DSS standard requirements[20].

Build and Maintain a Secure Network and Systems	1. Install and maintain a firewall configuration to protect cardholder data 2. Do not use vendor-supplied defaults for system passwords and other security parameters
Protect Cardholder Data	3. Protect stored cardholder data 4. Encrypt transmission of cardholder data across open, public networks
Maintain a Vulnerability Management Program	5. Protect all systems against malware and regularly update anti-virus software or programs 6. Develop and maintain secure systems and applications
Implement Strong Access Control Measures	7. Restrict access to cardholder data by business need to know 8. Identify and authenticate access to system components 9. Restrict physical access to cardholder data
Regularly Monitor and Test Networks	10. Track and monitor all access to network resources and cardholder data 11. Regularly test security systems and processes
Maintain an Information Security Policy	12. Maintain a policy that addresses information security for all personnel

Figure 2.1e: PCI Data Security Standard – High-Level Overview

---

[19] www.pcisecuritystandards.org

[20] Payment Card Industry (PCI) Data Security Standard, Requirements and Security Assessment Procedures, Version 3.2.1, May 2018

The specific requirements within PCI-DSS for WLANs are found in requirement 1.2.3, 2.1.1, and 4.1.1. They state the following:

- **1.2.3:** Install perimeter firewalls between all wireless networks and the cardholder data environment, and configure these firewalls to deny or, if traffic is necessary for business purposes, permit only authorized traffic between the wireless environment and the cardholder data environment.
- **2.1.1:** For wireless environments connected to the cardholder data environment or transmitting cardholder data, change ALL wireless vendor defaults at installation, including but not limited to default wireless encryption keys, passwords, and SNMP community strings.
- **4.1.1:** Ensure wireless networks transmitting cardholder data or connected to the cardholder data environment, use industry best practices to implement strong encryption for authentication and transmission.

The PCI-DSS standard also suggests testing procedures and provides guidance for each of these policies.

Because CWNP is an international professional wireless certification organization, we do not cover nor test on governmental regulations; however, as a WLAN designer, you should become familiar with such laws and industry best practices in your region and ensure that your designs are in compliance with them. This includes radio spectrum regulations, financial regulations, security regulations, safety regulations, and possibly more.

2.5: Collecting, Defining, and Using Technical Requirements

Technical requirements fall under the category of meeting the technical objectives of a project. Remember, the technical objectives are intended to build and maintain an IT infrastructure that is designed to help reach the business objectives. Working backward from the business objectives to the technical requirements, we get the following:

- **Company Mission or business objective** is to provide A–Z service to its customers.
  - **Need**: An IT network for the business that will support the business objective to provide A–Z service to customers.

- **Company Technical objective** is to provide the IT network that will adequately help the business reach the business objective.
    - **Need**: Business technical analysis that will gather enough intelligence, so the proper IT network can be designed.
- **Company Technical requirements** are a combination of the information gathered from the technical analysis and vendor research to find the best IT products to fulfill the technical objective.
    - **Need**: Professionals that are capable of designing, configuring, and implementing the IT network based on the objectives and requirements.

This is where the CWDP comes into play. As a seasoned professional, gathering and/or defining the technical requirements, and then designing and deploying the WLAN from those requirements is your responsibility. More times than not, the customer will not know the technical requirements, and you will have to collect or provide this information yourself. The technical requirements are derived from the business and stakeholder requirements. This section explains some of the common technologies and technical requirements that may be considered in this process.

Location Services

Using the Apple iPhone's Location Services in your WLAN has some specific requirements that need to be addressed to function properly. One main difference between a Location Services design and other designs for voice and data is that APs are required in the corners and possibly along the outlying walls of the building. This provides a more accurate location of the device or tag you are trying to locate using the WLAN. In smaller structures, placing APs along the outlying walls may not be necessary, and this depends on the RF propagation levels within the building that are measured during the survey. One thing to note is that the more APs you have in the area, the more accurate and precise the location of devices can be.

Location Services works better with proper placement of the APs, so the survey must be performed in a manner that is best for Location Services. AP placement should be designed so that a minimum of three or preferably four APs (in some cases five are required for accuracy) can detect and report the received signal strength of the device you are locating. These APs should detect the device's signal strength at -75 dBm or higher for best results.

This signal and AP requirement mean that the design cannot be engineered in the same manner as voice and data WLANs, because most of these designs do not have multiple APs overlapping at -75 dBm or greater throughout the facility. Instead, it is common practice to have a 25 percent overlap at the cell edges at -67 dBm for proper coverage and roaming capabilities, but this doesn't always provide multiple AP coverage throughout as Location Services demands. In some cases, WLANs are designed for Location Services with several of the design's APs in monitor mode. These APs are there only to assist in location of devices and will not service any clients for network connectivity. This approach is used for multiple reasons, including lowering the possibility of co-channel interference and providing more precise accuracy in pinpointing devices within the coverage area. The more radios you have reporting the device, the better the accuracy is in location placement.

How does it work? Location Services, also known as RTLS, work off several technologies that allow the RF to have more precise location abilities. One is *Received Signal Strength* (RSS). Because signals weaken over distance, the RSS can be used to determine the location of a transmitting client device. Here the transmit power, antenna gain, and path loss (based on the frequency being used) are all values known to the system. Path loss is determined by measuring the transmit signal of the AP or client device and then subtracting the received signal to get the path loss value. Estimated location calculations are then determined by the points of intersection of the RF cells from multiple APs and their estimated distances to the client device as shown in Figure 2.2.

In Figure 2.2, you see that the green AP is detecting the laptop's signal strength at a strong -43 dBm while the others are at a -68 dBm or lower. Another simple technique for locating the laptop is called *cell of origin*. The cell of origin can also be called the *associated AP*. It is the RF cell the laptop is closest to and in most cases the AP the laptop is associated with.

At times when proper surveys and designs are not in place, along with appropriate RF profiles, the roaming laptop can become a sticky client and roam past two or more APs while holding its connection to its original associated AP and cell of origin is not as dependable for Location Services.

Another technique used in RTLS is *time of arrival* (ToA). Time of arrival is a distance-based technique and requires precise synchronization of time from all APs, including the device being located. Because four of these APs have the device within -75dBm signal strength or better, all four of them can use ToA to calculate the location of the laptop. The distance between sender and receiver is found by using the time of the propagated signal transmission and the velocity of that signal. All four APs use the synchronized time to do the calculations and now have a distance between themselves and the laptop. This distance is plotted in intersecting lines, similar to what is displayed in Figure 2.2, and the intersection will mark the location of the laptop.

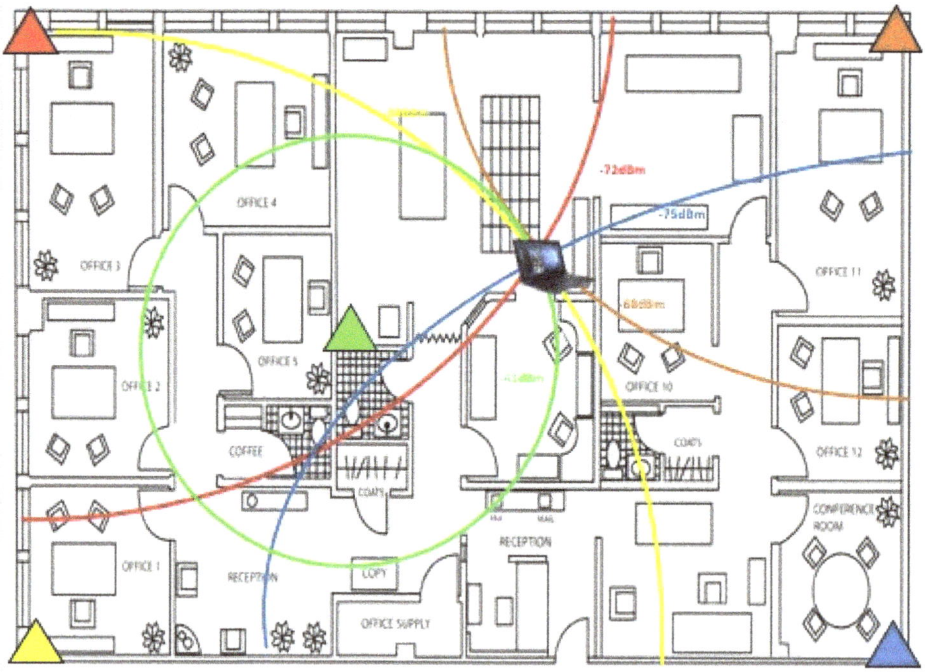

Figure 2.2: Estimated Location Calculations

The slightest discrepancy in time synchronization between the locating APs and laptop can result in large miscalculations in the location. For example, location calculations can be off by thirty meters or more with only a 100-nanosecond mismatch in time synchronization. Multipath, interferers, and noise can also create serious challenges to ToA location abilities.

RTLS can also utilize *Time Difference of Arrival* (TDoA) to help calculate a more accurate location of the laptop within the area surrounded by the four APs. TDoA uses relative time measurements at each receiving AP instead of using absolute time measurements. Because of this, synchronized time is not required from the laptop at the time of its transmission. This takes place when a signal, with an unknown start time, is transmitted from a source (the laptop), that signal is received from three or more APs (with synchronized time), and then shifted in time to find a position or alignment, then the time of each received signal is multiplied by the speed of light to determine the distance the laptop is located from each of the APs. These distances are now plotted in hyperbolic lines, similar to what you see in Figure 2.2, and the intersection of these lines indicates the location of the laptop. This whole process is based on a mathematical concept known as *hyperbolic lateration.*

One more technique used to increase the accuracy of an RTLS is *angle of arrival* (AoA). In using AoA, a single AP functions somewhat like a human's hearing ability. Using two or more antennas, the AP utilizes the spatial separation between the antennas to determine the arrival time of the signal. The human ear performs the same way. We hear a sound, and the time difference of arrival of that sound is used to determine what direction it came from; then we immediately turn our heads to look in the direction of the sound. AoA can greatly assist Location Services in achieving better accuracy. The concept of AoA is represented in Figure 2.3.

As you see, angle-of-arrival calculations can determine orientation within a few degrees just as the human hearing ability can. What allows us to find the object making the sound is the use of other senses such as our eyesight, which is what we immediately use after we turn our heads to look toward the sound.

In RTLS, designs commonly use multiple techniques to achieve optimal performance. For example:

- Ranging the laptop with received signal strength (RSS) helps in finding the distance from each of the four APs.
- Time of arrival (ToA) calculates the distance between transmitter and receiver.
- Time difference of arrival (TDoA) also provides accurate calculations of distance between transmitter and receiver within the area.
- Angle of arrival (AoA) utilizes spatial separation between antennas on each AP to calculate the angle from which the transmitted signal came.

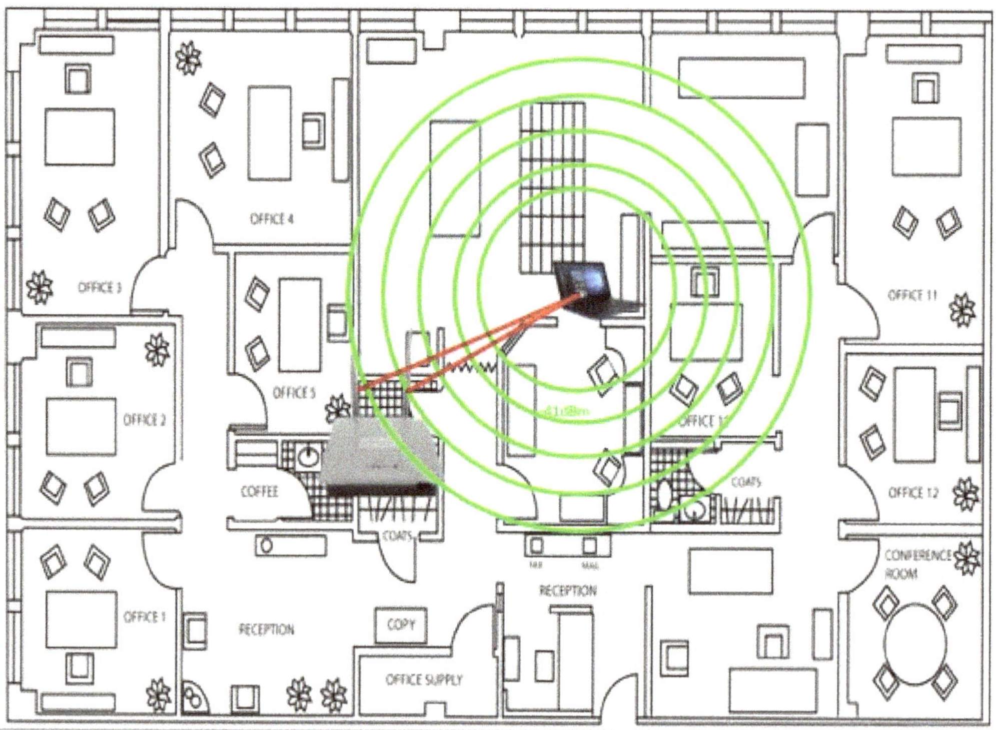

Figure 2.3: Angle of Arrival (AoA) Location Services

By utilizing two or more of these techniques in your design, RTLS accuracy will become more precise.

Bluetooth Low Energy (BLE) allows battery powered transmitters to work more efficiently for longer periods. This protocol is much more efficient at saving energy than other Bluetooth protocols. The battery life will depend on which vendor manufactures the product[21].

The location of devices using BLE is determined by the placement of several of the Bluetooth Low Energy Beacons in an area so that multiple beacons can detect a single device and provide a location within 1–2 meters. WLAN vendors are placing BLE

[21] Some wireless vendors now provide BLE integrated with their WLAN APs. This allows for Wi-Fi-based location algorithms and BLE location algorithms to be used and supports the modern movement of BLE use for IoT locationing.

beacons in new APs to provide the location ability found in Bluetooth Low Energy. Other uses for BLE include proximity tracking of wireless devices by retail stores to send sale price notifications to potential customers. BLE can be used to monitor and track activity. This is becoming popular with healthcare facilities and in-home health care.

Latency Requirements

Latency is the delay between the input time of a packet and the time that packet reaches its destination or it is the round-trip-time (RTT) for a higher layer communication to reach the destination and the acknowledgement to be received by the original transmitter. Controlling latency within a WLAN depends heavily on several things within the network:

- Routing and switching
- Applications used
- Number of users
- Volume of data transmitted and received by users
- Devices used by users
- Cabling and fiber infrastructure
- QoS and configurations
- WLAN design

There are many elements of a network that can cause latency in data transmissions, and each one must be considered when you know you are going to be using latency-sensitive applications. Voice and video applications are most commonly affected by latency. For voice applications, it is best to have latency at or below 150ms from mouth to ear and less than 1% packet loss. Video applications can have up to five seconds of latency and 2% packet loss, depending on buffering capabilities; however, real-time video communications are closer to the voice requirements with the exception of packet loss. Best case for voice is to get latency at or below 50ms, and best case for video is to keep latency below one or two seconds. How is this achieved? The answer to this question has multiple parts.

First, all the Layer 1 infrastructure needs to be designed correctly. The practice of daisy-chaining network switches from one to the other is a path to high latency in the network. Following a best-practices solution of having a distribution switch through which all access switches connect to the network is a great start. Next, all the infrastructure cabling should be installed in compliance with ANSI/TIA/EIA 568 standards.

Second, an important factor in avoiding latency is proper QoS settings across the network – end-to-end. Setting voice applications at highest priority will ensure that all voice packets will be delivered fastest, while video and data packets follow, respectively. Even with all of those elements in place, there still is the consideration of applications, number of users, traffic, and device types. The WLAN will be part of the network infrastructure, and when designing it, AP limitations come into play. A simple example of this would be an AP that can provide 400 Mbps placed in an office area where twenty-five employees will be constantly utilizing the WLAN to email CAD drawings, place VoIP calls to customers, and watch training videos for their safety awareness classes. This scenario is extreme, but it is not unrealistic. Similar scenarios exist in businesses everywhere, and all twenty-five employees want to throw their laptops or tablets out the window because of the poor performance of the WLAN.

The WLAN designer has to take into account all of the applications, users, and throughput demands within each given area (requirement area) to provide a design sufficient to avoid high-latency nightmares like the one just described. Ask the questions about applications, do research on buffering capability, and study the vendor's product you intend to use. Apply all your findings to the design decisions to avoid latency issues. Finally, be sure to configure QoS priorities from end-to-end to provide the best performance.

Signal Strength Requirements

In WLANs, *signal strength* can be explained as the magnitude of the transmitted signal measured at a distance or at a specific reference point out from the transmitter. This measurement, referred to as the *receive signal strength indicator (RSSI)* colloquially (although RSSI has a different meaning in the 802.11 standard), is in dBm (decibels as related to a milliwatt) and is shown as a negative number such as –67 dBm. Along with the received signal strength, there is another factor that has to be considered, and that is called the noise floor. The *noise floor* of a WLAN is made up of those signals and incidentally radiated energy that cause constant interference on the transmitted signal's channel.

These interfering signals and energy are not always a Wi-Fi related signal or a signal at all. At times, the interferer is electrical such as EMI (electrical magnetic interference). Whatever it turns out to be, the interference shows up as a constant dBm level on the channel you are using. That constant interference is the noise floor, and it, too, is measured in dBm. To find the accurate, usable signal in a WLAN, you must find the difference in dBm between the RSSI and the noise floor. This number is always shown as a positive number value and is referred to as the signal to noise ratio (SNR); it is found by determining the difference between the noise floor and the RSSI. This value is explained as follows:

Noise floor	-93 dBm
RSSI	-67dBm
SNR	26dB

The difference in numerical value between 93 and 67 is 26. These three numbers help to determine if RF signals will be efficient for voice and data traffic. They are also used in troubleshooting packet loss and retries. If you find an area of the WLAN has a noise floor value that is close to the RSSI value, then you know the SNR will be low with the possibility of packet loss and retries in that area. The higher the SNR, the better the throughput and RF quality[22].

Coverage Requirements

Coverage requirements define the physical or geographical spaces that must have WLAN access. They are often best determined using a floor plan or map of the facility or area. This section provides guidance on planning for coverage in common, industrial, outdoor, service, and guest areas. It is not a complicated process, as you are simply identifying the

[22] As a best practice, design for SNR and not signal strength. While you may indeed design for signal strength in the end, it is only after you have determined the noise floor that you can really do this. If you blindly design for a signal strength of -65 dBm everywhere, but you do not consider that the noise floor is -82 on the plant floor and -93 in the offices, you will not really be designing to the same specification. Indeed, you may not be able to design for a minimum SNR of 28 dB, because it may not be realistic in some spaces. But this reality should be discovered and communicated to the acquirer so that proper decisions can be made. For example, the organization may choose to implement all 6 GHz 802.11ax devices on the plant floor because no significant interference or high noise floor is detected in that band within that physical location.

locations that require coverage. In the Design phase, you must ensure that adequate coverage for the desired services and applications is provided in the design.

Common areas include the following locations:

- Workspace areas
- Hallways
- Lunch/break rooms
- Conference rooms
- Restrooms
- Stairwells
- Elevators
- Hotel rooms
- Lobbies
- Storage areas

Industrial areas are usually areas that have large equipment, shelving, and other items that cause unusual RF propagation results. These areas could be manufacturing floors, assembly lines, or even power generation facilities. Understanding the objects in the area and how they affect RF propagation is part of a successful survey.

Outdoor areas can typically be covered more easily than indoor areas because there are few if any obstacles causing attenuation of the signal.

The facility itself will have a significant impact on the design. Several factors must be considered, including these:

- Building materials
- Square footage
- Ceiling heights
- Multi-floor structures
- Wiring and power limitations
- Blueprints (floorplans)
- Facility access

Part of gathering requirements is learning about the facility where you will perform and then deploy the design. The more you know about the structure and building materials, the better prepared you will be to perform the site survey. Older buildings with thick block and concrete walls greatly reduce the range of wireless signals, leaving no option but to place an AP in each room. Metal structures can cause an increase in signal reflections, resulting in little or no RF penetration through the walls. In the end, knowing the building materials can help in the requirements analysis by revealing the impact the materials will have on the number of required APs in the design.

The square footage of a facility is an important factor because as RF signals travel away from the transmitter, they experience a natural reduction in signal called *free space path loss*. This factor sets limit to how far a signal can travel and still be useful. Knowing the total square footage to be covered, and the type of building materials the facility is built with, will help in estimating the number of required APs needed based on coverage and capacity demands.

Ceiling heights also become a factor when selecting specific APs, as higher ceilings impact RF coverage footprints. This challenge can require you to include AP models with external antennas, so that you can select among the variations of antenna models for the best coverage from the higher mounting positions.

Capacity Requirements

WLAN capacity requirements tell us how many user devices will be in a specified area utilizing the WLAN and what applications will be demanding throughput. The design of the WLAN is approached differently when a higher capacity of user devices and applications is required.

For example, if you have an area of space in a building that is 50ft × 50ft and you want to install Wi-Fi in this area, it's safe to say that a single AP will more than cover this. If you are going to perform a wireless site survey with only RF coverage in mind, this space will require one AP.

As we see in Figure 2.4, by just providing RF coverage for this area, a single AP can provide the desired coverage. If this same area is an open office with 40 to 50 employees who each have a laptop, smartphone, a tablet and a wireless VoIP phone, as in Figure 2.5, then you are looking at that single AP potentially servicing 200 wireless devices.

This scenario will not provide these employees with adequate throughput, and the WLAN will get a bad reputation because of poor foresight in the design process.

Capacity requirements are calculated by the design engineer based on the client device count, applications running on the client devices, and the size of the area in which these devices will or now reside. Obtaining these calculations will require the design engineer to research APs, client devices, applications and their amount of use.

Example: The fifty users tell you that they are regularly transferring files that are an average of 8 - 10 MB in size, and this traffic is all taking place internally. Also, they have noted that this type of file activity happens regularly throughout the work day, so it is necessary to have each file move very quickly and off the network so that the next one can be sent. An average of 1-minute transmission times, from start to finish, per file would be an acceptable rate. With this added information, you estimate 40 to 50 percent of the employees will be actively moving large data across the WLAN at any given time. This results in the following equation:

```
10MB × (50×.5) = 250 MB per minute

(file size) × (total users × 50%) = transfer rate
```

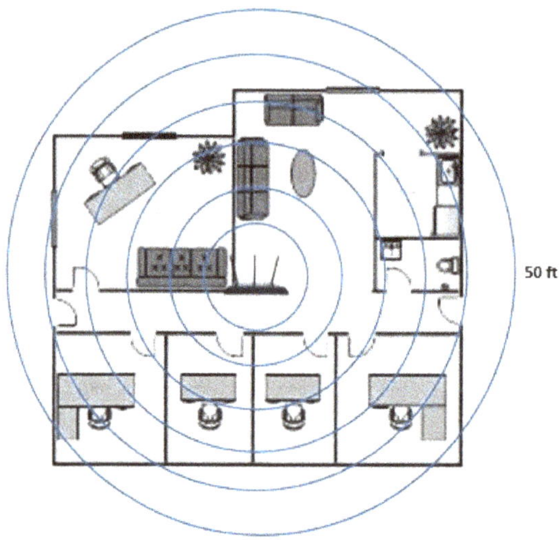

Figure 2.4: A single AP can provide the coverage

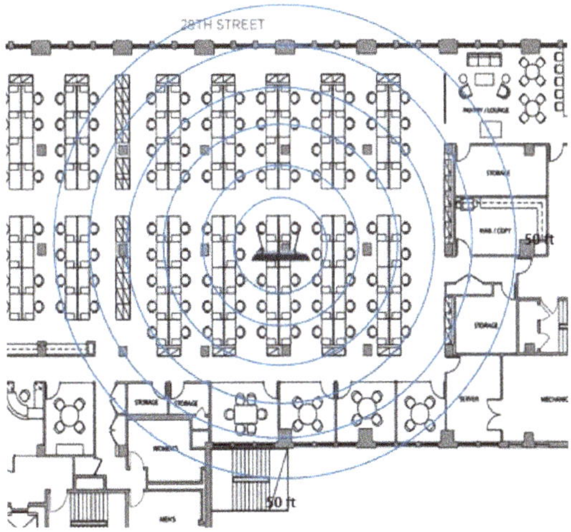

Figure 2.5: A single AP may not be able to provide the needed capacity

Now that you know the potential of regular data flow, you can design the WLAN to accommodate a 250 MB per minute demand. You should also buffer this for other network traffic not included in the specifics above for things like email use, smaller file transfers, and regular Web browsing. This can be over-engineered slightly to accommodate future growth of employee and mobile device count. It would be safe to target a 300MB per minute transfer rate for this, and now the design can begin. With 802.11ac standards claiming 1.3 gigabits per second or 166 MB per second, you can convert this number to a per-minute rate by multiplying it by sixty. 166MB per second × 60 = 9,960 MB per minute. It is important to mention here that these data rate theories are just that; theories. Different rates exist in AP models, deployment environments, and user demands.

You should not assume that a single 802.11ac AP will provide 9,960 MB per minute just because white papers and standards mention these calculations. These are perfect-world/perfect-scenario assumptions and do not represent the real-world findings. You should familiarize yourself with a product and perform tests on that product to find actual performance rates in similar environments before designing for capacity demands. In other words, it is a good practice to order the model of AP you intend to use for designs in plenty of time to learn the product's performance details. Also, other wired LAN traffic can alter network performance, causing network congestion and reduction of

WLAN performance. These factors must be considered in your equation to get more accurate calculations for everyday use.

Once you have a grasp on capacity and how to calculate for it, there are a few ways to achieve the capacity requirements:

- ***Co-existing APs***: When the need for more throughput than a single AP can provide becomes obvious, placing multiple APs in the same area, on different channels, will provide the increased throughput so that capacity requirements can be met.
- ***Smaller coverage cells***: When a large amount of throughput is needed in a congested facility, adjusting the design to more APs with lower power settings will produce smaller coverage cells, allowing client devices to connect within closer proximity of the AP. This spreads client devices out over more APs while providing fewer client connections per AP and higher data rates per client.
- ***Utilizing higher throughput protocols:*** 802.11 protocols have evolved over the last few years, and the biggest reason for this is capacity and throughput. One benefit from these protocols is the use of 802.11ac APs, even if your client devices are only capable of 802.11n. The 802.11ac protocol provides improved connectivity and performance for 802.11n devices.

When designing for capacity requirements, you should always survey the environment by performing spectrum analysis first to find, remove, or design around potential interferers that can reduce WLAN performance in critical areas.

Security Requirements

Security is a crucial component of any WLAN design. When designing security solutions, a CWDP considers all avenues of performance as well as threats. Important details such as fast secure roaming, network monitoring, Bring Your Own Device (BYOD)/Mobile Device Management (MDM), and guest access must be implemented properly to prevent attackers from infiltrating the network through one of many attack options.

Selecting the right authentication and encryption solution can be achieved from one of two perspectives—hard requirements or client support. When the selection is based on hard requirements, the organization dictates specific authentication and encryption that must be supported by devices attempting to join the network or the client device will not be allowed on the network. This mode is often enforced using Network Access Control (NAC) solutions. For example, many government implementations have minimum security requirements such as the DoD Instruction 8420, which requires:

- EAP-TLS support
- WPA2-Enterprise certification
- Implemented encryption must comply with FIPS-140

This is just one example, but it is an important example to consider. Additional examples include healthcare and payment processing systems. In environments like these, where security of communications is mission-critical, it is important to have hard requirements and only allow clients onto the network that meet these requirements.

In less secure environments, it is common to select a security solution that is based mainly on client support. For example, a WLAN might be designed around a product that can support only WPA-Personal and not WPA2-Personal or Enterprise. In this scenario, the client device drives the security solution. However, even in such cases, the security should at least meet some minimally accepted standard. For example, the WPA-Personal passphrase should be complex and longer (15–20 characters) to avoid simple attack methods.

Fast Secure Roaming

When VoIP or other highly mobile applications are utilized, roaming becomes a very important factor. In early WLANs, because roaming took significant time and could even result in a change of IP configuration, many environments were implemented with all wireless APs and their clients on the same VLANs, and no security or only passphrase-based security was used. As more wireless clients came onto the networks, it was obvious that a single large IP domain wouldn't work well. Additionally, early passphrase-based security solutions (such as WEP) proved to be very weak. Today, fast, secure roaming can be implemented in several ways, but two primary methods are used—proprietary and standards-based.

Opportunistic key caching (OKC) is a method used by many vendors but implemented with a bit of mystery behind the scenes. The 802.11 standard now has secure roaming defined within it, and the standards methods is becoming more and more common in modern WLAN hardware. The 802.11r and 802.11k amendments made it possible to begin implementing secure roaming in a standards-based method. The details of the roaming are not essential for a WLAN designer to understand, but one should be aware that three basic options are available for implementing secure roaming, depending on client requirements:

- OKC using proprietary solutions that may not work well with all devices
- 802.11-2012 fast secure roaming that may not yet be supported with all devices
- WPA-Personal or WPA2-Personal, which has the widest support but increases management overhead

It is important to note that an additional option is available with some vendors, which provide a per-use pre-shared key (PPSK). This solution allows each device to have its own PSK rather than sharing a PSK with every device on the WLAN. The result is an increase in security.

Devices that require fast, secure roaming include VoIP handsets, some one-to-one communication devices, and some industry-specific devices. Laptops and tablets may require it depending on the technologies used; however, when only non–real-time applications are used, slow roaming is often effective—particularly if the IP address is not lost. This is due to the recoverability in many protocols.

> Layer 2 roaming is acceptable for most applications as it does not break the network connections of higher layers. Layer 3 roaming may cause many applications to fail or require a reset because it changes the IP address and results in lost connections for higher layers.

Network Monitoring

WLANs have traditionally been considered vulnerable to all the following:

- Hijacking
- RF jamming (DoS)
- Protocol attacks
- Eavesdropping
- Spoofing

- Man-in-the middle attacks
- Management interface exploits
- Encryption cracking
- Authentication attacks
- Peer-to-peer attacks
- KRACK (Key Reinstallation Attack)

These attacks are common, and a solution exists to discover and mitigate many of them. Wireless Intrusion Prevention System (WIPS) solutions consist of a centralized server appliance with distributed sensors that are used for monitoring and sometimes interacting with the WLAN environment. Two kinds of WIPS system exist—integrated and overlay. Integrated WIPS are found as part of a WLAN infrastructure system where the functionality of a WIPS is built into the controller. An overlay WIP is a standalone system with its own sensors. Each type presents advantages and disadvantages.

When designing a WLAN, consider network monitoring or WIPS solutions. When recommending a solution, the choices are between the integrated and overlay models. The customer should understand both options and be able to make an educated decision as to which model is best for them.

In many cases, an existing WIPS solution may be in place with an older WLAN, but it is important to remember that such a system may require an upgrade itself to support full detection operations of the modern PHYs.

BYOD/MDM

Bring Your Own Device (BYOD) refers to the use of personal devices on the organization's network. BYOD can be considered a problem, but many organizations instead consider it a benefit.

These benefits may include:

- **Enhanced productivity:** Users can do more work on the same devices that hold their personal information, which often enhances productivity. For example, rather than using one mobile device for corporate email and another for personal email, the user may simply use two email apps on a single device. In some cases, multiple accounts may even be aggregated into a single email app. This use case provides for enhanced productivity through reduced device or app switching.

- **Decreased costs:** Organizations have lower equipment cost and maintenance by allowing employees to do business work on personal devices. In this case, the cost benefit is measurable. Rather than purchasing a device that the employee will use for personal and business uses, the organization simply allows the employee the option of using their personal device for business uses. Some businesses will give a reimbursement for employee participation in a BYOD program. Consider a company with 1000 employees who agree to use their own personal mobile phone for all business processes and accept a $50 per month allowance for this participation. For many organizations, this can be a savings of $20-100 per month per employee, totaling a savings of between $20,000 and $100,000.
- **Increased satisfaction:** Employees are thankful to have the option to select the devices they desire for work activities. Some people like Apple products, others like Android, other like Microsoft, and others may prefer another provider. Rather than forcing users to learn and use a different solution, many organizations allow the use of selected products that work with their mobile device management (MDM) solution so that users can operate with devices they know well.

BYOD is not the same as MDM. MDM is used to manage BYOD and, therefore, BYOD is sometimes said to be a subset of MDM. Strictly speaking, BYOD is defined as users utilizing their own devices for organizational work (and possibly personal uses) in the workplace. These personal devices include laptops, tablets, and mobile phones, but in rare cases today also include desktop computers. In most cases, desktop computers are used by telecommuters, and while this use does not qualify for the BYOD category, it resembles BYOD as the telecommuters are using their own devices (Use Your Own Device).

MDM is used to manage mobile devices that perform business functions—both organization-owned devices and BYOD systems. MDM solutions may be integrated with a WLAN solution, or they may be selected as an overlay solution.

When considering how to address BYOD in an organization, two basic options are available:

- Container - Containerization can be described as an implementation that secures and restricts access to other devices and enterprise applications in the workplace.

With containerization, many organizations are simply allowing access to the WLAN and therefore the Internet, but they are not allowing enterprise apps on these personal devices, a combination that might allow for data leakage and other security concerns even with the use of MDM solutions. Containerization may also be defined as the process used to ensure that enterprise data on personal devices are encrypted and secured—hence, in a container or workspace.

- MDM - The basic process of MDM for BYOD devices is outlined as follows:
    - Device detection determines that the device is not an enterprise device and needs to be processed by the MDM solution.
    - Device approval may involve several actions, including registration or enrollment, certificate installation, profile installation, and more.
    - Device management is the ongoing management of the device once access is granted and may include updates to apps, configurations, and the profile settings.

Another movement developing momentum is bring your own application (BYOA). Driven by many commonalities with BYOD, BYOA builds on the many users who already have user licenses to Microsoft Office and other applications that can be used to perform enterprise work. Use of these licenses can save on software expenses when implementing business computer systems for employees.

When planning BYOD requirements, consider the capabilities of your MDM for these specifics:

- Does it allow you to restrict settings when connected to the enterprise network that are released when connected to other networks?
- Does it provide restrictions for apps or only for hardware/OS?
- Does it work with the devices you must support?
    - Android
    - iOS
    - Windows
    - BlackBerry
    - Other
- Do you have the existing infrastructure dependencies?
    - PKI
    - Authentication service
    - Traditional network features (DNS, DHCP, etc.)

In some networks, Network Access Control (NAC), called Network Access Protection (NAP) by some vendors, is used to analyze a connecting device, determine if it meets minimum network access requirements, and then either grant or deny access accordingly.

The basic NAC process is as follows:

1. The device connects to the network.
2. The NAC infrastructure analyzes the device and processes the results.

If the device complies, network access is granted. If the device is not in compliance, network access is denied, or the device is placed onto a quarantine network. If placed into quarantine, the device may be updated with antivirus software or definitions and other required components.

Guest Access

In building a guest network, VLANs are used to separate guest users from the corporate network. One may choose to make the guest network open so that all users can access the network as if they were at a public hot spot. Or you can redirect them to a captive portal to simply agree to the use of the guest network. It may be decided that the guest network should be secured as well. In a secured guest network, users are required to know WPA-Personal or WPA2-Personal keys or a Web-based authentication code, or they are required to go through a guest registration process. Regardless of the steps, the result is that they are connected with strong, modern Wi-Fi security.

At the WLAN level, authentication methods include both personal and enterprise options depending on the method used to onboard users. At the user level, authentication methods are usually Web-based.

When implementing a WLAN specifically for guests, it is a best practice to separate it from the enterprise network using VLANs. Using a completely different VLAN and IP network allows for better security when separating guest users from the corporate network. Other measures can be administered on the guest VLAN, such as utilizing an anchor controller so that all guest VLAN traffic is routed through a DMZ and stays behind a series of firewalls for better corporate security. Guest access has become a common phenomenon in the corporate world, so it is important to build a good design that includes security and separation for protection of secured corporate data.

Wireless Device Roaming

Roaming is a matter that tends to cause many issues in corporate networks. If devices are not roaming from one AP to the other fast enough, complaints are submitted. If devices are dropping connection when moving between buildings on campuses, complaints are submitted. So many scenarios concerning the WLAN and roaming devices can haunt a network administrator if the design has not been correctly developed. Here are some helpful points to place on your checklist when concerned with roaming devices:

- ***Proper Cell Overlap***: One of the basic foundations to ensure that devices are roaming correctly is to make sure the survey was performed with a good 20–30 percent overlap of neighboring RF cells. This is needed for clients to determine when they will roam, and it gives the clients consistent overlapping coverage at every cell so keep roaming decisions from becoming inconsistent due to a bad survey.
- ***RF Profiles***: It is always a good practice to include RF Profiles in your WLAN design to set connection restriction on the APs. When designing the WLAN, you perform a survey based on a specific minimum signal strength at the cell edge such as -67dBm. When you read your AP's white papers, you can determine data rates accomplished at various dBm levels, and this will help you when building RF profiles for your controllers. For example, suppose you have surveyed for -67dBm at the cell edge for VoIP devices, and reading up on your device white papers you find it has a great connection rate at 36Mbs up to 54Mbs. Depending on other devices you may have using the network, it could create better roaming and connection rates for your VoIP devices if you set the minimum data rate to 24Mbs in your RF Profiles. This will keep devices from connecting at lower rates and ensure your VoIP phones will be looking for an AP to roam to when its connection rate approaches 24Mbs. Other RF Profile thresholds can be set to assist in this as well.
- ***Fast Secure Roaming Protocols***: As mentioned earlier in this chapter, making proper use of protocols such as 802.11r and 802.11k and implementing opportunistic key caching (OKC) will help speed up device roaming between cells.
- ***Layer 3 Roaming:*** Most roaming of wireless devices takes place in a Layer 2 broadcast domain, or more specifically in the same subnet and SSID. When a campus is designed with multiple VLANs, Layer 3 roaming provides a consistent

connectivity between each VLAN while the device keeps its original IP address. This basically becomes ***Roaming between***

- ***Roaming Domains***. In order to do this successfully, you must have a way to allow the roaming client device to maintain its original Layer 3 IP address while roaming across multi-VLANs. The way to do this is through ***Mobile IP***. A mobile IP-enabled network will have several key components:
    - **Mobile node (MN)** - This is the client device doing the roaming across the multi-VLANs.
    - **Home agent (HA)** - The HA is on the network router and/or Layer 3 switch, and it guarantees that the roaming client or Mobile Node receives its IP packets.
    - **Foreign agent (FA)** - This is also on the network router and/or Layer 3 switch. The job of the foreign agent is to help the roaming client by telling the home agent about the new location of the roaming client or HA. It does this by receiving the IP packets from the home agent that are to be received from the roaming client or mobile node.
    - **Care-of address (CoA)** - The CoA is on the local network router that gets packets that are sent by the HA to the MN.
    - **Co-located care-of address (CCoA)** - This is a care-of-address that is on the roaming client.

In a Mobile IP-enabled network, a client device is within its home subnet if its IP address came from the home router or home agent (HA). When the client device roams to anther subnet (considered a foreign subnet), it detects the foreign agent and registers with it or the mobile node (MN) CCoA. Next, either the HA or MN CCoA will set up a tunnel to the home agent for the roaming client. Once this is in place, packets sent to the roaming client device will be sent to the home agent first—just as if the client were still in its original subnet. The home agent then uses the tunnel to forward the IP packets to the roaming client. If the roaming client sends IP packets, they are sent through the foreign agent.

By utilizing Mobile IP, client devices can roam across Layer 3 boundaries without dropping their connection or their original IP address. This will allow VoIP calls to continue without reassociating on the new subnet and having to dial the call again.

Wireless Network Monitoring

Monitoring your WLAN is an important part of maintaining a healthy network. There's a multitude of reasons your WLAN may suffer in productivity, even after implementing a terrific design. Some of them are as follows:

- ***Rogue APs***: Even after a new WLAN has been implemented for employee use, you will still find that someone within the company will bring in their own wireless AP and plug it into the network. The WLAN has to be designed with this in mind and from the outset have a method in place to locate and mitigate these issues. A well-implemented intrusion prevention system can notify a network administrator of a rogue AP by using a nearby AP on the corporate network to attempt association to the rogue AP. This allows the intrusion prevention system to determine where the rogue is so that the administrator can go to its location and remove it. Rogue APs are bad for the RF environment as they cause both a higher noise floor and co-channel interference and, most of all, they become a possible unprotected portal into the corporate network.
- ***Neighboring networks:*** It is always possible that another business nearby will install a WLAN as well. The issue here is that they may not run a spectrum analysis and design around the channel usage of your existing WLAN. Instead, they may install APs close to yours (if you are in a multi-tenant building), and this can cause a noise floor issue or affect your employee devices attempting to associate to their network. Other issues such as co-channel interference can become a problem. In a perfect world, you would love for neighboring businesses to get together with your IT staff and get help building around your channel use, and placement of APs, but this isn't a realistic option in most cases. Hopefully, the WLC (WLAN controller) is configured for auto channelization with proper thresholds set so that the APs can detect the neighboring interference and change to clean channels to avoid the conflict. If not, that is the next step recommended for this type of environment.
- ***New microwaves or cameras:*** Other items such as wireless monitors, microwave ovens and security cameras can be installed after your WLAN is deployed. These things can cause serious interference to Wi-Fi devices and a stack of trouble tickets called in.

All in all, it is a great habit to set up some type of monitoring of your network to prevent service interruptions and security breaches. Study your options and find the best one to fit the needs of your environment.

Authentication and Encryption

A corporate network's first line of defense is the way client devices are authenticated to the network. Many options exist, but few are secure enough to protect sensitive data from being compromised. In the days of WEP keys, it was simply enough to have the key that everyone in the company uses and you had access to everything. But this way made it much too easy for the key to be detected by social engineering. Many employees would write the key on a sticky note and post it to their computer monitor to keep from forgetting it. There are many other means of discovering very weak passwords:

- Dictionary attacks
- Brute-force attacks
- Phishing
- Malware
- Keyloggers
- Rainbow tables
- Free software for cracking passwords

Authentication is the method or series of methods used to ensure the network is not being accessed by someone not allowed on it. To accomplish this, you first should use WPA2 Enterprise, also known as 802.1X, which utilizes some form of RADIUS server for authentication purposes. This method of authentication can be used by the server to authenticate a client, and by the client to authenticate the server. Various options of authentication can be used in 802.1X depending on which variant of authentication protocols is used. 802.1X also defines three specific elements in the authentication process:

- The Supplicant: A client device such as a tablet or smartphone and can also be the software found on the client device that provides credentials to the authenticator.
- The Authenticator: This can be a network switch, but in most cases, especially with a WLAN, it is a wireless AP or a WLAN controller.
- The Authentication Server: This is a host of some kind that RADIUS software is running on and can also be on the authenticator.

When a client device decides to join the WLAN secured with RADIUS authentication, it first starts with open authentication to become associated to an AP. Once open authentication is successful, the client device is connected to the WLAN in something like the way a computer's Ethernet cable is plugged into a network switch port. You will see the lights come on to show a connection, but it doesn't mean that access to anything has been granted.

After open authentication, the RADIUS process begins. It usually starts with an EAPoL frame, and then the process of requesting and providing credentials begins. Depending on the level of security and authentication a company is looking for, you may choose to use any of many versions of EAP to modify the process to the desired taste. The outcome is a set of keys used to enact secure encryption for the exchange of data across the WLAN.

CWDPs should familiarize themselves with both 802.1X and the EAP options along with what is required for the use of each one to provide customers with details on equipment and software needed to implement it.

WPA3 Enterprise offers a better 192-bit cryptographic strength and is gaining ground quickly. It also offers Wi-Fi Certified EASY CONNECT, which cuts down on the complexities of onboarding wireless clients with limited options on small displays. WPA3 maintains interoperability with all WPA2 client devices through a transitional mode of operation. To learn more specifics about RADIUS authentication and the 802.1X processes, read the *Certified Wireless Security Professional Official Study Guide.*

Wireless Bridges

Bridge links are used to connect buildings over distance so that wired links are not required. They may also be used as redundant connections along with wired connections. This section introduces the issues faced when planning for bridge links and mesh connections.

During the requirements analysis, you should determine whether bridge links will be needed and what LANs will be bridged as a result. This will help you determine if the bridge links will be point-to-point or point-to-multipoint.

Wireless bridges support two major configuration settings:

- Root
- Non-root

Bridges work in a parent/child relationship, so think of the root bridge as the parent and the non-root bridge as the child.

Wireless bridges, also referred to as bridge links, can provide many benefits. One main benefit from a wireless bridge is cost savings. When a company has multiple buildings on a campus and does not want to pay a monthly fee to a service provider for Internet service for each location, a wireless bridge is a great solution. The company can have the service provider install the Internet connection at one location and then utilize a wireless bridge to share the bandwidth between buildings. In order to do this, you first must collect some information:

- *How many buildings will you be distributing the service to?* If it is a single structure, then you will utilize a *point-to-point bridge*. If you have two or more buildings you will be distributing to, you will utilize a *point-to-multi-point* bridge.
- *How far away are the buildings you are distributing to?* If the structures are all within a close proximity of your main building, *omni-directional antennas* may be used for the bridge links. If the buildings are large distances away from the main building (root), you may need to utilize more *directional antennas* with higher gain in order to obtain proper connectivity across the bridge link.
- *Do you have radio line of sight between buildings?* Simply having a line of sight that allows you to see the other structure is not always satisfactory for a wireless bridge. A wireless bridge depends on connecting signal lobes that are dependent on what is called the *Fresnel zone*. Also known as *radio line of sight*, a Fresnel zone is a cylindrical ellipse drawn between transmitter and receiver. The size of the ellipse is determined by the frequency of operation and the distance between the two sites. As you see in Figure 2.6, the radio line of sight broadens out over distance, which could allow for partial obstruction from structures or trees along the path of the signal.

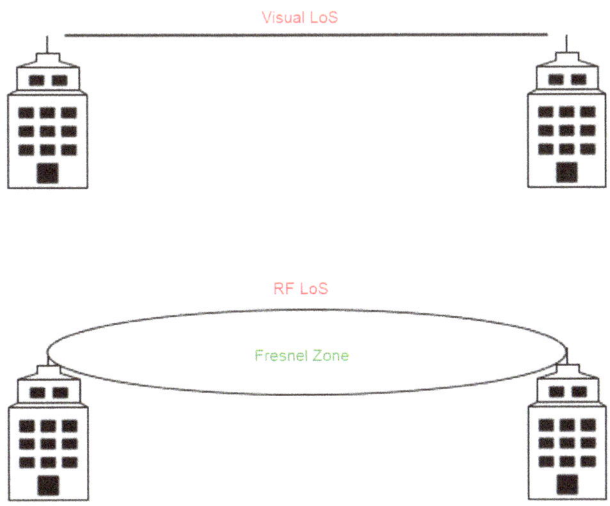

Figure 2.6: Visual LoS vs. RF LoS

When designing a wireless bridge, you should always check for good clearance for the signal. The Fresnel zone cannot be obstructed more than 40% without causing serious degradation of the signal as shown in Figure 2.7.

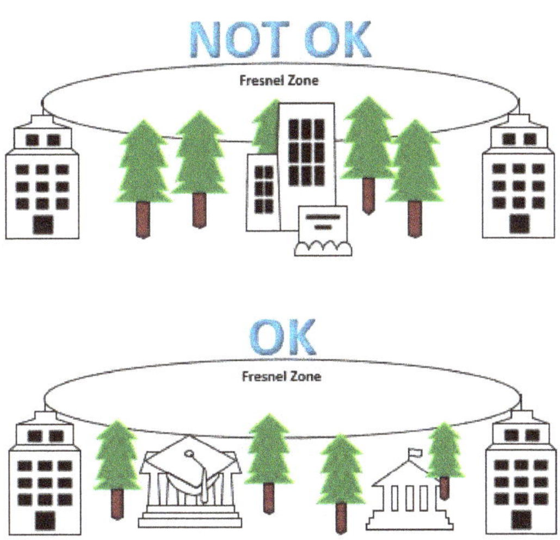

Figure 2.7: Fresnel Zone Blockage\Clearance (PtP)

The same applies to point-to-multi-point bridges as well. You may choose to have an omnidirectional antenna at the main building (root) and any combination of directional antennas at other sites (non-root), but the Fresnel zone needs to be calculated into the success of the bridge.

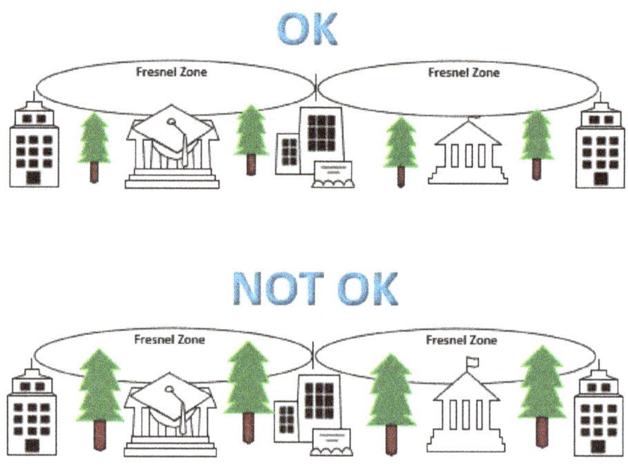

Figure 2.8: Fresnel Zone Clearance\Blockage (PtMP)

It is essential to understand that the Fresnel zone will be different depending on the distance of the bridge, the frequency being used, and the type of antennas being used. Most Fresnel zone calculators solve for a generic zone based on distance and frequency alone.

A bridge link budget is important for a successful wireless bridge. It needs to include the following:

- Radio transmit power
- Cable loss
- Antenna gain
- dB loss from lightning arrestors and connectors
- Free space path loss
- Curvature of the earth
- EIRP

Any time you are designing a bridge link, those items affect the bridge performance. Both sides of the bridge (root and non-root) should be factored into the calculation to find the proper solution.

Mesh deployments are similar to a bridge with a few exceptions. The root is connected to the LAN, and the non-root is a remote AP that allows clients to associate to it on one radio while a second radio performs as a back haul to the root AP to gain network connectivity for its associated clients. Mesh designs can be deployed both indoors and outdoors where needed. Non-root deployments are limited by a lack of available power sources at needed locations. This restriction is commonly overcome using solar power.

Bridge links and mesh networks are discussed in further detail in Chapter 6.

Client Devices

All client devices (laptops, tablets, mobile phones, and so on) should include a wireless radio. Without this radio, a device cannot associate to the WLAN. More specifically, it must be an 802.11-compliant radio. Each WLAN client device is composed of a similar set of hardware components and software elements. The hardware components include chipsets for radio control and management, antennas for RF transmission and reception, and interfaces for connectivity to the device intended to communicate on the wireless network.

The chipset and antenna are points of differentiation. Chipsets provide the actual implementation of the 802.11 PHYs that are supported by the client device and, therefore, the radio. For example, a chipset may support only the transmission of 2.4 GHz signals and support the DSSS, HR/DSSS, ERP, HT, and HE PHYs, or it may support the 2.4 GHz signals and the 5 GHz signals as well, which allows for support for the OFDM, VHT and HE PHYs in addition to the PHYs operating in the 2.4 GHz band. The newest of devices may even support the 6 GHz band, supporting only HE PHYs there.

A device that supports both the 2.4 GHz and 5 GHz PHYs is often described as an 802.11a/b/g/n/ac/ax adapter. These devices cannot operate both bands at the same time but must switch between them or operate on only one of them and this is also true when the 6 GHz band is included. The same is true for most HT (802.11n), VHT (802.11ac), and HE (802.11ax) client adapters. The clients can usually operate only in one frequency band at a time and some of them support only a single frequency band.

Laptop computers are among the most common Wi-Fi clients. They usually have very few issues pertaining to WLANs because they typically have stronger wireless adapters and better antennas than those of other wireless devices. However, they also present some challenges.

Laptop vendors manufacture their products incorporating many types of wireless adapters. Some are single stream, while others are multi-stream. Some still come with 802.11n adapters, while most new ones are equipped with 802.11ac or 802.11ax. Even if all laptops within a company came with 802.11ac or 802.11ax adapters, they will still differ in features of the standard. For example, they may or may not support transmit beamforming. They may or may not support multi-user MIMO. And they will support different numbers of spatial streams.

One advantage of most laptops is the powerful batteries they use. This allows for better Wi-Fi connectivity in many instances thanks to higher output power for the Wi-Fi radio.

When designing a WLAN, selecting a good laptop and adapter combination is essential if this task falls on you. For example, choosing a MacBook Pro with 802.11ac 3×3 chipsets allows for packet capture without the need for odd adapters and external antenna kits. Figure 2.9 shows a MacBook Pro laptop with 802.11ac support.

Figure 2.9: MacBook Pro laptop

Tablets, like the iPad shown in Figure 2.10, can now perform practically any function that can be performed by a laptop computer. This is particularly true with some of the crossover tablets like the Surface Pro 3. It runs on the Windows operating system and can run almost any software application that a laptop or desktop can. However, tablets typically have less powerful processors and memory and may also use slower wireless chipsets to conserve battery power. For example, they may use a single-stream chipset instead of a two-stream chipset, which is more common in laptops.

Figure 2.10: iPad tablet

Mobile phones, like the Samsung Galaxy G6 (as well as newer models) shown in Figure 2.11, have evolved into small–form-factor tablets in that they can act as Wi-Fi clients and, in many ways, can be treated as tablets. In addition, mobile phones are devices that work on a provider network, so transitioning from the provider network to the WLAN should be planned. If the mobile phones are owned by the organization, it can provide a significant cost savings to have them automatically connect to the enterprise WLAN when it is available. This will allow for reduced-sized data plans for all employees with employer-owned phones[23].

[23] It is not the intent of this book or the CWDP exam to ensure your knowledge of all modern devices. That is the ongoing responsibility of the WLAN designer, at least in relation to the devices used on the networks designed. However, you should know the common characteristics of these devices that will impact their performance and functionality on a WLAN so that you can design a well-performing network for them.

Figure 2.11: Samsung Galaxy S6

802.11 VoIP phones include several components, such as the call manager, voice gateway, and the infrastructure routers. However, for the purposes of CWDP study, it is most important to familiarize yourself with the handsets, their capabilities and requirements.

Most VoIP handsets or phones have the following basic features:

- IP address assignment
- Firmware updatable
- VLAN support
- Security features like WPA/WPA2 pre-shared key (PSK)
- 802.11a/g/n wireless PHY support
- Transmit power of 25 mW or more

Wireless desktop computers are nonmobile devices but often use wireless NICs to connect to the corporate network. This trend is likely to continue now that 802.11ac is in its second wave and future standards will prove to be even faster. The wireless workspace provides these advantages:

- No need for cable runs to each workspace.
- Easy speed upgrades through replacement of Wi-Fi adapters.
- Easier moves from one workspace to another.

However, adding Wi-Fi-enabled desktops to the WLAN introduces a demand for greater density and faster pipes through the Wi-Fi connections.

Wireless desktops may be equipped with onboard Wi-Fi, internal Wi-Fi adapters, or USB adapters. In many cases, the supposed onboard Wi-Fi is really just mini-PCIe included with the motherboard. However, this configuration is advantageous as it means that the adapter can be replaced to gain faster or different Wi-Fi capabilities according to your needs. The internal Wi-Fi adapters are typically PCI or PCIe with antennas that protrude out the rear of the computer case.

Internet of Things (IoT) is a reference to devices that communicate with one another over the internetwork provided without necessarily requiring human interaction. These include medical devices, automation systems, tracking devices, and personal devices. The key concept is that they communicate regularly across the network. This communication may involve small amounts of information transmitted frequently, and many devices will communicate with each other all the time.

These rapidly growing technologies demand stable network access and communications. Some are Bluetooth, others are Wi-Fi, and some use proprietary wireless communications, but the focus here is on the devices that operate in the 2.4 GHz and 5 GHz frequency bands.

By the year 2020, it was estimated that more than 50 billion devices will be utilizing the Internet and many of them using Wi-Fi[24]. So, it is essential to consider these devices when designing WLANs.

Finally, when defining requirements for your mobile client devices, it is important to know the following:

- How many devices are expected to be used within the WLAN?
- What PHYs will be supported by the devices?
- What will the expected load generation be from the devices?
- What is the growth rate expected for such devices in your organization?

---

[24] In fact, thanks to IoT, by the beginning of 2021, we saw well beyond the forecasted number of IoT devices. It is likely that wee will surpass 70-80 billion devices by 2025 and 110-20 billion devices by 2030.

Client devices drive the need for better Wi-Fi. Be sure to know the devices that will be used in the environment you are designing for.

2.6: Design Constraints and Restrictions

Any time you are asked to perform a WLAN design, there are almost always restrictions that make for a difficult situation in one or more areas. These restrictions can cause limitations to the aesthetics of the installation. An example of this could be an office in an historical building that wants Wi-Fi but cannot allow you to drill holes in the structure. This limits your ability to run cabling as well as AP mounting, leaving you with the possibility of exposed Panduit runs on the ceilings. Such installations lack in good aesthetics but may be the only option you are left with.

The more common aesthetics requirement is simply related to appearance. The organization does not want the WLAN Aps to be seen or they don't want them to be obvious, but they want the WLAN to function well. Sometimes the solution is as simple as color matching and at other times custom skins can be acquired for the APs to help them blend with the environment. Additionally, wireless APs mounted in handrails, ballasts, and other items can coexist in the environment without obvious presence.

Another example of mounting restrictions is a medical facility that wants Wi-Fi coverage in patient rooms but does not want the APs themselves mounted in the patient rooms. This becomes extremely difficult when performing your survey for proper overlapping cell coverage, as many of the APs will have to be mounted in the open hallways with excessive overlapping signals.

Building architecture constraints can impact mounting, cable runs and other factors in the WLAN deployment as well. Older buildings may have thick internal concrete walls that heavily attenuate RF signals. A multi-floor building imposes different constraints on a WLAN design than a single floor building. Such architecture constraints must be identified and addressed in the requirements and then in the design.

Another building architectural constraint is having early access to the area you need to mount an AP or antenna to determine the feasibility of doing so. For example, if it turns out there is asbestos in the wall where you need to mount the AP, you are not going to get permission to drill into the wall. This happens often in older buildings, and it presents design constraints that are difficult to work around. Sometimes the only solution is to redesign an entire area so that a couple of APs can be moved from places of restriction.

Solution architectural constraints are different. The organization may have chosen to use cloud technologies for the management of all infrastructure and networking technologies. In such a case, the WLAN design will have to accommodate this constraint with the use of a cloud-managed WLAN solution. Contrariwise, the organization may have chosen to allow not management of networking devices from outside the facility, which would rule out the use of cloud-managed solutions. Identifying such solution architecture constraints early is essential.

Project budget is a major constraint on a design at times as a customer's needs and expectations may conflict with the budget allotted to fulfill the needs. When this becomes a factor, it is imperative to weigh out all options on how to complete the design. Here's an example.

A customer wants Wi-Fi coverage on power generation floor that is a quarter mile long. They also want Wi-Fi coverage outside the building on both ends, and there is only one network switch available in the middle of the floor. So, the distance out from the switch is over 650 feet in both directions. The customer's budget will not allow for the installation of new fiber and switches at each end of the building, so now you must design for mesh connectivity to the ends of the building. A difficulty here is that EMI (Electrical Magnetic Interference) is usually a serious issue in a power-generation plant, and the 2.4 GHz frequencies are affected by this, so utilizing a model of AP with dual 5 GHz radios would be the best option. In selecting a vendor, you are looking for a satisfactory AP model that will provide the solution your network needs. In most cases, every vendor can provide you with a solution, but it doesn't always fit the budget.

The project may begin with the budget already set before the plan has even been created. To make it less painful, this kind of budget is often aptly named a top-down budget. Of course, the opposite is a bottom-up budget, where you are asked to determine what it would cost to deploy a WLAN that meets all requirements. Either way, budget restrictions are a practical limit to how much money is available. This budget constraint can impact the wireless equipment you purchase and available throughput in a coverage area. The point is simple: what you can implement is directly linked to the demands of the users and the available resources.

Alongside the budget, time constraints are an essential consideration. The organization may have a hard deadline by which the WLAN must be deployed. Such a deadline will impact the actions taking along the way to complete deployment if the deadline is inflexible. For example, if you are asked to implement a new WLAN for a stadium where the next year's Super Bowl will be played, you have until some time before the game to complete the project. You cannot have extensions and therefore some things you might prefer to do cannot be done. Schedules are sometimes the highest priority.

Building and safety codes will sometimes cause limitations to a design. Just as the historical building cannot have holes drilled into it, other building codes may restrict items from being mounted on ceilings at all. Some safety codes require low-smoke, zero-halogen network cabling, which results in a higher cost for cabling and restricts the budget when trying to design within the budget.

Physical access to all areas of the facility requiring coverage can become a constraint. It is important to bring this up during the initial meeting with the customer to ensure that this access clearance is granted. If open access is not available because of the environment or security, then have plans for an escort during the duration of the design so that time delays do not become a problem when access is needed to specific areas.

2.7: Documentation

The final topic of this chapter is a crucial part of the requirements analysis. It is essential that all information gathered is documented in an understandable and acceptable format. The IEEE 29148-2018 standard provide guidance on the structure of requirement statements and the quality or validity of a requirements set for a system. These guidelines are applicable to WLANs and other wireless networks and should be considered for your WLAN deployment projects. This topic will be discussed at the end of this section. First, let's explore the documents that may already exist that can be helpful in the requirements engineering process.

Gathering Existing Documentation

In WLAN designs, there are multiple areas of documentation involved. It is essential to the design process to have access to many of these documents to provide optimum designs, upgrades, and expansions. When a facility has no WLAN at all and you are starting out with nothing, it is much easier to design the WLAN. When any part of the facility has an existing WLAN in operation, the task of expanding becomes more complex. Also, when you are upgrading an existing WLAN to new equipment, there is more work to be done besides just changing out APs. No matter which stage you come in on, documentation is key to success.

Existing network equipment should always be surveyed before starting any type of WLAN design, to answer questions like the following:

- Are existing switches up-to-date and compatible with new standards?
- Are there enough switch ports available on existing switches for the WLAN to be implemented without purchasing new switches?
- If more ports are needed, are the existing switches stackable?
- Do the existing switches have the proper PoE budget overall and per port for the APs being used for the new WLAN?
- Are the existing network closets within standard distances of potential AP locations, or will new locations be needed?
- Is there existing fiber that can be used to add more network closets?
- Are the existing VLANs on the switches big enough to support the new WLAN?

All these questions, and sometimes more, will need to be answered before designing a WLAN at any stage, whether it be a new design or an expansion of existing wireless infrastructure. Creating good documentation for these questions is a great start to your design. These answers become a part of your deliverables when it comes time to give a design proposal to a customer. The last thing you need to hit you in the middle of the implementation is lack of switch ports or noncompliant PoE on existing switches. Document these things before starting your design process so that your deliverables can include new switch hardware where needed. If none is needed, then you have a document that lists the inventory of each existing network closet.

Existing cabling is another element one should document. In many cases, there are several available network drops near the locations you intend to place APs. If these drops were put in for extras at the time the wired LAN was installed, it is very possible that they were intended for network expansion and possibly for a WLAN. If several of these existing cable drops are found, and can be utilized for the WLAN, it could save the project budget thousands of dollars and help out with any time constraints the installation may encounter.

It is also a good practice to gather documentation on existing network services such as DNS, DHCP, NTP and authentication servers. When the design comes down to what type of security will be implemented, knowing what security is already in place can help with the decision on securing the WLAN.

Also, almost every business has some form of a DHCP server, whether it be on a single router, multiple Layer 3 switches, or a service such as Info Blocks. Finding which form of IP address hosting is in use will help answer a couple of questions:

- Is the current system best suited for the size of WLAN that will be added?
- Are the current subnets big enough to support the WLAN and the client devices that will utilize it?

Just one basic example: If a company has a /24 network and over 200 IP addresses are already taken up by network devices and end-user stations, then adding a WLAN consisting of a WLC and 30 APs will deplete the IP pool, leaving few if any addresses for client devices that will use the WLAN.

In most companies, the IP pools are bigger than a /24 and several pools exist, but at these companies there are also hundreds of workstations and IP Phones pulling addresses. In designing a WLAN, you should never assume that a plentiful reservoir of IP addresses will be available. Always document what is used and what is available. Then you can decide, prior to starting the wireless design, if a wireless VLAN exists or needs to be created. In many cases multiple VLANs exist just for implementing a WLAN, but plenty of cases come up short of IP addresses after implementation. Do the legwork up front and save the headache of change management latter.

Accurate and validated floor plans will come in very useful for all the scenarios just discussed. The best way to document available network cables is to place them on the facility's floor plans and name them. All network closet locations along with the number of switches in each closet should also be documented on a valid facility floor plan. Any information you find during your spectrum analysis should also be placed on a floor plan to document what the interferer or existing wireless device is and where it is located within the facility. Having good floor plans that are up to date is key to a good design. These floor plans can be used to document necessary information and be used for the wireless survey for documenting new AP locations.

Finally, if you do find existing wireless APs, not only should you document their location or gather existing documentation of their location on a floor plan, but you should also collect as much data about this system as you can possibly find. Ask for the documentation from the previous survey. This can help you identify the capabilities of the existing WLAN and help you locate existing cabling that may be used for new APs. The bottom line of documentation is that you would rather have too much than not enough.

Environmental Factors Documentation

Environmental factors should be documented as you gather information for the design. Some of this information will be discovered during requirements engineering and some may have to be gathered and documented before the design can begin. Following are some key pieces of the environmental information that are usually required:

- The results of
- The results of wireless environment scans.
  - Spectrum analysis: Spectrum analysis can provide important insights into the RF noise floor and other potential interferers of the intended WLAN. This information is most commonly shown on a floor plan that gives the type of interferer and where it is located as well as the information recorded from the spectrum analysis.
  - Packet captures: Using a protocol analyzer to capture WLAN communications from existing networks can easily reveal the varied types of devices used in and around the target installation area. Existing channels, PHYs, and other technologies in use can be discovered and used for the design process.
  - Wi-Fi scanners: In some cases, a full packet capture is not required and using a Wi-Fi scanner, like Acrylic Wi-Fi tools or Wi-Fi Explorer Pro will provide all of the information you need about existing wireless networks, clients, PHYs, and so forth.
- An accurate floor plan that shows the locations of all new APs to be installed with the WLAN.
- The location of all existing switches, ports available, existing PoE budget per switch, and any switches to be added in the design.
- The location of existing cable drops that could be utilized for the new WLAN.

- The location of existing WLAN equipment in the event this design is an upgrade or an expansion of an existing WLAN.
- Existing network services (DHCP, RADIUS) that are sufficient for the new WLAN implementation, and/or changes that need to be made to existing network services.
- Pictures: Be sure to include pictures of existing switch locations and potential new switch locations, existing wireless APs (if any), any interferers, existing routers, existing cable drops that can be used, and pictures of new AP locations for installers to use for installation purposes.
- Heat maps: Survey coverage maps are often called heat maps, and many customers want them included in the deliverables for their future use. Others have asked for copies of Air Magnet and ESS files to verify coverage areas and proper overlap. Save these files in a manner that they can be included in the deliverables if asked.
- Materials list: The deliverables should include the AP models to be used and how many, any antennas, switches, controllers, and so on. This list should be something the customer can place an order by, so it should include everything down to the licenses needed for the system.
- A description of the design and how it meets and exceeds the customer requirements in detail. This should provide details about throughput expectations from the customer and what is included in the design to meet those expectations.
- Attenuation factors for building materials that can be used to fine tune propagation modeling in WLAN design software.
- Ceiling heights, if not provided with the floor plans, so that you can accurately configure them in the WLAN design software.

This list is just a start, and it could be different for every customer. During the initial meeting with the customer, expectations and deliverables should be well defined to the point that both you and the customer agree upon the terms. Remember to treat the design itself as a project and go through the proper stages of project management to ensure that everyone involved stays informed of all progress, risks, and potential delays. The more communication you have, the better the documentation. The better the documentation, the better the implementation will be. As a CWDP, you may wear many hats at times. No matter which hat you are wearing, there is no substitute for good documentation.

Categorizing Requirements

While many divide requirements into the simple functional and non-functional requirements, the 29148-2018 specification provides six examples of requirement types or categories. These types include[25]:

- **Functional/Performance:** Describe the system or system element functions or tasks to be performed by the system. Performance is an attribute of function. A performance requirement alone is an incomplete requirement. Performance should be stated quantitatively. More than one performance requirement may be associated with a single functional requirement.
- **Interface:** The definition of how the system is required to interact with external systems (external interface), or how system elements within the system, including human elements, interact with each other (internal interface).
- **Process:** These are stakeholder, usually acquirer or user, requirements imposed through the contract or statement of work. They include compliance with national, state, or local laws, including environmental laws, administrative requirements, acquirer/supplier relationship requirements, and specific work directives. Corporate policy or practice may also impose process requirements.
- **Quality (Non-Functional):** Include a number of 'ilities; in requirements to include, for example, transportability, survivability, flexibility, portability, reusability, reliability, maintainability, and security.
- **Usability/Quality-in-Use:** (for user performance and satisfaction) provide the basis for the design and evaluation of systems to meet the user needs. Developed in conjunction with, and form part of, the overall requirements specification of a system.
- **Human Factors Requirement:** State required characteristics for the outcomes of interaction with human users (and other stakeholders affected by use) in terms of safety, performance, effectiveness, efficiency, reliability, maintainability, health, well-being, and satisfaction. These include characteristics such as measures of usability, including effectiveness, efficiency, and satisfaction; human reliability; freedom from adverse health effects.

---

[25] IEEE 29148-2018, clause 5.2.8.3, Examples of the requirements type attribute

When considering requirements in the two common categories of functional and non-functional requirements and placing all requirements in one of these two categories, consider the following definitions:

- **Functional Requirements:** what the users need for the system to work. Functional requirements are functions and features.
- **Non-Functional Requirements:** requirements that support the functions in performance, quality, or compliance, but do not describe a function of the system.

Examples of functional requirements include:

- Within the manufacturing plant, when a device attempts to locate and connect to the WLAN, an AP shall be available to provide connectivity at all times.
- Devices shall maintain connectivity when moving throughout the facility within constraints that support VoIP communications.
- Push-to-talk devices shall provide voice communications with the push of a button and require no additional user actions.

Examples of non-functional requirements include:

- The WLAN system shall support 99.99% uptime.
- A monitoring solution shall be implemented that reports on uptime, interference issues, security violations, and performance metrics based on defined policies for security and performance.
- The WLAN solution shall separate guest users from internal users and limit guest capabilities based on security policy 2.3.12.

Deciding into which category to place a requirement is sometimes a challenge as a requirement can seem to include both functional and non-functional characteristics. A functional requirement may be constrained by a non-functional performance metric or security metric within a single requirement statement. For example, the requirement statement, "Devices shall maintain connectivity when moving throughout the facility," is a functional requirement. The constraint of "within constraints that support VoIP communications" is a non-functional performance constraint by some interpretations. This is where the soft side of requirements engineering comes into play. It is far more important that the requirement be documented than that it be properly categorized, in most cases.

IEEE 29148-2018 Recommended Requirements Writing Guidelines

As we finish this important chapter on requirements engineering, it is important to consider how you may choose to write requirements statements. The 29148-2018 standard provides solid guidance for this process and the recommendations are addressed here as they apply to WLAN solution requirements.

First, you should understand the different between a requirement and a requirement set. A requirement is a single capability, characteristic, constraint, or quality factor related to the WLAN system. It should be stated in an unambiguous manner so that it can be interpreted only in the way intended. It should also be complete so that additional information is not required to understand it. Additionally, the requirement should be feasible, which simply means that it can be implemented within known constraints. Finally, it should be necessary. If it is not necessary it is not required, though it may be desired.

A requirement set is a collection of requirements. Requirement sets can be considered hierarchically. For example, you may have a large requirement set that includes all requirements for a WLAN design; however, this large set may be divided into specific requirement sets related to security, aesthetics, performance, reliability, and so on. How you structure the full-scale requirement set will likely be dependent on the size of the requirements set generated. If you have a total of thirty requirement statements, you may choose to place them all in one requirement set. If you have a total of three hundred requirement statements, you may choose to organize them into a component- or feature-based hierarchy for easy management and verification.

Each requirement statement should have the following characteristics[26]:

- **Necessary:** The requirement defines an essential capability, characteristic, constraint, and/or quality factor.
- **Appropriate:** The specific intent and amount of detail of the requirement is appropriate to the level of the entity to which it refers. The level of abstraction is appropriate to the level of the entity.
- **Unambiguous:** The requirement is stated so that it can be interpreted in only one way.

---

[26] IEEE 29148-2018, clause 5.2.5, Characteristics of individual requirements

- **Complete:** The requirement sufficiently describes the necessary capability, characteristic, constraint, or quality factor to mee the entity need without needing other information to understand the requirement.
- **Singular:** The requirement states a single capability, characteristic, constraint, or quality factor. Multiple conditions may be stated under which it is to be met, but the requirement is singular.
- **Feasible:** The requirement can be realized within constraints with acceptable risk.
- **Verifiable:** The requirement is structured and worded such that its realization can be proven to the customer's satisfaction. In optimal state, the requirement is measurable or observable.
- **Correct:** The requirement is an accurate representation of the need from which it was derived.
- **Conforming:** The individual items conform to an approved standard template and style for writing requirement statements, when applicable.

The entire requirement set should have the following characteristics[27]:

- **Complete:** The requirement set stands alone such that it sufficiently describes the necessary capabilities, characteristics, constraints, or quality factors to meet the describe system or system component needs without needing further information[28].
- **Consistent:** The set of requirements contains individual requirements that are unique, do not conflict with or overlap with other requirements in the set, and the units and measurement systems are homogeneous. The terminology used within the set of requirements is consistent, i.e. the same term is used throughout the set to mean the same thing.
- **Feasible:** The complete set of requirements can be realized within entity constraints (e.g., cost, schedule, technical) with acceptable risk[29].

---

[27] IEEE 29148-2018, clause 5.2.6, Characteristics of a set of requirement

[28] TBx items (To Be Defined [TBD], To Be Specified [TBS], To Be Resolved [TBR]) should not exist in finalized requirement sets. However, TBx items may be included early in the requirements discovery process so that they are not forgotten. Through system analysis and requirements analysis, they should be resolved before the requirement set is considered complete.

[29] Feasibility includes the concept of 'affordable.'

- **Comprehensible:** The set of requirements is written such that it is clear as to what is expected by the entity and its relation to the system of which it is a part.
- **Able to Be Validated:** It is practicable that satisfaction of the requirement set will lead to the achievement of the entity needs within constraints (e.g., cost, schedule, technical, legal and regulatory compliance).

When writing requirement statements, it is useful to have a defined requirement construct. Think of this construct like a template for each statement. While you may use other constructs from time-to-time, most requirement statements can follow the construct recommended in 29148-2018. Having a defined construct and language specification helps to bring clarity when later interpreting the requirements. The standard recommends the following common approach for language specification in requirement statements[30]:

- Requirements are mandatory binding provisions and use 'shall'.
- Non-requirements, such as descriptive text, use verbs such as ‘are’, ‘is’, and ‘was’. It is best to avoid using the term ‘must’, due to potential misinterpretation as a requirement.
- Statements of fact, futurity, or a declaration of purpose are non-mandatory, non-binding provisions and use 'will'. 'Will' can also be used to establish context or limitations of use.
- Preferences or goals are desired, non-mandatory, non-binding provisions and use 'should'. They are not requirements.
- Suggestions or allowances are non-mandatory, non-binding provisions and use 'may'.
- Use positive statements and avoid negative requirements such as ‘shall not’.
- Use active voice: avoid using passive voice, such as ’it is required that’.
- Avoid using terms such as ‘shall be able to’.

Additionally, the recommendation for a well-formed requirement statement is that it contain one or more of the following:

- It shall be met or possessed by a system to solve a problem, achieve an objective, or address a stakeholder concern.
- It is qualitied by measurable conditions.

---

[30] IEEE 29148-2018, clause 5.2.4, Requirements construct

- It is bounded by constraints.
- It defined the performance of the system when used by a specific stakeholder or the corresponding capability of the system, but not a capability of the user, operator or other stakeholder[31].
- It can be verified (e.g., the realization of the requirement in the system can be demonstrated).

The template for a requirement statement, at least a functional requirement statement, is defined in the 29148-2018 standard as:

[Condition][Subject][Action][Object][Constraint of Action]

For example:

When a station attempts to locate and connect to the WLAN **[Condition]**, an AP **[Subject]** shall be available **[Action]** to create an association **[Object]** within all areas of the manufacturing plant **[Constraint of Action]**.

Whether you choose to use this model or another, creating a consistent structure for requirement statements allows for increased understanding and more accurate implementation of those requirements. When you have a template or construct for a requirement statement and use consistent language with all documented requirements, you have a set of requirements that can be read, understood, implemented, and validated against. If the requirements are vague and use inconsistent language, understanding is significantly reduced, implementation is often erroneous, and validation is nearly impossible.

[31] To understand the phrase, "but not a capability of the user, operator or other stakeholder," consider the following example. The organization uses an application where the user must select four different items on a screen, click next, select two items on the screen, click next, and then enter some textual information. A system requirement statement would not depend on the user's ability to select type the information quickly. These abilities are beyond the capability of the system designer to address. While the system designer may be able to implement checks to ensure proper or possibly proper items were checked or typed, the system designer is not expected to impact the speed at which the user can make decisions and take actions.

Requirements Engineering in the Real World

In the book *Mastering the Requirements Process*[32], the authors, Suzanne Robertson and James Robertson, begin chapter 1 by stating:

> Requirements are what the software product, or hardware product, or service, or whatever you intend to build, is meant to do and to be. **Requirements exist whether you discover them or not, and whether you write them down or not.** Obviously, your product will never be right unless it conforms to the requirements, so in this way you can think of the requirements as some kind of natural law, and it is up to you to discover them.

Indeed, requirements do exist whether you discover them or not. In fact, while the processes discussed in this chapter may seem daunting, they can usually be achieved for a WLAN in a few hours of meetings and a day or two on location for the average WLAN project. Certainly, some projects require more time, and some require less time; however, the increased iterations required during design and, again, during implementation and, again, during validation and optimization often takes far more time than a good requirements engineering effort itself.

Some organizations will not budget for the time and cost of performing requirements engineering. These organizations still pay for requirements engineering, it just takes place iteratively during the design and deployment process (and sadly often into operations as well). It's not a line item in the budget, but they still pay for it, and I would argue that they often pay more in the end. The CWDP should always recommend a proper requirements engineering effort, but if it is not provided, the CWDP will simply have to assume the best requirements possible and work through the lengthy iterative requirements discovery process along the way[33].

Can you imagine someone building a house without a set of blueprints? I can, because I watched my ancestors do it when I was growing up and it looked like they built them

---

[32] Robertson, Suzzanne and Robertson, James. *Mastering the Requirements Process: Getting Requirements Right, Third Edition.* 2013.

[33] To be clear, iteration nearly always exist in WLAN projects. However, it can be greatly reduced by performing proper requirements engineering in the define phase. A good requirements engineering plan will always allow for some iteration, but it will also prevent as much of it as possible in WLAN projects.

without blueprints. A WLAN design without requirements is like building a house without blueprints. You may select good equipment and even install it "correctly" with the right mounting kits and cable runs, but that doesn't mean it will work as the users require. Only requirements can tell you that.

The good news is that as you design and implement more and more WLANs, you will gain expertise that allows you to better gauge the needs of a particular deployment. However, nothing replaces proper inputs (requirements) in the accomplishment of proper outputs (design). To quote Charles Babbage, "On two occasions I have been asked, 'Pray, Mr. Babbage, if you put into the machine wrong figures, will the right answers come out?' I am not able rightly to apprehend the kind of confusion of ideas that could provoke such a question."

2.8: Chapter Summary

In this chapter, you learned about the importance of requirements engineering and the IEEE 29148-2018 and supportive standards. You learned about the concepts related to requirements engineering and the items that should be considered in WLAN projects. With proper requirements analysis you have the right foundation on which to build an effective WLAN design. Without requirements analysis, it is effectively impossible to implement solutions that meet the needs of an organization. In the next chapter, you will explore an area of requirements engineering unique to wireless networks (both Wi-Fi and non-Wi-Fi IoT wireless networks) known as requirement areas.

2.9: Facts to Remember

Be sure to remember the following facts as you prepare for the CWDP certification and be sure that you can explain the details related to them:

- The IEEE 29148-2018, IEEE 15288-2015, and IEEE 15289-2019 standards define the requirements engineering processes, outcomes, and documentation.
- Requirements engineering is the process of discovering a project's requirements.
- A requirement is anything that must be provided by the product or outcome of the project.
- Business requirements are known as the objectives of the organization.
- User requirements include the goals or tasks that users must perform when using the WLAN.
- Technical requirements, or functional requirements, are those technical capabilities that allow the WLAN to meet user and business requirements.
- Regulatory requirements or constraints are those imposed by policies or government agencies.
- Client devices must be considered when planning a WLAN, including PHYs they support and range of connectivity.
- Applications must be considered related to their characteristics, such as throughput requirements, data load, and more.
- Security requirements should be defined including authentication, encryption, and monitoring.
- Physical coverage requirements are essential as the WLAN must serve the users where they require service.
- Building factors should be discovered, including building materials and square footage. Blueprints should also be gathered and facility access for the site survey acquired when needed.
- Planning bridge links includes selecting point-to-point and point-to-multipoint deployments
- At each phase of the WLAN design process, the right documents should be created or maintained.

2.10: Review Questions

1. A Project Charter is a document that officially recognizes a project and gives the project manager________________________.

    a. Exclusive authority over the project and its budget.

    b. Written permission to begin work on the project.

    c. Permission to access resources for project details.

    d. Access to funding to begin hiring resources.

2. A home agent and foreign agent are components in what is known as a __________________ enabled network.

    a. Layer 3

    b. Mobile Node

    c. Tunneled IP

    d. Mobile IP

3. Obtaining validated floor plans is important because they can be used to document __________________ for the deliverables.

    a. AP locations

    b. PHYs in use

    c. Network security type

    d. Network performance

4. In wireless bridge deployments, the elliptical open area between the root bridge and the non-root bridge where the transmit and receive signals form a stable connection with each other is referred to as the ________________________.

   a. Free Space zone

   b. Line of sight zone

   c. Link budget

   d. Fresnel zone

5. When deploying a new WLAN across existing switches, it is important to document every switch and to ensure the proper _______ output is provided for the model of APs being deployed in the design.

   a. Port

   b. Ethernet

   c. PoE

   d. IoT

6. A ____________________ consists of a centralized server appliance with distributed sensors that are used for monitoring and sometimes interacting with the WLAN environment.

   a. Wireless Intervention & Prevention System

   b. Wireless Interaction Prevention System

   c. Wireless Intrusive Prevention System

   d. Wireless Intrusion Prevention System

7. Which one of the following represents the construct recommended by IEEE 29148-2018 for a requirement statement?

    a. [Condition][Action][Subject][Object][Constraint of Action]

    b. [Condition][Subject][Action][Object][Constraint of Action]

    c. [Condition][Action][Subject][Object]

    d. [Condition][Object][Subject][Action][Constraint of Action]

8. What term is used to describe a requirement that defines what a system must allow a user to do?

    a. Non-Functional

    b. Functional

    c. Human Factors

    d. Quality

9. WPA3 offers Wi-Fi Certified EASY CONNECT for clients with__________________.

    a. 192-bit cryptographic strength

    b. Complexities of authenticating

    c. Limited options on small displays

    d. Interoperability with WPA2 clients

10. The main purpose of Opportunistic Key Caching (OKC) is to provide ______________________ for client devices on the enterprise WLAN.

    a. Protocol compliance

    b. Private Key exchange

    c. Secure Roaming opportunities

    d. Fast Secure Roaming

2.11: Review Answers

1. **The correct answer is B.** A project charter is a formal document that officially recognizes a project. This document is also official permission granted to the project manager to begin work on the project.

2. **The correct answer is C.** The home agent and foreign agent are both components in a mobile IP enabled network. A mobile IP enabled network provides the ability for client devices (mobile nodes) to roam across Layer 3 boundaries while keeping their original IP address. The foreign agent establishes a tunnel with the home agent and receives IP packets for the mobile node from the home agent. Because of this tunnel, the home agent receives and forwards packets for the mobile node just as if it was still in the HA domain.

3. **The correct answer is A.** Validated floor plans can be used to document new and existing AP locations, interferers, switch locations and available ports, existing cable drops, new switch locations, and many other things. By using a validated floor plan for documenting these types of data, it is much easier to provide accurate locations for these items.

4. **The correct answer is D.** The open space between the root and non-root bridge is free space. The specific area where the transmit and receive signals come together to form a stable connection with each other is the Fresnel zone and is referred to as radio line of-sight. The Fresnel zone is elliptical and must have at least 60% of this zone unobstructed in order for the bridge link to be an acceptable and stable connection. The size of the Fresnel zone depends on the distance between the root and non-root, the frequency being used, and the type of antennas.

5. **The correct answer is C.** When documenting existing switch locations for a new WLAN design, it is important to determine whether or not the switch provides the correct type of PoE output. Many APs require 30w of power to function while others require much less. Switches also have a power budget that will allow for 15w per port if all ports are being utilized for PoE, or 30w per port if half of the ports are utilized for PoE. Know the power requirements of the APs being used as well as the power outputs provided from the switches you intend to use.

6. **The correct answer is D.** A Wireless Intrusion Prevention System monitors the LAN environment by listening for transmissions from devices not part of the enterprise network. If a rogue AP is found, the sensor will interact with it as a client device would and attempt to associate to it. If the sensor has success associating to the rogue AP, the WIPS system will discover a path back to the centralized server appliance on the network through the rogue AP. If this is successful, the WIPS begins counter measures to prevent enterprise clients from associating to this device.

7. **The correct answer is B.** [Condition][Action][Subject][Object][Constraint of Action] is the construct recommended.

8. **The correct answer is B.** Functional requirements define what a user of a system should be able to do or what tasks they must perform. Non-functional requirements define the 'ilities' such as reliability, survivability, and maintainability.

9. **The correct answer is C.** WPA3 does provide for a stronger 192-bit cryptographic strength and it does have interoperability with WPA2 clients. Where Wi-Fi Certified EASY CONNECT specifically helps is with client devices that have limited options on small displays such as VoIP handsets.

10. **The correct answer is D.** Opportunistic Key Cashing serves a specific purpose of aiding in fast secure roaming of client devices. Using OKC allows for a client device to share its PMK ID with all APs on the same Layer 2 network. By doing this, even though a client hasn't roamed to a specific AP yet, it can skip the 802.1X EAP exchange process and go straight to the 4-way handshake, eliminating a lot of latency while roaming.

Chapter 3: Defining Requirement Areas

Objectives Covered:

1.4 Define requirement areas including essential metrics for each requirement

After you've discovered the customer requirements for the new WLAN, there are other specific requirements that can present themselves before, during, and after a site survey. It is always best to catch these requirements before performing the site survey so that it doesn't have to be repeated with new requirements in mind, but sometimes they slide under the radar when initial interviews and discoveries do not reveal these specifics up front. In this chapter, you will learn of specific requirements relating to facility areas (requirement areas) that are impacted by client device types, security settings, application specifics, SSIDs and WLAN profiles, along with the challenges faced in many vertical markets. We reference these areas as *requirement areas* because each area can have unique requirements. When defining requirements, do not think of just the entire WLAN for an entire organization. Instead, think of the areas requiring wireless access and what each area requires; this is the concept of requirement areas.

3.1: Requirement Areas Explained

WLANs are implemented in buildings, on campuses, and across multi-site locations (though each location implements an individual WLAN that may be managed from a central location). At each location, the WLAN may have varying requirements in different areas. Some such areas stand out, like warehouses vs. front offices, or student classrooms vs. administration areas, or attendee areas in stadiums vs. management offices. Defining requirement areas allows you to clearly describe the requirements of these different areas.

A requirement area is not necessarily bounded by walls or floors but by use cases, device types, application requirements, and other metrics that cause different requirements to exist from one area to another. The requirements of a requirement area are defined by the most stringent device requirements in that area. For example, if one set of devices requires 15 Mbps throughput and other devices in the same area require only 2 Mbps throughput, the higher throughput requirement will become the metric for the requirement area. It can become a bit more complex in that you can specify that 15 Mbps must be available for the more demanding devices and 2 Mbps be available for the less demanding devices, but if they use the same SSID, it is more likely that the device capabilities will drive this variation more than something specified in the requirements. Of course, if dynamic VLAN assignment is available (typically through RADIUS authentication) and if the WLAN management solution provides for traffic shaping for

VLANs, the network may be configured so that the devices requiring 15 Mbps get more of the network resources and the devices requiring 2 Mbps get less. This configuration requires the selection of appropriate vendor hardware and software for implementation, which could be part of the requirements and design process.

Factors that are often considered when defining requirement areas are capacity, quality requirements, device types, applications, and security[34]. These varying factors that may drive the need for different requirement areas are discussed in the ensuing sections of this chapter.

3.2: Capacity

Capacity is a quality requirement. It is defined as the *degree to which the maximum limits of a product or system parameter meet requirements*[35]. Parameters may include the number of concurrent connections, the amount of bandwidth, the throughput and latency of transactions, and so on. Capacity is a sub-characteristic of the Performance Efficiency characteristic in the Product Quality Model.

In WLANs, capacity is determined by bandwidth and latency[36] from the wireless medium perspective and processing capabilities from the supporting services perspective. Capacity must be considered in relation to the wireless medium and the services that support the WLAN. We will explore supporting services first as the needs are similar to wired networking systems and include IP addressing, infrastructure

---

[34] While these are the most common factors that cause the need for different requirement areas, they are not the only factors. As a WLAN designer, it is important that you evaluate each WLAN project independently to determine if specific requirements may be needed in some areas that are not needed in others. While the entire network could be designed to meet the most demanding requirements, in most cases, the cost of the network would increase significantly. By defining requirement areas, you can reduce costs and increase user satisfaction. In some cases, the differences from one requirement area to the next is not a cost factor as the only difference is in the configuration of the APs and not the number of APs. In other cases, the differences will demand more APs, different antennas, or other cost-based factors.

[35] ISO/IEC 25010-2011. Systems and Software Engineering – Systems and Software Quality Requirements Evaluation (SQuaRE) – System and Software Quality Models

[36] For a good overview of the technical underpinnings of these issues in WLANs, see Andrew Von Nagy presentation on Vimeo at: vimeo.com/118975822

capacity (routers and switches) and network services capacity (Internet connectivity and IoT device service requirements). Then we will explore the wireless medium.

Capacity for Supporting Services

An area of capacity often missed, much to administers' irritation, is IP address availability. In today's world of enterprise networks, office spaces are becoming more congested daily, with smaller workspaces and more employees occupying the area as well as more devices being carried and used by each of these employees. In scenarios like these, the first thing that takes a hit is not always network performance. Instead, it is the IP address pool. Designing a corporate WLAN usually means it will connect to an existing wired LAN, which means the existing VLANs will more than likely pick up the responsibility to issue IP addresses to the APs connected to switchports on the existing LAN. That may not seem to be much to become alarmed about, but in the case where 300 APs are deployed it can be.

The next place where IP pools take a hit is underestimating the client count. An example would be when a company has 150 employees that all have desktop computers. The LAN designer may have set up a /24 network[37] for this scenario, believing there would be at least 75 addresses left for new employees, added servers and printers, and a guest or two over the years. Then along comes Wi-Fi. A couple of controllers, thirty or forty APs, a few added switches and the spare IP addresses are gone. Now, suddenly, there are 150 employees with two and three devices each that connect to the WLAN and pull an IP address even if they are not actually using the device. This is not the only concern you should have about capacity, but it is an issue and should be part of the conversation during the discovery stages of the project. VLAN expansion has to be dealt with often and adding a WLAN to an existing network is a project that will consume IP addresses as well as require more VLANs to be created. We will discuss this further later in this chapter.

Infrastructure capacity is also important. Some WLANs tunnel all traffic through the wired network to a WLAN controller. Others direct traffic from the APs directly to the intended destination. In both cases, data from wireless stations is ultimately traversing

---

[37] A /24 network is a reference to a network using a 24-bit subnet mask for the IP addresses on the network or a subnet mask of 255.255.255.0. To learn more, explore any good Network+ or CCNA study guide.

the wired network to reach some end point or data is coming from the wired network to the wireless stations. In an environment with no WLAN or with insignificant WLAN usage and new plans will introduce or increase WLAN usage, the capacity of the wired network must be prepared to handle the load from the WLAN as well. It cannot be assumed that the wired network will simply keep up. Switches and routers should be evaluated to ensure that they can handle the load. This task may not be the responsibility of the WLAN designer, and often it is not, but the WLAN designer should be careful to point out this consideration to the appropriate group within the organization.

Internet access demands attention as well. When adding a WLAN the users are likely to increase Internet access demands, particularly those accessing the WLAN from their mobile devices like phones and tablets. They will use them for personal and business Internet access in restrooms, breakrooms, meeting rooms, their offices, and as they roam the halls. Many organizations allow employees to access a guest WLAN from their personal devices and some even allow access to the corporate WLAN. Either way, the increase Internet access capacity must be considered.

It's not just about their active actions, but the inactive actions performed by the mobile devices must be considered. These devices will check social media for updated, email updates and dozens of apps could be installed on the phone that check for updates as well. These actions from hundreds or thousands of personal devices can significantly increase Internet consumption. Therefore, in many WLAN deployments, it is important to ensure that the organization has proper bandwidth[38] at the Internet connection point(s).

---

[38] The term bandwidth is used in multiple ways, and this can become particularly confusing in literature related to wireless communications. In wireless communications, bandwidth typically references the total frequency space used in a communication. It is not even safe to use it for the total frequency space used in a channel. Within cellular and Wi-Fi systems that use OFDMA, a communication may use far less bandwidth that that available in the channel. Another way bandwidth is used, and equally valid, is to reference the transmission capacity of a computer network in general. Context well help to understand which use is intended. By the way, the third use of the term is equally valid, given that dictionaries are our only source of validity for word usage. The third use has to do with workload capacity and is mostly used to refer to humans and their ability or lack thereof to take on more work but has been used to a lesser degree in reference to the bandwidth of a system. For example, "the server lacks the bandwidth," as a reference to the computing capacity of the system rather than the communications capacity.

IoT-related services should also be considered if Wi-Fi is used to implement IoT devices. Many IoT devices require special network servers, such as MQTT servers that act as brokers between the IoT devices and other devices or systems and database servers that provide the storage services for the IoT devices.

Technically speaking, IoT devices on the WLAN would simply be another use case; however, historically when performing WLAN design, most designers have focused only on the application bandwidth requirements for the various applications running on the client devices. These devices, however, have been used mostly by direct human interaction (though exceptions have certainly existed). A delay in application responsiveness simply meant that the human waited a little longer for the response. What was acceptable was based on the experience of the user with the same or similar applications in the past. If the current system is as responsive or more responsive, the users were typically happy.

The same cannot be said for many IoT devices. The device is not emotionally attached to the responsiveness of the network based on past experiences. The device needs the response in a given window of time to perform the action required and the action may be safety critical. This has been true for many Wi-Fi applications in healthcare and military deployments for years but is now true for a large majority of IoT devices that act as actuators[39] through communications across the WLAN. Figure 3.1a illustrates the difference.

Consider an actuator that turns off a machine if it reaches dangerous or near-dangerous temperature levels. Random delays, like those often experienced when browsing the Internet, are unacceptable in such IoT scenarios. Two options exist: perform on device processing to shut the machine down or implement network capacity requirements that ensure the machine will receive a timely response to indicate the action that should be taken. A failsafe mechanism may also be required. For example, if the IoT device does not "hear" back from the analysis system within 70 ms, it may shut down the machine. The only purpose of the analysis system (on the other end of the network link), in this

[39] An actuator is a component of an IoT device that allows the device to interact with the physical world. It is a component of a cyber physical system. Some IoT devices contain only sensors, some contain only actuators, and some include both. For example, an actuator can turn a machine on or off, increase or decrease flow in a pipe, increase or decrease temperature in a boiler, speed up or slow down the rate of movement in a robot arm, and so on.

case, is to prevent shutting down the machine if some other factor makes the current temperature acceptable. Otherwise, the actuator can shut down the machine to maintain safety for nearby humans and to protect the functional health of the machine and machines around it.

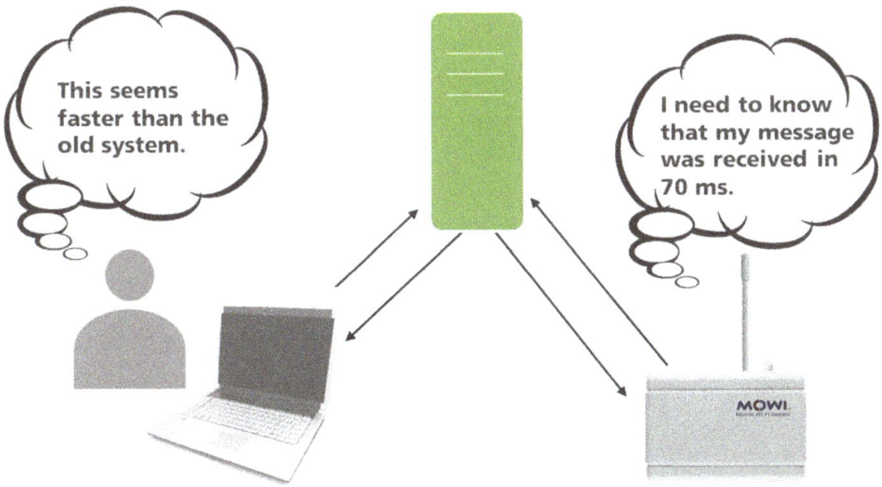

Figure 3.1a: Delay Tolerance of Humans vs. IoT Devices

Capacity on the Wireless Medium

In a WLAN, capacity is a characteristic of a channel and not of a BSS. In other words, if you have multiple BSSs in the same channel, the capacity of the channel is decreased by the existence of more BSSs (and therefore APs, whether virtual or physical) in that channel. Each AP must attempt to transmit a Beacon frame, by default, about 9.76 times per second. If you implement four BSSs in the channel, the requirement is now the transmitting of Beacon frames at a rate of about 39 times per second. The overhead from the SSID-based Beacons alone can consume from 10-13% of the channel capacity when running four SSIDs. Additionally, if you run multiple physical APs in the channel with the same SSID, similar capacity loss results.

Now, when you factor in all the clients attempting to communicate through the channel, you reduce the capacity even more. Given that multiple BSSs are running in the same channel and multiple clients are associated with each BSS, the total channel capacity consumption is the total of all client station and AP station transmissions in the channel. The point is that capacity is a factor of the total devices in the channel and not of the total devices in a BSS. BSSs that use the same channel in the same contention domain (coverage area) impact each other's capacity.

This kind of medium capacity demand can be seen by exploring airtime utilization in a spectrum analyzer. What percentage of time is the channel "busy" with RF energy above the noise floor? Each device has a duty cycle – the amount of time the device is communicating on the RF medium. If each device must be active .5 percent of the time and the APs are consuming 8 percent of the time, then the channel can only handle 184 such devices. Using the full capacity of the channel, 184 devices, would leave no room for spikes in activity. At the same time, it would leave no allowance for some times of inactivity for whatever reason. Sadly, the real world isn't even as simple as this. You will have some devices that require less than .1 percent of the time and others that require varied amounts of time up to some maximum, say 3 percent of the time. Significant information must be gathered about the client device types so that you can properly calculate capacity demands.

Client count and type is an important consideration in a requirement area. Many businesses have different use cases throughout the facility and some of them are simply storage areas that require little to no wireless coverage. Others may be warehouse space where scanner guns are used for inventory, and then there is the busy congested area packed with as many employees possible to save space and utilities. Out of the three areas mentioned here, they all require different coverage and capacity to support the demand for signal and throughput. This is where designing for capacity becomes important.

Let's say that AP model A is going to be used in this facility and it will support 20 devices with 10 Mbps throughput each. The small warehouse only has about fifteen scanner guns being used. It will be fine to deploy one AP model A on each end of the warehouse to provide adequate coverage throughout the warehouse as well as more than enough throughput for the fifteen scanner guns and then some.

Let's look at the office area condensed with employees and hundreds of mobile devices. During your site survey, you found that two of the model A APs provided great coverage for the office area, and not a single spot anywhere was found to have a signal level less than VoIP quality. The issue now is capacity and it has conflict in multiple places. For starters, remember that AP model A can support 20 devices with 10 Mbps throughput each, so, according to the survey, you have the ability to support 40 devices with 10 Mbps each in the office area now. If you simply look at the heat map of your survey software, the coverage is great and you're on to the next area, but your survey is incorrect because capacity was not considered.

Let's toss in some more hypothetical numbers and say there are 75 employees in this two–access-point survey area, and each of those employees has a laptop, a tablet, a smart phone, and a wireless company VoIP phone. Quickly the scenario has changed due to the number of devices in the coverage area. Instead of the assumed 40 employees, there are 75 employees. Instead of a single device per employee, there are four devices per employee. This makes the device count go from the assumed 40 to 300. With these numbers, let's go back to the specifications of AP model A that is being installed. Forty devices can utilize 10 Mbps throughput from a single AP, and we have two of them in the office area, providing a total of 800 Mbps throughput there. Just doing the simple math and not taking into consideration any rate shifting due to distance from the AP, 800 divided by 300 = 2.67 Mbps per device. This would probably be acceptable in the warehouse with scanner guns only, but it will not be sufficient for anything in the office work environment. Something worth mentioning again is the fact that the heat map on your survey software shows –67 dBm or higher throughout the entire office area, but the design is still inadequate because capacity wasn't taken into consideration.

The next factor to be considered for capacity calculations is the requirements of the devices being used by the employees. If you performed the site survey utilizing a dual-antenna wireless NIC on a laptop in order to provide the –67 dBm coverage throughout the area on the heat map, then you are going to have some signal issues if the wireless VoIP phones being used have only a single antenna. The point here is that your WLAN should be designed for the weakest/lowest powered device being used. The laptop would be the device that has the best transmit power out of the four devices mentioned, so you would not want to base your design on its receive signal strength.

Finally, you want to look at the applications that will be functioning on each device and what the throughput demands are for each application as well as the duty cycle of these applications. In other words, if all employees will be using an application on their laptops that demands 20 Mbps and all employees will be utilizing this application for one hundred percent of the workday, then you have a 1500 Mbps demand on the WLAN all day long just from a single application on laptops alone. Based on the ability already known about AP model A, it would take four of them just to supply enough throughput for the single application being utilized on the laptops, without even considering other applications or the other devices. For this reason, capacity considerations must be factored into the design before you perform a survey for highly congested areas.

3.3: Quality Requirements and WLANs (ISO/IEC 25030-2019)

The IEEE 29148-2018 standard, explained in the preceding chapter, references the IEEE 15288-2015 standard, which in turn references the ISO/IEC 25030-2019 standard in relation to quality as a requirement in systems. Quality is an important requirement in WLAN systems from several perspectives and this section will explain the 25030-2019 standard and apply it to a WLAN solution Figure 3.1b provides a high-level overview of the processes involved in establishing quality requirements based on these standards[40].

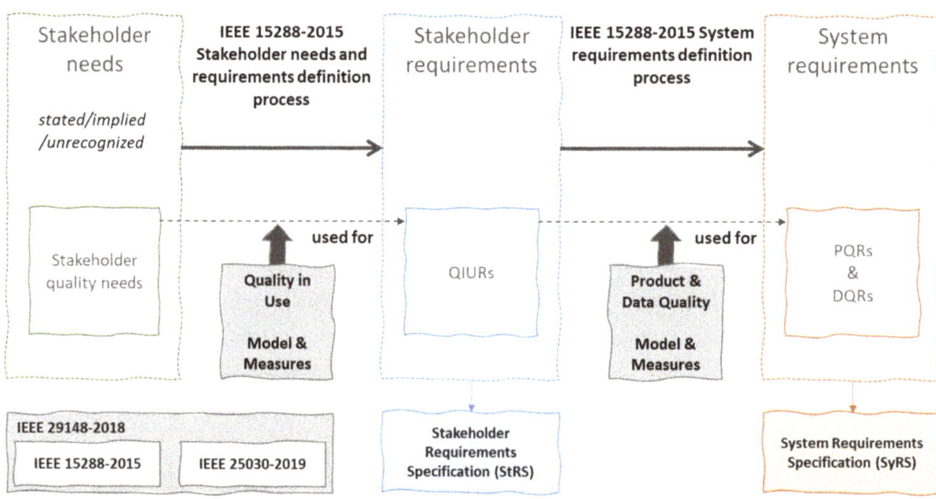

Figure 3.1b: IEEE 25030-2019 Relation to Other Requirements Engineering Standards by the IEEE, ISO, and IEC (Reproduced from ISO/IEC 25030-2019)

---

[40] It is highly recommended, as a network design professional, that you become intimately aware of the requirements standards available from the IEEE and ISO/IEC. They are based on decades of experience and bring tremendous value to the knowledge base of any network or systems engineer.

Some terms in Figure 3.1b require an explanation:

- **Quality in Use Model:** A quality model defining five characteristics related to the outcomes of interaction with a system: effectiveness, efficiency, satisfaction, freedom from risk, and context coverage.
- **Product Quality Model:** A quality that categorizes system quality properties into eight characteristics: functional suitability, performance efficiency, compatibility, usability, reliability, security, maintainability, and portability.
- **Data Quality Model:** A quality model that defines fifteen characteristics for data quality requirements and that is beyond the scope of this material.
- **QIURs:** Quality in Use Requirements
- **PQRs:** Product Quality Requirements
- **DQRs:** Data Quality Requirements

From figure 3.1b, you can determine that the stakeholder needs are used to derive stakeholder requirements. These requirements are based on the Quality in Use Model and result in Quality in Use Requirements (QIURs). From the QIURs we derive the Product Quality Requirements (PQRs based on the Product Quality Model) and the Data Quality Requirements (DQRs based on the Data Quality Model). The PQRs are of most importance to us and, in fact, the DQRs are defined in a completely different standard from the QIURs and PQRs. Within the IoT space, DQRs are equally important, and this can be true in Wi-Fi-based IoT as well. For more information on that standard, see the CWIDP certification in the CWNP wireless IoT track.

To further illustrate how quality requirements are derived from processes with users and stakeholders as the source of needs leading to requirements, and from other information and communication technologies (ICT) systems as constraints, consider Figure 3.1c. This figure illustrates the derivation of quality requirements from different sources at different levels. Users and other stakeholders, such as regulators, managers, and industry standards groups, lead to QIURs or stakeholder requirements. Then, within the information system solution, multiple ICTs may exist. For example, the wired network infrastructure could be considered an ICT and the individual services running on that network, such as DHCP, NTP, and Internet access, could be considered ICTs; however, they are not the target ICT under the scope of requirements development for the WLAN.

If we take the WLAN to be the inner-most ICT on the left side of the image, PQRs (our primary concern with a WLAN) will be derived from the QIURs and the environment in which the WLAN will operate. Our greatest concern will be the requirements for the WLAN in the context of the existing systems, which are in the context of the user-based, which provided the QIURs. You can begin to see the hierarchy of derivation. At the same time, the non-target ICTs, for example, the wired network and network services previously mentioned, certainly give requirements as constraints to the WLAN. For example, if we cannot upgrade the wired network, the WLAN will be limited to the current capabilities of the network. If we cannot add new services, the WLAN will be limited to the capabilities of the existing services. At the same time, we should not assume that the wired LAN cannot be upgraded or that the services cannot be upgraded. Known what is available on the wired network, in relation to supporting services for the WLAN is important both for recommendation upgrades and for designing within constraints[41].

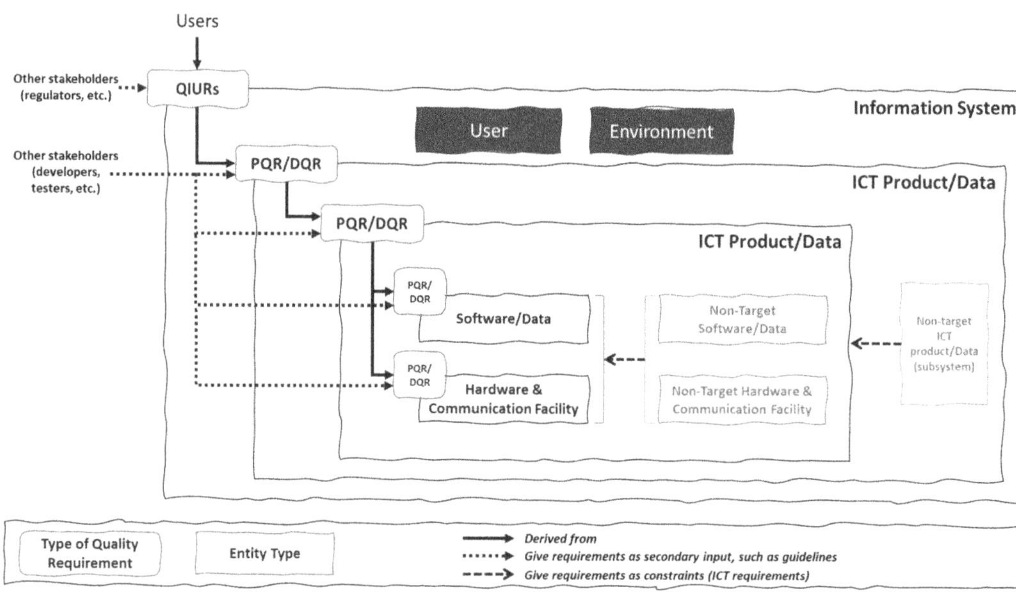

Figure 3.1c: Derivation of Quality Requirements (Reproduced from 25030-2019)

---

[41] At this point, I hope you're beginning to see the value of the standards-based requirements engineering processes that we began exploring in Chapter 2 and continue to explore in this chapter. They provide an excellent model and, possibly of greater importance, they provide a strong argument to go to management and plead for effective requirements engineering in the design process. The first time or two that you use these models in a design project, they may seem overwhelming, but just like mastery of the 802.11 protocols, they quickly become second nature.

As stated previously, the Quality in Use Model defines five characteristics. These characteristics are represented in Figure 3.1d.

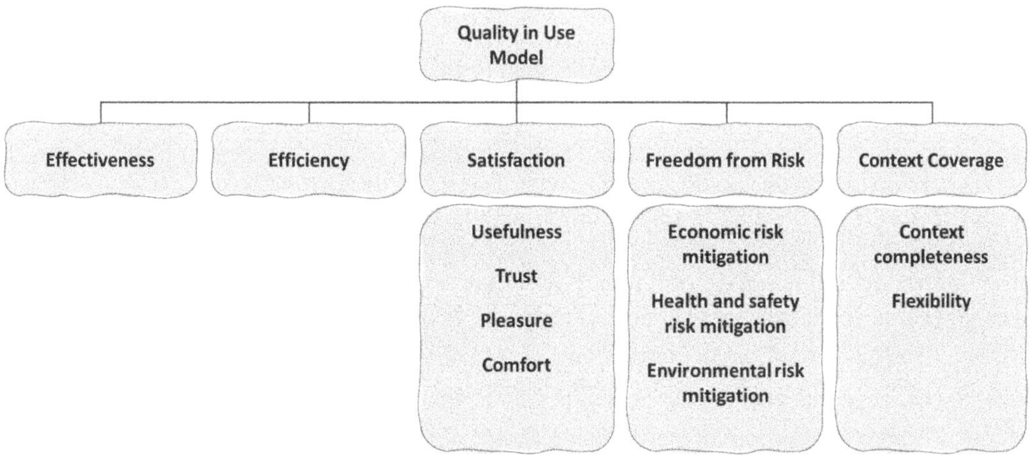

Figure 3.1d: Quality in Use Model with Characteristics and Sub-Characteristics

Effectiveness is the accuracy and completeness with which users can achieve specific goals with the system. In a WLAN, this will be most impacted by connectivity and consistent connectivity. The focus is on completeness through the ability to maintain a stable connection throughout the activity performed. In some cases, this stakeholder need may demand that the device in use can maintain connectivity while the user moves throughout the facility performing a task. If the connection is lost for a significant amount of time, the application may reset causing a loss of previously performed actions, or a state of incompleteness. A stakeholder requirement statement coming out of this quality characteristic might be: *The WLAN must provide connectivity in the warehouse as users move from place to place with roaming actions that do not result in the loss of connectivity at the application layer.*

Efficiency is a quality characteristic that encompasses the resources expended in relation to the accuracy and completeness with which users achieve specific goals with the system. The focus is on accomplishing effectiveness with only the required resources and while avoiding overengineering. A stakeholder requirement statement coming out of this quality characteristic might be: *The WLAN shall be implemented using the required number of network connection points to achieve effectiveness and satisfaction and no more.*

Satisfaction is the first quality characteristic in the model to have sub-characteristics. Satisfaction is the degree to which user needs are satisfied when a product or system is used in a specified context of use. It is a factor of usefulness, trust, pleasure, and comfort. *Usefulness* is a measure of the perceived ability to achieve goals using the system. *Trust* is a measure of the degree to which the user can expect the system to function as intended. *Pleasure* is a measure of the degree to which a used fulfills personal needs through the use of the system, including acquiring new skills and knowledge, communicating personal identity, and provoking pleasant memories[42]. Finally, *comfort* is a measure of the degree to which the user is satisfied with physical comfort when using the system.

As an illustration of the last sub-characteristic, think of the commercials and movies where people are standing on chairs, rooftops, or other uncomfortable locations in an attempt to gain a signal for their cellphones. This is an example of poor quality in relation to the comfort sub-characteristics. A stakeholder requirement statement coming out of this quality characteristic (satisfaction) might be: *The WLAN shall function in all areas of the facility such that users are never required to change locations to gain access and that function shall provide stable and consistent connections for the achievement of lengthy processes which require state endurance over an extended period*[43].

Freedom from risk is a quality characteristic that defines the degree to which the product or system mitigates the potential risk to economic status, human life, health, or the environment. It includes three sub-characteristics: economic, health and safety, and environment risk mitigation. A stakeholder requirement statement coming out of this quality characteristic might be: *The WLAN shall be implemented within {regulatory agency} guidelines to avoid exposure to harmful RF energy and shall be powered through energy harvesting solutions that minimize energy consumption from the power grid*. Another example statement might be: *The WLAN shall be implement with proper security to protect against the*

---

[42] Provide me with a Commodore 64 keyboard interface to the system and you will achieve a significant level of pleasure-related quality for me.

[43] This is a good place for a point of clarification. At this stage, you are not defining system requirements. You are defining stakeholder requirements. From the stakeholder requirements you will derive system requirements. As you design the WLAN, the system requirements specification will guide your design. As long as they were developed with proper links to stakeholder/user requirements and business requirements, you should not be required to recursively evaluate the source of the system requirements during the design process.

theft of personal privacy information or corporate intellectual property that could cause financial harm to the individual or organization.

Context coverage is the final quality characteristic in the model and is a measure of the degree to which the system can be used with the other four characteristics in the context or environment in which it will operate. This characteristic includes two sub-characteristics: context completeness and flexibility. *Context completeness* is a measure of the ability of the system to be used with effectiveness, efficiency, satisfaction, and freedom from risk in all specified contexts of use. Stated differently, you could say that the system can be used in all requirement areas while meeting the effectiveness, efficiency, satisfaction, and freedom from risk requirements in those area. *Flexibility* is a measure of the degree to which the system can be used in contexts beyond those initially specified in the requirements. That is, how challenging will it be to adapt the selected WLAN solution to new requirement area definitions or completely new locations?

As you can see, the Quality in Use Model surfaces many important considerations in the design of a WLAN. Such quality requirements should be considered in the requirements engineering process and for each requirement area, when needed.

The Product Quality Model is used to define quality requirements at the system level. It will not be treated as exhaustively here but is well-documented in the ISO/IEC 25010-2011 standard and brief definitions of the characteristics and sub-characteristics follow. Figure 3.1e illustrates the Product Quality Model used for system quality requirements.

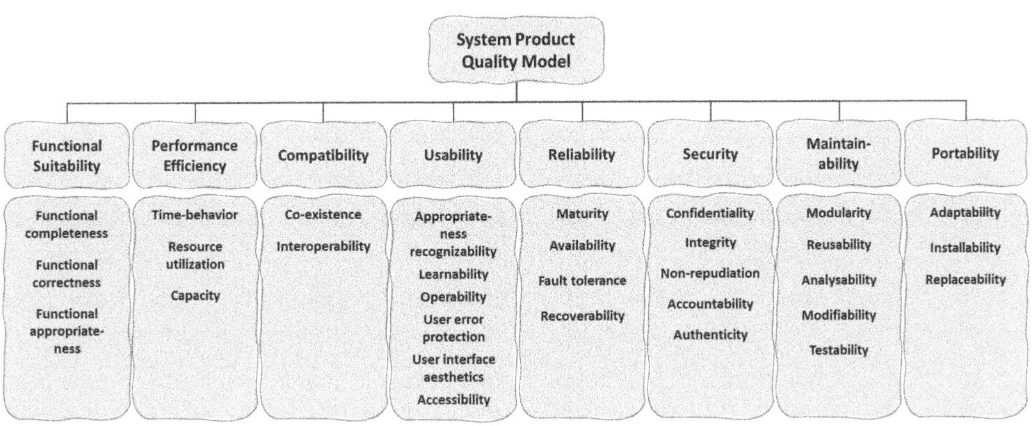

Figure 3.1e: Product Quality Model Characteristics and Sub-Characteristics

The characteristics and sub-characteristics are defined as[44] (those most important to WLAN requirements are indicated in color, though all may apply in some contexts):

- **Functional Suitability:** degree to which a system provides functions that meet stated and implied needs when used under specified conditions
    - *Functional completeness:* degree to which the set of functions covers all the specified tasks and user objectives
    - *Functional correctness:* degree to which a system provides the correct results with the needed degree of precision
    - *Functional appropriateness:* degree to which the functions facilitate the accomplishment of specified tasks and objectives
- **Performance Efficiency:** performance relative to the amount of resources used under stated conditions (spectrum can be considered as a resource in wireless networking system)
    - *Time-behavior:* degree to which the response and processing times and throughput rates of a system, when performing its functions, meet requirements
    - *Resource utilization:* degree to which the amounts and types of resources used by a system, when performing its functions, meets requirements (consider spectrum for Wi-Fi)
    - *Capacity:* degree to which the maximum limits of a system parameter meet requirements (parameters can include number of concurrent users, communication bandwidth, throughput of transactions, etc.)
- **Compatibility:** degree to which a system can exchange information with other products, systems, or components, and/or perform its required functions, while sharing the same hardware or software (or spectrum) environment
    - *Co-existence:* degree to which a system can perform its required functions efficiently while sharing a common environment and resources with other products (or systems), without detrimental impact on any other product (or system)
    - *Interoperability:* degree to which two or more systems can exchange information and use the information that has been exchanged (in the context of Wi-Fi: roaming capabilities)

[44] ISO/IEC 25010-2011, clause 4.2, Product Quality Model

- **Usability:** degree to which a system can be used by specified users to achieve specified goals with effectiveness, efficiency, and satisfaction in a specified context of use
    - *Appropriateness recognizability:* degree to which users can recognize whether a system is appropriate for their needs
    - *Learnability:* degree to which a system can be used by specified users to achieve specified goals of learning to use the system with effectiveness, efficiency, freedom from risk, and satisfaction in a specified contest of use
    - *Operability:* degree to which a system has attributes that make it easy to operate and control
    - *User error protection:* degree to which a system protects users against making errors
    - *User interface aesthetics:* degree to which a user interface enables pleasing and satisfying interaction for the user
    - *Accessibility:* degree to which a system can be used by people with the widest range of characteristics and capabilities to achieve a specified goal in a specified context of use
- **Reliability:** degree to which a system performs specified functions under specified conditions for a specified period of time
    - *Maturity*: degree to which a system meets needs for reliability under normal operations
    - *Availability*: degree to which a system is operational and accessible when required for use
    - *Fault tolerance*: degree to which a system operates as intended despite the presence of hardware or software faults
    - *Recoverability*: degree to which, in the event of an interruption or a failure, a system can re-establish the desired state of the system
- **Security:** degree to which a system protects information and data so that persons or other systems have the degree of data access appropriate to their types and levels of authorization
    - *Confidentiality*: degree to which a system ensures that data are accessible only to those authorized to have access
    - *Integrity*: degree to which a system prevents unauthorized access to, or modification of, computer programs or data

- *Non-repudiation*: degree to which actions or events can be proven to have taken place, so that the events or actions cannot be repudiated later
- *Accountability*: degree to which the actions of an entity can be traced uniquely to the entity
- *Authenticity*: degree to which the identity of a subject or resource can be proved to be the one claimed

- **Maintainability:** degree of effectiveness and efficiency with which a system can be modified by the intended maintainers
    - *Modularity*: degree to which a system is composed of discrete components such that a change to one component has minimal impact on other components
    - *Reusability*: degree to which an asset can be used in more than on system, or in building other assets
    - *Analysability*: degree of effectiveness and efficiency with which it is possible to assess the impact on a system of an intended change to one or more of its parts, or to diagnose a system for deficiencies or causes of failures, or to identify parts to be modified (monitoring in WLANs)
    - *Modifiability*: degree to which a system can be effectively and efficiently modified without introducing defects or degrading existing product quality
    - *Testability*: degree of effectiveness and efficiency with which test criteria can be established for a system and tests can be performed to determine whether those criteria have been met
- **Portability**: degree of effectiveness and efficiency with which a system can be transferred from one hardware, software or other operational or usage environment to another (from cloud management to on-premises, from autonomous to controller-based, etc.)
    - *Adaptability*: degree to which a system can effectively and efficiently be adapted for different or evolving hardware or other operational usage environments (adding 802.11ax, incorporating 6 GHz, etc.)
    - *Installability*: degree of effectiveness and efficiency with which a system can be successfully installed and/or uninstalled in a specified environment (short-term WLANs for mobile operations, etc.)
    - *Replaceability*: degree to which a system can replace another system for the same purpose in the same environment (WLAN upgrades)

As you can see, this standard provides an exhaustive set of characteristics and sub-characteristics to consider for system quality requirements. The WLAN you are planning is the product, and the system quality requirements define the quality-related capabilities that should exist in the system.

3.4: Client Device Types

As mentioned in a previous topic, when designing a WLAN, providing for the needs of client devices is a useful design driver. In many cases, designing for the least capable client (typically a mobile phone or barcode scanner) is a best practice. This section introduces common capabilities of various client devices, including laptops, tablets, mobile phones, and stationary devices. It is a good practice to collect information about the intended functions of each device and their throughput demands so that capacity limitations do not become a negative factor in the final design.

> No matter how little impact you may think a given client device has, the number of these devices used can impact WLAN performance, especially when factored in with all other wireless client devices and the collective duty cycle.

Laptops

Not too long ago, laptops were the most popular Wi-Fi–capable client device. Today, although laptops are still quite popular, tablets and mobile phones have surpassed them as the most commonly used mobile devices. Laptops are still popular and used in most places, and they are manufactured by many vendors. The components within laptops vary from one model to another, so it is important to do your research on them before assuming capabilities.

When considering client devices, you must ask the following questions:

- What PHYs do they support?
- What data rates are supported?
- What power level do they transmit at?

- What channels are allowed?
- What channel widths are supported?
- Do they have DFS support?
- What are the receive sensitivity levels?

Most laptops manufactured today will support 802.11n and 802.11ac. However, support for 802.11n does not always mean the 5 GHz band. Several chipsets support 802.11n in the 2.4 GHz band only. On the other hand, it is important to note that when a device supports 802.11ac, it may be a 5 GHz chipset only, and connecting to a 2.4 GHz network will not be possible. Also, just because a laptop supports the 802.11ac (VHT) PHY doesn't mean that all data rates of that PHY are supported. Therefore, though the PHY supports very high data rates, client devices will support a limited subset of these data rates depending on implemented features of the PHY. The number of spatial streams, guard interval, and channel width all come together to impact data rates.

Some clients support a limited subset of channels in 5 GHz. For example, it is not uncommon for a 5 GHz client to support only channels 36, 40, 44, 48, 149, 153, 157, and 161. These are the channels that are available in nearly all parts of the world. If you have a large number of clients supporting only these channels, the WLAN design should include at least one of these channels in all of the required coverage areas. Additional cells can be created using other channels for capacity, but support for the client base must be provided.

Most laptops today support both 20 and 40 MHz channels for 802.11n and many 802.11ac laptops now support the 80 MHz and 160 MHz channels even though support for these channel widths is not likely to be seen in enterprise WLANs.

DFS channels may or may not be supported in laptop clients so it is also important to discover this information before designing channel use to include these channels when a majority of your client devices may not be able to use them.

Finally, it is important to understand receive sensitivity ratings. Laptops tend to use chipsets and antennas that provide for better sensitivity ratings than tablets or mobile phones. This is why it was mentioned earlier in this chapter that the WLAN should be designed around the sensitivity of the least capable devices. If the low-powered tablet or mobile phone's connectivity is working as it should with a less capable receive sensitivity, then the more capable, higher-powered laptops or stationary devices will also have great connectivity given that their receive sensitivity ratings are much better. The receive sensitivity rating defines the lowest signal strength at which the receiver can demodulate a given data rate signal. Receive sensitivity is impacted by channel width as well. That is, when using 40 MHz channels, a better signal across the channel is required to get the same MCS as when using a 20 MHz channel.

At the time of writing, several laptops have come to market including 802.11ax support. However, Wi-Fi 6E (802.11ax in 6 GHz) has not come in the laptop market at the time of writing, but they will begin to come into the market in the next 3-6 months from the time of writing. If you are reading this in early-to-middle 2022, they are likely available. If you have plans to deploy a 6 GHz network, purchasing laptops with Wi-Fi 6E capabilities going forward will be an important requirement.

Tablets

Most tablets today support 802.11n and 802.11ac PHYs. However, it is important to remember that most of them are only one- or two-stream clients. In other words, most tablets are not capable of performing at data rates available for laptops, because laptops have more data streams. Most tablets also have lower receive sensitivity ratings than laptops.

Wi-Fi 6 (802.11ax) tablets are available in the market at the time of writing and Wi-Fi 6E tablets should be coming to market soon. During the lifetime of this version of CWDP, we will see the introduction of more common use of the 6 GHz band for Wi-Fi as more and more regulatory domains approve its use. WLAN designers must be prepared to plan for it. The RF design will be very similar to 5 GHz (the frequency bands are not significantly higher), but they must be prepared to address the new security constraints imposed in that band, which are discussed later in this book.

Mobile Phones

Most mobile phones support only one stream, and very few of them support two streams. Therefore, with mobile phones, it should be assumed that only single-stream data rates will be available for most of these devices as of 2021. In fact, it is unlikely that multiple-stream mobile phones will gain majority market share because the battery life is significantly reduced, unless the device associated to the AP as a single stream device when batter drops below a certain level. However, most users would then have poor quality of experience and become frustrated. Time will tell how mobile phone vendors deal with this issue.

Stationary Devices

Stationary devices fall into two categories: desktop computers and specialty devices. Desktop computers provide exceptional flexibility to the WLAN design as they can be easily upgraded with PCI adapters or USB adapters.

Specialty devices are a different story altogether. These devices, often found in manufacturing and healthcare, may use USB adapters to connect to WLANs, but in many cases are restricted to a specific adapter, and many of these adapters are still 802.11b/g or 802.11a devices. The reason is simple. These specialty devices often cost many thousands of dollars. They are expected to be used for 5–10 years. Therefore, as they near the end of their life, they still utilize older adapters. Providing for such devices is an essential part of the design process because they are often beyond support agreements and drivers cannot be acquired for the custom operating systems to use newer USB adapters.

Barcode Scanners

Barcode scanners are used for inventory management in many industries. The devices have gone from being computer-connected devices that depend on the computer to do all the processing to devices with localized technology for entering locations, quantities, and more. This information is then passed back to a server using 802.11 WLAN connections.

Such barcode scanners often use 802.11b/g technologies, with a few supporting 802.11a. This older technology will constrain your design in areas requiring use of barcode scanners, and even though some can connect to 802.11n APs (or 802.11ac APs in 5 GHz), they will not benefit from the newer technologies. However, given that their throughput demands are very low, this is not a significant concern if 802.11b data rates are not utilized.

Healthcare Devices

Since the evolution to paperless record-keeping, WLANs have become an essential part of the healthcare industry. The primary concern with healthcare devices is availability and durability. The network must be available to the devices, as the information they communicate is critical. This means coverage and capacity become important at all new levels. Every location must be covered by at least two APs, and the throughput capacity must be sufficient such that critical communication packets are not lost.

Most devices used in healthcare are either stationary or nomadic. The nomadic devices are indeed portable but are not typically used while in motion. Instead they are transported to a room or treatment area and are then utilized in that location. Such equipment includes mobile x-ray units, pulse oximeters, intravenous pumps, and vital sign monitors.

Highly mobile devices that require connectivity during motion are Wi-Fi phones, Vocera communication badges, and real-time location tracking devices. The same kinds of demands are placed on these highly mobile devices in healthcare as in other environments; however, in healthcare, as in emergency response, the stakes are higher.

ID Badges

ID badges have traditionally been magnetic strip-based; however, newer badges are using RFID, Bluetooth, and other near-field communications (NFC) technologies. Figure 3.1 shows a communications badge from Vocera. This particular badge is an excellent example of what might be used in healthcare settings. The badge is dual-band so it can connect to both 2.4 GHz and 5 GHz networks. It also provides hands-free operation with its high-quality voice recognition capabilities. It supports the use of many authentication protocols, which makes its operation in healthcare more popular.

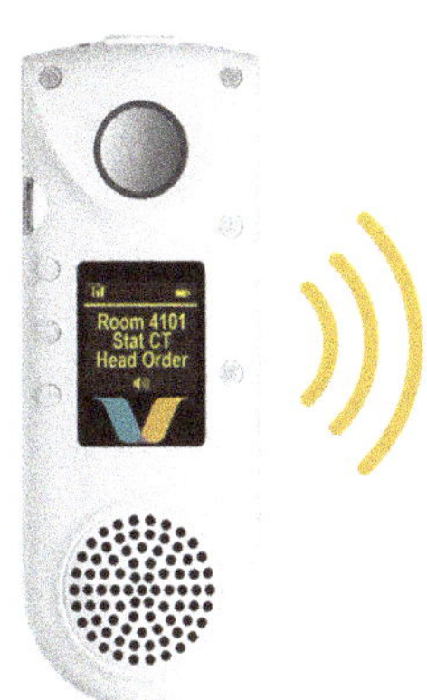

Figure 3.1: Vocera Communications Badge

Motion or Vibration Sensors

This type of wireless device is almost guaranteed to have only one stream and a low sensitivity rating. It is commonly found in manufacturing and power production facilities. Used to keep a continuous log of vibrations within large motors or power generating equipment, it is more efficient to allow software to capture continuous data and measure tolerant or intolerant vibrations in equipment. Many of these devices function only in 2.4 GHhz and require only a very low amount of throughput to perform data collection. Others function through BLE and 900 MHz.

Location Tracking Systems

The Ekahau A4+ is an example of an asset-tracking tag. It is shown in Figure 3.2 and is also a 2.4 GHz-only device. Running off a USB power source with built-in battery backup, the tag continues to operate even if unplugged temporarily. The range of this device is not specified in the documentation, but it runs at 11.5 dBm output power and should have a range of 50–100 meters in free space.

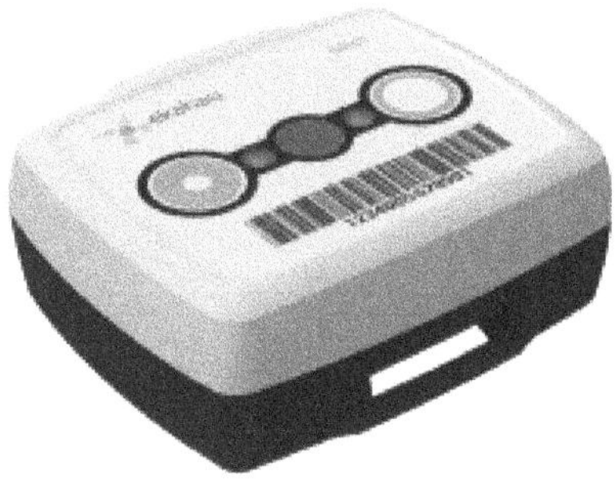

Figure 3.2—Ekahau A4+ location tag with 802.11g

Wireless Cameras

Wireless cameras are IP cameras that rely on 802.11 connections to the network rather than Ethernet connections. They transmit live video streams, or they store video streams on local storage (SD cards, for example) to be accessed and transmitted on demand. Most devices transmit the video streams to a central server that then records the video locally. The throughput demands of the devices will depend on the video codecs and quality implemented. In most cases, the quality can be adjusted and, when lowered, will reduce the throughput demands.

Internet of Things (IoT) Devices

IoT devices are covered lightly in the CWNP Wi-Fi certification track (CWNA, CWAP, CWSP, and CWDP, leading to CWNE)[45], but it is important to consider a few unique characteristics of these devices. IoT devices that use Wi-Fi are still Wi-Fi client devices, but they may have a few unique constraints:

- IoT devices are often single-stream devices to conserve battery power, if powered in this way.
- IoT devices often use 802.11n chipsets instead of 802.11ac or 802.11ax, though we may see this change with the new enhancements in 802.11ax, specifically Target Wake Time (TWT) and OFDMA, which can provide significant value to IoT devices.
- IoT devices most frequently transmit small amounts of information frequently or infrequently, but rarely transmit large payloads. The exceptions are video-related and augmented-reality-related IoT solutions.
- Many IoT devices use non-Wi-Fi protocols or support Wi-Fi and other protocols, such as BLE (Bluetooth Low Energy).

3.5: Applications and their Requirements

Designing for applications is about designing for application behavior. Application behaviors impact roaming and network requirements like latency, throughput, and availability. This section provides information to assist you in planning for various application requirements and several specific application types.

---

[45] For more detailed information on IoT devices related to Wi-Fi, see the CWISA Study and Reference Guide. For more detailed information on IoT devices that use other protocols, see the CWICP Study and Reference Guide. For more detailed information on additional and important concepts related to wireless IoT, see the CWIIP and CWIDP study and reference guides.

Roaming Requirements

All mobile devices will at some time be required to roam from one BSSID to another. For example, a laptop may need to roam from one AP to another, but depending on the application running on the laptop, it may not require continual Layer 3 (IP addressing) connectivity. However, if that same laptop is running softphone software and a call is placed on it, the user may desire to maintain a call while roaming.

The following devices have demanding roaming requirements:

- Mobile phones acting as VoIP phones
- 802.11 VoIP handsets
- Some push-to-talk devices

The following devices have less demanding requirements:

- Tablets
- Laptops
- Mobile phones not acting as VoIP phones
- Barcode scanners
- Mobile healthcare devices

Planning for Roaming in the Design

Client device roaming works in one of several ways:

- Layer 2 roaming across APs within a single controller or without a controller
- Layer 2 roaming across APs connected to separate controllers
- Layer 3 roaming

When Layer 2 roaming occurs, the client device maintains the same IP address, and roaming times average 40ms or less. Layer 2 roaming across APs managed on a single controller is called *intra-controller roaming,* while Layer 2 roaming across APs on separate controllers is called *inter-controller roaming*.

The 802.11 standards define what should take place as a client roams from one AP to another, but they do not provide specifics on how communications must occur. This flexibility allows vendors to provide competitive features in this area. The good news is that the major vendors all have intra-controller and inter-controller Layer 2 roaming solutions that provide sufficient roaming speeds for wireless VoIP.

Layer 3 roaming takes place as the client roams to an AP that cannot provide the same IP configuration, because it is connected to a different subnet. In this scenario, the IP address must be reallocated from the DHCP server, and the client is placed on a new subnet. This action causes the client's Layer 3 connections to be lost, and if data was in transition at the time of the relocation, the process will have to be started again on the new subnet. With VoIP applications, the same holds true. The call will drop and have to be re-established after the move is complete and the phone receives an IP address on the new subnet. To solve this issue, we have 802.11 protocols that define fast, secure roaming.

Implementing Layer 2 roaming without the loss of the client's IP configuration isn't difficult in unsecured wireless networks. The problem is, before 802.11i and 802.11r, it was difficult to implement a standards-based wireless network that offered both a secure connection and fast roaming. For this to happen, client devices need the ability to roam without being required to complete another full 802.1X authentication process. This is accomplished through key caching.

802.11 Key Hierarchy

The following list shows the key hierarchy derived from the 802.1X authentication process you should become familiar with to understand the key caching process:

- **Master Session Key (MSK):** This key is generated from the 802.1X EAP exchange process and is unique to the client session. No other client will receive the same MSK. This key is then used to create the PMK.
- **Pairwise Master Key (PMK):** The PMK is created from the MSK. The Authentication server then sends this PMK to the Authenticator or WLC. The PMK is now at both the client device (Supplicant) and the WLC (Authenticator).
- **Pairwise Transient Key (PTK):** This key is generated during the four-way handshake from the PMK, Authenticator (AP) address (AA), Supplicant (client) address (SPA), Authenticator nonce (ANonce), and Supplicant nonce (SNonce). This key is calculated on both the Supplicant and the Authenticator and is never sent across the air. This is the key used by the client to encrypt data frames.

- **Group Master Key (GMK):** A supporting key that may be used to generate a group temporal key. The GMK may be regenerated within the AP periodically to reduce the exposure of the group temporal key.
- **Group Temporal Key (GTK):** The key used to protect broadcast or multicast MPDUs on a wireless link.

Now that you have a basic understanding of the key hierarchy, next we will look into some protocols and processes that help make roaming successful while using real-time applications.

- **802.11k** includes the use of neighbor reports on nearby APs so that client devices can know their surroundings ahead of time resulting in reduced roaming time.
- **802.11r** (also called fast transition roaming) provides fast, secure handoffs for wireless devices in motion by caching keys already established in the original 802.1X process.

The following key definitions from the 802.11r standard will help in understanding the enhancements to the roaming process:

- **Fast basic service set (BSS) transition:** A station (STA) movement that is from one BSS in one extended service set (ESS) to another BSS within the same ESS and that minimizes the amount of time that data connectivity is lost between the STA and the distribution system (DS).
- **Fast basic service set (BSS) transition (FT) four-way handshake:** A pairwise key management protocol used during FT initial mobility domain association. This handshake confirms mutual possession of a pairwise master key, the PMK-R1, by two parties and distributes a group temporal key (GTK).
- **Fast basic service set (BSS) transition (FT) initial mobility domain association:** The first association or first reassociation procedure within a mobility domain, during which a station (STA) indicates its intention to use the FT procedures.
- **Mobility domain:** A set of basic service sets (BSSs), within the same extended service set (ESS), that support fast BSS transitions between themselves and that are identified by the set's mobility domain identifier (MDID).
- **Over-the-air fast basic service set (BSS) transition (FT):** An FT method in which the station (STA) communicates over a direct IEEE 802.11 link to the target AP.

- **Over-the-DS (distribution system) fast basic service set (BSS) transition (FT):** An FT method in which the station (STA) communicates with the target AP via the current AP.

Pre-authentication is the first way of removing delay from the roaming process. It is used by a wireless station that hears, during the scanning process, other APs to which it may choose to connect. The full 802.1X/EAP authentication is performed over the Ethernet infrastructure for the purpose of remaining on-channel with its current AP while preparing for connectivity with another AP. Pre-authentication support is optional and is not provided by all manufacturers. If it is to be used, it must be enabled on all APs and client devices on the WLAN in order to function correctly.

Pairwise master key (PMK) caching eliminates the entire 802.1X EAP process for roaming purposes once a client has already completed that process during its association to the first AP. Once a client device has associated and authenticated to the wireless network, a PMK secure association (PMKSA) exists between the authentication server and the client device. A PTK secure association (PTKSA) exists between the AP and the client device once the four-way handshake is completed. It is important to note that accomplishing a PMKSA and PTKSA takes time—sometimes too much time for real-time applications, unless time-consuming steps can be removed from the process. The solution to the delay caused by establishing a PMKSA is PMK caching.

In this process, APs cache a PMK identifier from the original session. This identifier verifies that the client has completed EAP authentication and has a unique pairwise master key. At this point, the client device immediately goes into the four-way handshake, skipping the lengthy 802.1X process. No wireless traffic overhead is introduced through PMK caching making it ideal for VoIP applications.

Opportunistic Key Caching (OKC) is an extension of PMK caching not defined in the 802.11 standard, but necessary to enable optimized Layer 2 roaming. With OKC, all APs on the same Layer 2 network will receive a copy of the client device's PMKID. Any of these APs the client decides to roam to will immediately recognize the client's PMKID and go directly into the four-way handshake making the roam from AP to AP much quicker.

Latency Requirements

Latency, as previously defined, is equal to delay. It is a measurement of the time required for bits to be moved from the originating device to the destination device. It is measured in milliseconds. Some applications have stringent requirements on latency and VoIP is the prime example. Figure 3.3 shows the steady flow of VoIP packets from one client to another, and also displays a latency issue from packet congestion at the network router.

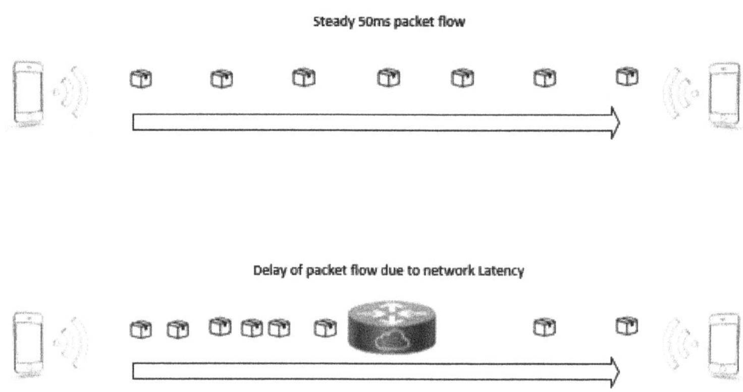

Figure 3.3: VoIP Packet Flow is Steady

The general requirement of VoIP is 150 ms latency or less, unidirectionally. This demand indicates that the round-trip time (RTT) cannot be more than 300ms, with less than 1% packet loss. However, it cannot be assumed that a 300 ms RTT is acceptable. While rare, it is possible that the communication could have a latency value of 120 ms one direction and then 180ms the other direction. The RTT is still 300 ms, but the 180 ms direction is unacceptable for normal VoIP communications. This can be caused by a poor design, in which the AP has a higher transmit power than the client device, causing mismatched send/receive data rates and mismatched modulation types between the AP and client. Communication from the AP to the client device has a much better throughput to the client, but the client device ends up with less throughput, and the VoIP packets must be sent to the AP at a slower rate, resulting in more latency in that direction.

Latency can be impacted in three areas:

- The wireless connectivity

- The wired network
- The processing devices

Interestingly, the processing devices often cause significant latency and are not always considered in the equation. For example, an overloaded router may be delayed in forwarding packets and begin causing extra delay in the network. Additionally, if a VoIP handset uses encryption, the encryption results in extra delay within the handset; however, this is not part of network communications and is typically factored into the voice communications process.

You should test the network for latency, and this should include the full communications path and not just the wireless link. For example, test an actual communication from a wireless client through the AP, to the switch and on through the routers, into another switch and out through another AP which transmits and receives the VoIP packets with a second wireless client. This will provide you with a much better picture of latency for VoIP calls.

Data-Throughput Demands

In addition to roaming and latency demands, most applications have specific throughput demands to function properly. Throughput is the amount of useful data that the network can transfer. It is typically measured in megabits per second (Mbps) or gigabits per second (Gbps).

Determining throughput demands is easy. For example, you can look up values for specific video or audio codecs to determine their throughput requirements. Most applications provide a minimum requirement list for proper functionality, but in some cases, you will have to perform tests to get this information. Such tests can be performed by connecting the client and the application server to a gigabit switch while performing various functions in the application and measuring the throughput. This simple test can give you a close estimate of the demands of the application and allow you to better plan for it in scale. In other words, if this test resulted in a demand of 250 Kbps for a single employee's laptop to use the application, then 50 employees using the application at the same time would need 12.5 Mbps of throughput for all of them to effectively utilize the application simultaneously.

Voice/Video

Let's look at the preceding requirements when applied to voice and video applications. From a design perspective, latency and throughput issues must be addressed up front when designing for voice and video applications across a WLAN. For voice communications, latency is the most important factor, but throughput is also a concern, especially when there is a lot of potential traffic on the network. Because voice packets are so small, if latency is resolved, throughput is rarely an issue.

For video, latency and throughput must be provided for, especially if the video is live streaming and not stored files such as training videos. Latency still becomes an issue in both live and stored video, but the two differ significantly. Live video must have adequate throughput with little latency to be delivered without interruption and/or lost sections of video. Stored video requires the same metrics to be delivered undisturbed, but it can be buffered and pick up where it left off. The big difference is the delivery protocols TCP and UDP. Live video is delivered via UDP, which is a "best effort" type of delivery just like voice communications. If a packet is dropped in a UDP transmittal, then it is just lost. In stored video such training videos, the video content consists of a file, just like any other data that is accessible and can be called up. The TCP protocol can detect lost packets and resend them so that the entire video can be seen. Throughput, latency, and capacity calculations still play a big role in this environment because training videos that must continuously buffer and/or resend dropped or lost packets tell us that one, two, or all three of these important metrics are insufficient for this application.

Table 3.1 shows the basic requirements for voice, video, and data communications. In this table you will notice that data communications do not have real issues with latency or jitter. These factors have the most impact on real-time applications like voice and video. You can also see that video is not as demanding as voice, partly because video services may buffer some of the data before playing it even for real-time. This is the reason that you may be watching a live stream at the same time as someone else, but their live stream is ahead of or behind yours by a few seconds. Voice packets cannot utilize a lot of buffering without causing silence in the conversation while the remaining portions of the audio are being buffered.

Voice over IP (VoIP)

Besides actual VoIP handsets, there are many Wi-Fi client devices and types that can utilize software to function as a VoIP phones, such as laptops, tablets, or even desktops.

When using any device for VoIP communications, there are always minimum requirements to adhere to in order to get the best performance out of VoIP applications on that device. Practically all of these devices follow a common standard of a –67dBm signal strength and an SNR of 25dBm in order to provide proper data rates to the device. Also, it is important to provide adequate overlap of individual coverage cells for proper roaming of the VoIP device while in use. This coverage overlap should be a consistent 20 – 25% throughout the coverage area. An important factor in VoIP design is the power settings of the APs. They should be set as close as possible to the output power of the lowest-powered device being used on the WLAN. In other words, if your VoIP handsets have a maximum transmit power level of 30mW, then you should design your WLAN to have all APs transmit at or below that level to ensure continuous two-way communication. Some design for 3dB below the VoIP handset's power level so that RRM can raise the transmit power of specific APs 3dB in the event an AP goes down and leaves a coverage hole. Once RRM kicks in and fills the coverage hole, the transmit level of the APs are still at the level of the VoIP handsets and not higher.

Property	Voice	Video	Data
Latency	≤ 150ms	≤ 200–400ms	N/A
Jitter	≤ 30ms	≤ 30–50ms	N/A
Loss	≤ 1%	≤ .1–1%	N/A but may benefit from less than 2-5% loss for performance reasons
Bandwidth	30–128Kbps	384–20+Mbps	Variable

Table 3.1: Voice, Video and Data Compared

Many VoIP devices support push-to-talk features so that dialing a number is not required and faster communications can be achieved. This feature usually requires a form of multicast packet delivery to be configured.

If deploying an outdoor WLAN, it is important to check channelization of the APs along walk paths and not assume that auto features will provide proper channel settings. These APs see each other across rooftops and airways. They do not always provide proper channel settings based on ground-level roaming. Many VoIP devices will not roam across same-channel cells without dropping the call. It may be necessary to manually set channels for these scenarios.

Video

Video applications over Wi-Fi can become complex, depending on several factors as video is a high-bandwidth application. One scenario that comes to mind is video conferencing. In this application of video over Wi-Fi, the video itself is live and there are usually many users involved in the conference. This case has to comply with real-time requirements and could also require a design for capacity if many users are in the same area using the same WLAN for access. For these reasons, the WLAN design has to build from knowledge obtained about the facility and use cases for the WLAN throughout. Here is an example from a past project:

The location was an amphitheater with copper-lined walls and concrete floors. This amphitheater seated over 400 university students who would gather for specific lectures. The dean wanted to provide Wi-Fi coverage so that all 400 students could use their laptops to participate in a live video conference lecture while also recording the session to their personal laptops for playback later. The IT department placed an AP in the center of the room and tried to set up a conference, and it failed miserably. It is important to point out here that this took place when autonomous APs were our only option and 802.11b/g was our limitation for frequencies.

As it turns out, the application being used for the video conference would allow for 22 client devices to connect and participate in the video conferencing without bandwidth issues. With that information, it was going to take 20 APs to provide for the demands in this area. The big issue to overcome was obviously co-channel interference, since 2.4 GHz was the only option to work with and the design was limited to three channels. The great thing about the amphitheater was the stair-step levels, which meant there had to be an area under the seats that could be accessed.

Figure 3.4 demonstrates how the design was completed to satisfy the demands for the application. Looking at Figure 3.4, you see 21 triangles representing APs. The thick black lines represent the 18-inch-thick concrete support walls underneath the amphitheater.

The floor was an 8-inch-thick concrete slab that caused just enough signal attenuation so that each AP would reach approximately twenty-five seats with adequate signal strength and then fall out rapidly from seat to seat. The concrete support walls were reinforced with steel rebar and would attenuate the RF signal over 50dB from one side to the other. With this arrangement, Both, real-time requirements for voice and capacity requirements were met. The university called in 400 students for a video conference to test the design. They were well pleased.

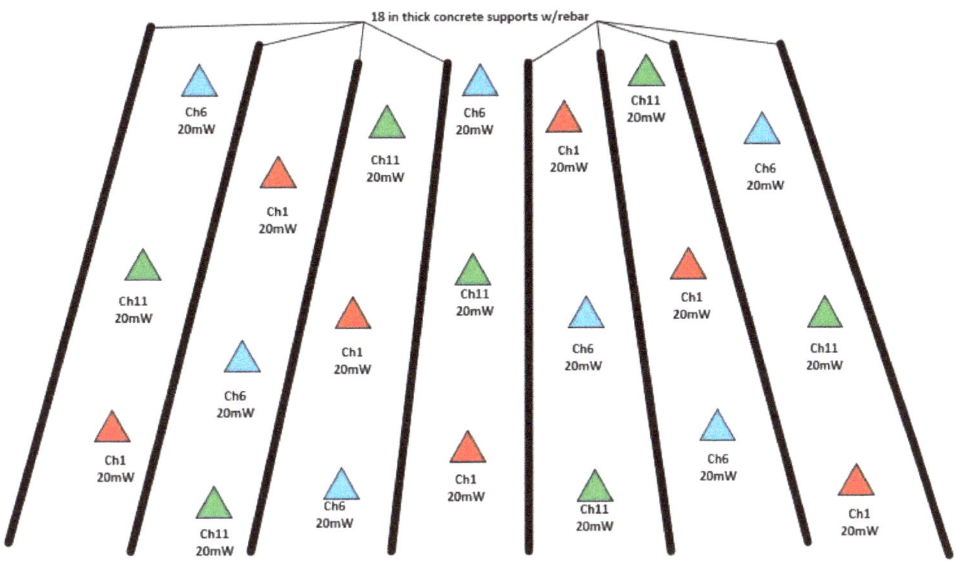

Figure 3.4: Design Meeting Requirements

The point of this example is to demonstrate the need for finding out specifics concerning the project requirements and the need to know limitations on equipment and applications. As we discussed for capacity at the beginning of this chapter, coverage doesn't always mean adequate coverage. Sometimes you will have to think outside the box and become creative with your design to provide the customer with the WLAN they desire, especially when the application is as bandwidth-demanding as voice.

The next video application we will discuss is streaming video. Streaming video typically uses buffering to reduce the latency demands on the application. For example, the player may buffer anywhere from five seconds to one minute of video so that network delays do not prevent smooth play of the video. When designing for video, be sure to consider the kind of video being implemented.

Both voice and video use codecs to encode and compress the data in order to reduce network throughput requirements. Table 3-2 lists common audio codecs and their characteristics. Table 3-3 lists common video codecs and their characteristics.

Figure 3.5—ASCOM MYCO Smartphone/VoIP Device

Codec	Bitrate	Compression (Yes/No)
G.711	64Kbps	No
G.722	64, 56, 48Kbps	Yes
G.723/723.1	5.4, 6.3Kbps	Yes
G.729/729A	8Kbps	Yes

Table 3.2: VoIP Codecs

Codec	Bitrate	Compression (Yes/No)
MPEG Basic	Varies by profile (96Kbps to 135Mbps)	Yes
H.263	384, 768Kbps	Yes
Uncompressed	1.5Gbps for 1080p	No

Table 3.3: Video Codecs

Standard Data

If a standard data network is all that's needed out of the WLAN, these requirements are much easier to design for than those of real-time data networks. However, unless you are implementing separate WLANs, one for real-time applications and one for standard data applications, you will need to accommodate QoS configuration in both WLANs because applications such as Skype and FaceTime will be used on either or both WLANs.

Also, when planning for standard data, you still must plan for the required throughput demands (capacity). For example, if the average user will require 2 Mbps and 25 users will be connected to each AP, the APs must provide 50 Mbps data throughput. Given that throughput is often 50 percent of the data rate, the users will require a data rate higher than 100 Mbps to ensure that throughput demands are met.

Location Services

In WLANs, location services are implemented in one of two ways:

- 802.11-based
- non-802.11

With non-802.11 systems, RFID or some other RF mechanism is used to track the device, and the system may use Wi-Fi to communicate with central servers, but the actual location tracking is performed using non-Wi-Fi tags or devices. 802.11-based location services utilize various triangulation techniques, such as those covered in detail back in Chapter 2. When designing for location services, it is important to ensure that at least three APs, four for best results, can see every location where a client device may be. These four APs should see the device at a minimum signal level of –75 dBm or higher for most accurate tracking results. Some of the APs detecting client devices may be set to listening mode only and will not allow client association. These listening APs will be used only for applications such as location tracking and intrusion detection.

Quality of Service (QoS)

QoS is essential for real-time communications. Voice, video, and a few other technologies are time-sensitive, and their packets must be delivered from sender to recipient quickly and with very minimal delay as they travel. This section provides detailed information on prioritizing network traffic using QoS.

It is important to configure and implement QoS correctly throughout the network when using applications such as voice and video. This helps to prioritize packets that are more latency-sensitive, so they are sent out first even if they didn't arrive first. In Figure 3.6, we see voice, video and data packets entering the router, all at random times. QoS settings prioritize the voice packets to be sent out first, followed by the video packets, and then the data packets. This order is used because the voice packets are more time-sensitive than the others, and too much delay can and will cause parts of the conversation to be unheard or portions of video traffic to freeze up or be dropped completely.

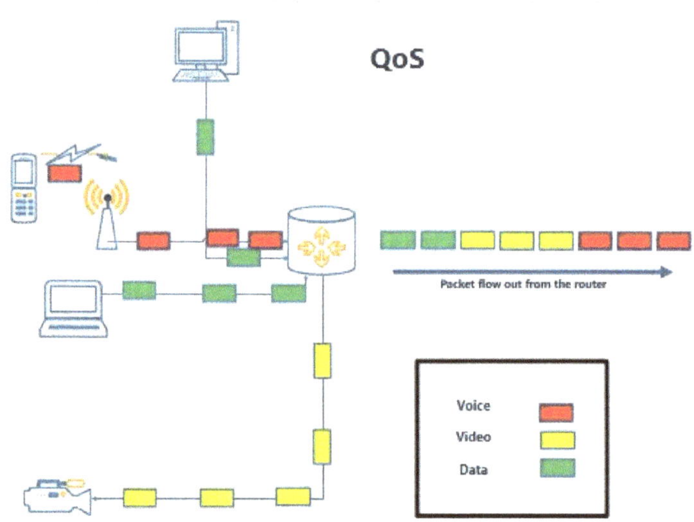

Figure 3.6: QoS and Its Impact

In configuring QoS, you first must determine the network performance requirements for the various traffic types: voice, video, and data. Next you must categorize the traffic types. Here are some examples of categories you might choose:

- Low latency
- Low priority

Or you can divide them this way:

- Realtime
- Critical
- Best effort

Once you have your applications grouped into categories, you can begin to prioritize each application. Next, apply QoS configurations that mark traffic on the network so that every device that a packet travels through will recognize the packet's priority and send it out accordingly. Figure 3.7 shows an example of QoS classifications and markings based on the application being used.

Application	L3 Classification			L2
	IPP	PHB	DSCP	CoS
Routing	6	CS6	48	6
Voice	5	EF	46	5
Video Conferencing	4	AF41	34	4
Streaming Video	4	CS4	32	4
Mission-Critical Data	3	AF31*	26	3
Call Signaling	3	CS3*	24	3
Transactional Data	2	AF21	18	2
Network Management	2	CS2	16	2
Bulk Data	1	AF11	10	1
Best Effort	0	0	0	0
Scavenger	1	CS1	8	1

Figure 3.7: QoS Tags/Markings/Classifications

QoS Queues

Traditional QoS also impacts wireless links, as QoS is an end-to-end solution, meaning that it must be implemented throughout the network on the wired and wireless sides to gain the most benefit. Traditional QoS uses queuing methods and frame markings.

Four basic types of queuing are commonly used:

- First-In-First-Out (FIFO)—packets leave the device (such as a switch or router) in the order they arrived.
- Fair Queuing—an equal amount of time is given to transmitting for each stream of packets.
- Class-based Queuing—packets belonging to higher-priority classes are transmitted first.
- Reservations—packets may belong to a reservation queue, granting them a specific percentage of the throughput available.

802.11 QoS, as originally defined in 802.11e, is effectively a class-based queuing system. Such systems use markings (shown in Figure 3.7 earlier) to identify the traffic and place it into classes. Wired networks use Class of Service (CoS), Type of Service (ToS), and Differentiated Services Code Point (DSCP) markings. CoS is a Layer 2 Ethernet marking. ToS and DSCP are Layer 3 IP markings. Today CoS and DSCP are the primary players in wired Ethernet QoS.

On the wired network, DSCP is Layer 3 QoS, and CoS is Layer 2 QoS, also referenced as 802.1p. Implementing end-to-end QoS requires configuration at Layer 2 and Layer 3 at a minimum.

DSCP uses a 6-bit value, but the leftmost three bits map the same as the CoS bits. That is, CoS voice is 5 (or 101 in binary) and the DSCP value for voice is 40 (or 101000 in binary). The reality is that some vendors have implemented their own numbers and classes, but the general concept to remember is that a device should map the CoS to a DSCP value for wired network communications and vice-versa. With this understanding, you can consider 802.11 QoS.

802.11 QoS

802.11e introduces QoS to the 802.11 family. It implements a Layer 2 QoS solution for the wireless link. It is the responsibility of the Ethernet-connected devices (APs or controllers) to convert the 802.11e markings to 802.1p and/or DSCP markings for communications on the wired side.

As described in the *CWNA Study Guide*, 802.11 uses the Distributed Coordination Function (DCF) to gain access to the medium and transmit a series of 1s and 0s (a frame). 802.11e introduced the Enhanced Distributed Channel Access (EDCA). The basic enhancement is that 802.11 frames are assigned an access category (AC) in one of four values:

- AC_BK: Background—lowest priority
- AC_BE: Best Effort—low priority
- AC_VI: Video—middle priority
- AC_VO: Voice—highest priority

The basic functionality is that higher-priority categories utilize lower numbers for their countdown timers while contending for the network medium. For example, AC_VO begins with a minimum value of 3 and a maximum value of 7, whereas AC_BK and AC_BE begin with a minimum value of 15 and a maximum value of 1023. As you can see, AC_BK and AC_BE frames (standard data) begin their contention window at roughly twice the value of the maximum that a voice frame (AC_VO) would have. Therefore, higher-priority traffic will gain access to the medium more frequently than lower-priority traffic.

Data traffic will sometimes win the contention battle over voice traffic, but at most, that would only happen once for every two voice wins. The real odds are far worse for data traffic, but remember, this is all measured in milliseconds, and data traffic will still get through while latency issues for voice traffic are removed.

The Wi-Fi Alliance created the Wireless Multimedia (WMM) certification to validate equipment that properly implements EDCA as defined in 802.11e. Devices that are WMM certified should be selected when QoS is essential.

When an AP tunnels to a WLAN controller with centralized data forwarding, the WLC converts DSCP values to 802.1Q user priority (UP) tags or 802.1p tags.

3.6: SSIDs and WLAN Profiles

When designing a WLAN, there will always be a need for traffic separation. This separation is required for keeping corporate data networks secured from guest users and is also used to assign bandwidth to priority traffic.

In wired networks, network traffic is separated using virtual LANs or VLANs. A VLAN is a logical subnetwork that isolates a group of network devices from other network devices operating on the same physical LAN.

On the WLAN the same is true. VLANs are set up for keeping corporate traffic separate from guest user traffic, and voice traffic separate from other broadcast domains. IP address pools (on separate subnets) are assigned to each VLAN and the wireless VLAN is given an SSID, better described as a name. Some examples of such VLANs might be

- ACME_CORP
- ACME_GUEST
- ACME_VOICE
- ACME_BYOD

The SSID is mapped to a specific VLAN so that wireless clients associated to a particular SSID can pull an IP address from the VLAN mapped to the SSID, providing a Layer 3 separation of networks.

As you can see by the example in Table 3.4, each SSID is assigned to a unique subnet for client devices to obtain an IP address from. Also note that the VLAN ID is also a unique identifier. Layer 3 routing must be set up for these four virtual networks to be able to communicate with each other even though all four of them are using the same physical medium.

Also, the data that is sent over WLANs includes the SSID in the packet so that information isn't sent to the wrong location when multiple SSIDs are present in the same service area.

SSID	Subnet	VLAN ID
ACME_CORP	192.168.0.0 255.255.255.0	VLAN300
ACME_GUEST	192.168.1.0 255.255.255.0	VLAN301
ACME_VOICE	192.168.2.0 255.255.255.0	VLAN302
ACME_BYOD	192.168.3.0 255.255.255.0	VLAN303

Table 3.4: SSIDs mapped to VLANs and subnets

A WLAN profile (which may include the RF Profile) is a set of configurations for the WLAN that will assign specific settings to APs the profile is assigned to. An administrator may want a few different profiles configured for various environments within the organization. Here are some profile examples:

- **Office areas**: This profile may be set up to accommodate a higher density of APs to meet requirements for capacity, VoIP, and specific software applications. The power settings may be configured to 12mW or lower for the density issue and available data rates set to a mandatory speed of 36Mbs or higher with all data rates below 36Mbs turned off. This profile can be configured for 5 GHz frequencies and/or 2.4 GHz frequencies.
- **Warehouse:** This profile may be strictly set up for scanner guns on 2.4 GHz to allow for warehouse management systems to function throughout the warehouse area. This profile would more than likely include 802.11b data rates for older scanner guns the company refuses to replace. If VoIP services are also needed in the warehouse, the 5 GHz radios would use a different profile to provide adequate settings for VoIP devices to function.
- **Outdoor Mesh**: If several APs are deployed outdoors in a mesh topology, you will want to set up profiles for both frequency bands. For example: if the 5 GHz band is used for bridging between APs, then a bridge profile can be set to have 80Mhz bonded channels for back haul throughput. The 2.4 GHz band would then be used for client access so 20Mhz channels will be set along with a lower power setting to accommodate the weakest client device on the network.
- **AP Model:** One other reason you might want to set up specific profiles would be for AP models. Different models of APs propagate their signals differently, have different receive sensitivity, use variations of antennas and end up leaving a

significantly different size of coverage footprint as well as sensing client devices differently. In the case that multiple models of APs must be used in your design, a profile may be warranted for each model of AP deployed.

Other configurations can be set in RF profiles to provide other settings that make the network function better. Profiles are designed and stored on the WLC and new APs can be assigned to a profile as they are added. This provides lots of flexibility in both WLAN design and AP deployment across multiple environments.

3.7: Security Settings

As technology grows, so do the threats to secure connections. This section provides guidance for designing security solutions from many layers of security options available for wired and WLANs.

Firewalls

Existing network firewalls will more than likely require changes in the configuration to allow the LWAPP and/or CAPWAP ports to pass wireless traffic through. By default, LWAPP uses UDP ports 12222 and 12223, while CAPWAP uses UDP ports 5246 and 5247

Additional firewall configuration settings may be required, so be sure to check vendor documentation for the ports (both UDP and TCP) used for various communications and ensure that the appropriate firewall rules are in place.

ACLs

Access Control Lists (ACLs) are used on network routers and WLCs to control network traffic. Like a firewall, a router must allow all required communications to pass through while denying unauthorized traffic. Having well-written ACLs in place helps to control and manage traffic. ACLs are often used to restrict users from accessing data that is off-limits to them. They are also a great tool for restricting access to certain web sites or even traffic utilizing certain protocols. For example, an ACL can be implemented that will deny ICMP traffic in order not to allow pinging addresses on the network.

When you understand how ACLs function, they are a great asset for controlling the traffic that comes and goes through the network.

RADIUS and LDAP

RADIUS is used for authentication (as mentioned earlier in the 802.1X EAP process), and LDAP is a protocol used to access a directory service. In many deployments, RADIUS servers are used for authentication, but they often depend on large-scale directories for account information such as Microsoft Active Directory. LDAP is used to make the connection from the RADIUS server to the Active Directory Domain Controller.

From a design perspective, you can recommend the use of a RADIUS server that fits well with the existing network solutions. For example, in a Microsoft Active Directory environment, using the Microsoft RADIUS server makes sense. In a Linux/Unix environment, FreeRADIUS or a commercial RADIUS product may be recommended, or the organization may choose to use a RADIUS server from the network infrastructure provider like Cisco.

RADIUS uses one or more variations of an EAP protocol in order to authenticate a user, server or both. RADIUS authentication also provides seeding material that eventually helps to generate keys used for encrypting data on the WLAN.

RADIUS Server Selection

When using the Extensible Authentication Protocol (EAP) for enterprise WLANs, a RADIUS server is required. Selecting the appropriate RADIUS server will be driven by these three important factors:

- The existing infrastructure solutions (Windows, Linux, Cisco, and so on)
- The existing directory services (Active Directory, LDAP, and so on)
- The EAP type required

Many variations of EAP exist, and the one to be used will depend a lot on what type of credentials will be used during authentication. Next, we will take a look at the different EAP types.

EAP Type Selection

Selecting the appropriate EAP type is a factor of desired security and client support. The more secure EAP types require both server and client certificates or at least server certificates, such as EAP-TLS and EAP-TTLS. PEAP is also a commonly used EAP type and as long as it is implemented appropriately, it should provide sufficient security. You can use Table 3.5 to compare EAP types based on features and requirements.

	EAP-MD5	LEAP	EAP-TLS	EAP-TTLS	PEAP	EAP-Fast
Mutual Authentication	No	Yes	Yes	Yes	Yes	No
Certificates Required	No	No	Client/Server	Server Only	Server Only	No
Dynamic Key Generation	No	Yes	Yes	Yes	Yes	Yes
Costs and Management Overhead	Low	Low	High	Low/Medium	Low/Medium	Low
Industry Support	Low	High	High	High	High	Medium
Credential Security	Weak	Weak	Strong	Strong	Strong	Strong

Table 3.5: Common EAP Types

The following list shows the most common EAP types used with wireless networks:

- EAP-TLS: Client and server certificates required
- TTLS (EAP-MSCHAP-v2): Only server certificates required
- PEAPv0 (EAP-MSCHAP-v2): Only server certificates required
- PEAPv0 (EAP-TLS): Client and server certificates required
- PEAPv1 (EAP-GTC): Used with token card and directory-based authentication systems; only server certificates required
- EAP-SIM: EAP for GSM Subscriber Identity Module; mobile communicators
- EAP-AKA: For use with the UMTS Subscriber Identity Module; mobile communications

Encryption Solutions

802.11 WLANs now have three major security protocols. Each option has personal and enterprise version of the protocol. Those protocol specifics are listed next:

1. WPA:

    a) **WPA – Personal:** WPA-PSK (Pre-Shared Key). Utilizes a 256-bit key, which is a huge security increase compared to the 64-bit and 128-bit keys used in the WEP system that WPA was created to replace.

    b) **WPA – Enterprise:** 802.1x EAP authentication with Temporal Key Integrity Protocol (TKIP) encryption and RC4 cypher with message integrity checks and a per-packet key system.

2. WPA2:

    a) **WPA2 – Personal:** WPA2-PSK (Pre-Shared Key). Utilizes a 256-bit Key. CCMP Encryption and AES cypher with TKIP and RC4 as an option.

    b) **WPA2 – Enterprise**: 802.1x EAP authentication with CCMP as the encryption method and AES cypher with TKIP and RC4 as an option.

3. WPA3 (Table 3.6 defines the acronyms used in describing WPA3):

    a) **WPA3 – SAE Mode (Personal):** WPA3 – Personal makes it much more difficult and time-consuming for an attacker to use brute-force dictionary attacks by forcing an attacker to interact with the network every single time a compiled guess at a password is attempted. This is accomplished by utilizing Simultaneous Authentication of Equals (SAE). When a Basic Service Set (BSS) is configured in WPA3-SAE mode, the Protected Management Frame (PMF) is set to Required. The Management Frame Protection Required (MFPR) bit in the Robust Security Network (RSN) Capabilities field is set to 1 in the RSN Element (RSNE) transmitted by the AP. A WPA3-SAE station will negotiate PMF when associating to an AP using WPA3-SAE Mode.

    b) **WPA3 – SAE Transition Mode:** When WPA2-PSK and WPA3-SAE are configured on the same BSS (mixed mode), PMF will be set to capable (MFPC bit is set to 1, and MFPR bit is to 0 in the RSN Capabilities field in the RSNE transmitted by the AP). When WPA2-PSK and WPA3-SAE are configured on the same BSS (mixed mode), the AP will reject an association for Simultaneous Authentication of Equals (SAE) if PMF is not negotiated for that association. A WPA3-SAE station will negotiate PMF when associating to an AP using WPA3-SAE Transition Mode.

c) **WPA3 – Enterprise 192-bit Mode:** When WPA3-Enterprise 192-bit Mode is used by an AP, PMF is set to Required (the MFPR bit in the RSN Capabilities field is set to 1 in the RSNE transmitted by the AP).

When WPA3-Enterprise 192-bit Mode is used by a station, PMF is set to Required (the MFPR bit in the RSN Capabilities field is set to 1 in the RSNE transmitted by the STA).

Acronyms	Definition
BSS	Basic Service Set
MFPC	Management Frame Protection Capable
MFPR	Management Frame Protection Required
PMF	Protected Management Frame
PSK	Pre-shared key
RSN	Robust Security Network
RSNE	RSN Element
SAE	Simultaneous Authentication of Equals
SSID	Service Set Identifier
WPA2 - PSK	Wi-Fi Protected Access 2 Pre-Shared Key
WPA3 - SAE	Wi-Fi Protected Access 3 – SAE
WPA3 - Enterprise	Wi-Fi Protected Access 3 – Enterprise

Table 3.6: Common Wi-Fi Security Acronyms

The EAP cipher suites for use with WPA3-Enterprise 192-bit Mode are:

- TLS_ECDHE_ECDSA_WITH_AES_256_GCM_SHA384
- ECDHE and ECDSA using the 384-bit prime modulus curve P-384

- TLS_ECDHE_RSA_WITH_AES_256_GCM_SHA384
- ECDHE using the 384-bit prime modulus curve P-384
- RSA ≥ 3072-bit modulus
- TLS_DHE_RSA_WITH_AES_256_GCM_SHA384
- RSA ≥ 3072-bit modulus
- DHE ≥ 3072-bit modulus

Wi-Fi Protected Access 3 has also made provisions for IoT devices to onboard more easily through *Easy Connect*. This new security protocol will see new IoT devices shipping with a QR code that can be scanned to introduce it to the network. This can be done through large batch deployments as well. Easy Connect is backward-compatible with WPA2.

Although WPA2 is still mandatory with WPA3 optional, it is expected that WPA3 will replace WPA2 as the mandatory standard by late 2019.

When selecting an encryption solution for your WLAN, client requirements help to drive this decision more than anything else (except for governmental and other highly-secure deployments). Some clients may still only support TKIP (WPA), and therefore you choose to utilize it rather than replace hundreds of client devices.

6 GHz 802.11ax Security

HE is the first, and currently only, PHY in the 802.11 standard to support the 6 GHz band. The channelization within 6 GHz was covered in earlier chapters, so we will not address it again here. Instead, we will explore the operational rules for Wi-Fi in the 6 GHz band – named Wi-Fi 6E by the Wi-Fi Alliance.

In this section, we will address the security requirements for Wi-Fi in 6 GHz. These requirements are direct and to the point. However, given that this will be the first mention of Opportunistic Wireless Encryption (OWE) in the book, we will also explore how this security solution works. The following are the requirements for 802.11 wireless connections in the 6 GHz band using 802.11ax (HE):

- All open networks must use OWE instead of just Open Systems authentication.
- WPA3 (SAE) is required for password-based/passphrase-based networks.

These rules are defined within the 802.11ax amendment in the negative. That is, the standard defines what cannot be used, which leaves only those listed above as allows.

The standard states that the following apply to a STA operating in the 6 GHz band[46] (numbering added):

- (1) The STA shall not use the following pre-RSNA security methods: WEP, Open System authentication without encryption, Shared Key authentication
- (2) The STA shall not use the following cipher suite selectors:
    - 00-0F-AC:0 (Use group cipher suite)
    - 00-0F-AC:1 (WEP-40)
    - 00-0F-AC:2 (TKIP)
    - 00-0F-AC:3 (WEP-104)
- (3) The STA should use Opportunistic Wireless Encryption, as specified in IETF RFC 8110, when connecting in an infrastructure BSS without authentication (as a replacement for Open System authentication without encryption)
- (4) The STA shall not use the following AKM suite selectors:
    - 00-0F-AC:2 (PSK)
    - 00-0F-AC:4 (FT authentication using PSK)
    - 00-0F-AC:6 (PSK with SHA-256)
    - 00-0F-AC:19 (FT authentication using PSK with SHA-384)
    - 00-0F-AC:20 (PSK with SHA-384)
- (5) The STA should use SAE (AKM suite selectors 00-0F-AC:8 and/or 00-0F-AC:9) if authenticating using a password where IEEE 802.1X is not used (as a replacement for PSK)
- (6) The STA shall use management frame protection (MFPR=1) when using RSN

Public Key Infrastructure (PKI)

Because many EAP types rely on certificates and some depend on client certificates, a PKI may be required. A PKI consists of certificate authorities (CAs) that either authorize other CAs to distribute certificates or generate and distribute (issue) certificates themselves. As part of your WLAN design, you may have to determine whether a PKI will be available so that you can select the appropriate EAP type. If a constraint is in place requiring an EAP type that must have client certificates, you should be sure to document the requirement of a PKI in the design plans, although the details of the PKI can be determined by those implementing the PKI project.

---

[46] IEEE 802.11ax-2021, clause 12.12.2 Security constraints in the 6 GHz band

Role Based Access Control (RBAC)

RBAC is a security solution still very widely used in the enterprise world. It organizes users into groups or roles and assigns capabilities or authorizes users based on those groups or roles rather than manually authorizing each of the users. Users can be assigned roles and then perform any function allowed by the role assignment.

BYOD and MDM Solutions

When designing for Bring Your Own Device (BYOD), Mobile Device Management (MDM) becomes an important factor. MDM is used to onboard, manage, and decommission mobile devices (both organization-owned devices and BYOD units).

MDM solutions may integrate with Network Access Control (NAC) solutions to perform onboarding with security. For example, when a new device connects that should have antivirus but does not, it can be denied network access and placed into a quarantine network until it complies with anti-virus requirements.

If required as part of your design, it is important to select an MDM solution that meets the needs of the organization. Consider the following questions:

- Will the MDM integrate with existing directory services?
- Is self-service onboarding required?
- What devices are supported and what operating systems and versions are used?
- Will containerization be required?
- Will features like remote wipe, remote lock, and encrypted storage be required?

Additional questions may be required in some installations, but the important thing is to ask the right questions and match the MDM solution to the organization's needs.

WPS

Wi-Fi Protected Setup (WPS) is a push-button security solution that should be avoided in enterprise deployments. The basic process is to place the AP/WLAN router into WPS mode and then do the same on the client device or software utility and have the AP/WLAN router and client negotiate security information between them. This information is then used in future connections. WPS has proven to have vulnerabilities when not implemented properly and should not be utilized in enterprise WLANs.

Virtual Private Networks (VPN)

In today's world of hackers, war drivers, and eavesdroppers, you should never assume that your wireless connection is safe from intruders. There is always someone attempting to circumvent security, if for no other reason than to see if they can do it. Most of these attempts take place at fast-food places, coffee shops, and public gatherings where hotspot wireless connections are plentiful. Too often the corporate employee needs to connect, check email, and send out a few documents, only to find out later that information on the laptop was compromised. A dependable security solution for hotspots is the use of a VPN every time you associate to any WLAN. Almost all public hotspots are provided with no encryption or security. They are simply open, unrestricted networks that remain targets of malicious evil doers. To protect sensitive data during the use of hotspots, users are usually told to use secure protocols like HTTPS and SFTP while browsing. However, such protocols are not always available while using a VPN, which secures all activity inside a protected tunnel.

If the customer you are designing for has a regular supply of remote employees who work from anywhere and everywhere as part of their daily routine, you should discuss and recommend a VPN solution as part of the security design plan. Many organizations already have devices or servers that can act as VPN endpoints, such as Cisco ISR routers or Windows servers, making a VPN solution a very affordable addition to the overall design.

Endpoint Security

Endpoint security is used to ensure that the client devices on the network are secure and do not introduce vulnerabilities to the network. Endpoint security can be as simple as running antivirus software or as complex as detecting whether a wireless link is encrypted or not. Some basic features of endpoint security software are:

- Antivirus
- Configuration checks
- WLAN security enforcement
- Local proxy features

- Content filtering
- Reporting to centralized servers

Wireless Intrusion Prevention Systems (WIPS)

A wireless intrusion prevention system (WIPS) is a strong, front-line defense against unwanted wireless visitors within the boundaries of the corporate facility. It is an option that can be implemented in different ways such as the following:

- **Integrated WIPS**: The integrated WIPS system is part of the existing WLAN, making the implementation cheaper for the overall budget. The client-servicing APs scan several channels, looking for intrusions, while they are not servicing clients. When a client device needs to associate to the part-time AP, the AP will stop scanning and begin servicing clients again as a normal functioning AP. The argument against this WIPS solution is that the WIPS is only part-time and there is a strong possibility that all channels will not be scanned before the AP goes back to work servicing clients. While this is true, other idle APs in the area continue to scan while the first AP switches back to client service. However, if the facility has a continuous flow of traffic on the WLAN throughout the wireless coverage area, there is a chance that WIPS scanning will be much less effective, because all APs have the priority to serve clients first. If this is the case, an integrated WIPS may not be right for your design.
- **Overlay WIPS:** An overlay WIPS is a system comprising APs that are separate from those that service clients. In fact, the overlay system is best designed after the WLAN design is complete. The overlay WIPS APs listen to and monitor the spectrum on a full-time basis, instead of part-time, providing a much better guard over the facility. This solution does end up being more expensive because of the extra expense added in hardware and labor.

A WIPS can be configured to recognize different classes and types of wireless devices that propagate signals within a service area, and then report rogue devices and/or other changes to the administrator through various options. Additionally, a WIPS can provide collected data for analyzing the overall security of the network.

WIPS solutions are available in three types:

- Software-based

- Hardware-based
- Cloud-managed

All can monitor the unbounded wireless medium using wireless sensors and all report/record captured information to a database. The WIPS can deploy countermeasures to prevent intrusions, as needed. These countermeasures are based on identifying the intrusion through a comparison of the captured information to its database of intrusions.

Some abilities of the WIPS include sending out alerts on many different threat conditions, which fall into the following categories:

- User authentication and traffic encryption
- Rogue and ad-hoc mode devices.
- Configuration vulnerabilities
- Intrusion detection on security penetration
- Intrusion detection on DoS attacks

This helps IT and security administrators with corporate security policy compliance when dealing with wireless client devices that belong to employees and guests.

WIPS solutions also provide a variety of useful features, including these:

- Hardware sensors for monitoring
- 24×7 monitoring
- Threat mitigations
- Detecting threats to the wireless infrastructure
- Built-in reporting systems
- Integrated RF spectrum analysis to monitor and view the RF spectrum
- Validating compliance with corporate security policy and legislative requirements
- Retaining collected data for further forensic investigation

The introduction of a rogue AP into the corporate network by an employee creates multiple layers of threats to the entire network. Most of these rogue APs go for months or even years before being found or questioned. With a WIPS in place, a WIPS-sensing AP will act as a client device and attempt to associate to the rogue AP. Once successful, the WIPS sensor will send a packet back to the WLC or WIPS server through the rogue AP and discover the switchport it is connected to. The port can then be disabled to prevent any further traffic to and from the rogue device. A WIPS coupled with location service will provide the ability to map the location of the rogue device on a floor plan so it can be found and removed quickly.

No matter the platform you choose, a WIPS implementation can prove to be very effective in preventing sensitive information from being accessed by unauthorized users.

3.8: Understanding the Challenges Introduced by Common Vertical Markets

As a CWDP, you will most likely face many challenges throughout your career. These challenges will have you thinking outside the box to find solutions, and the result will be that you develop valuable skills along the way that put you in higher demand. The following section provides some insight into challenges that you may face while designing WLANs for various markets.

Healthcare

The first concern in healthcare is information privacy. Federal regulations on patient privacy dictate strict policy when dealing with health records. Since the push for paperless records keeping in healthcare facilities, WLANs have become the most popular means of facilitating the collection, storing, and retrieving of healthcare data. This places a burden of security compliance on you, the designer, if you are involved in designing the total turnkey solution. It is important to become fluent in WLAN security options, including the functions of each EAP type used in 802.1X authentication. Beyond this, remember that security is always best when layered. Utilize firewalls where applicable and some form of intrusion prevention. Discuss security policy with the customer and help them develop one if needed. Most importantly, if you haven't already, study the *CWSP Study Guide* and get a deep understanding of what it takes to secure a WLAN.

A challenge faced by everyone who has ever taken part in a WLAN design at a hospital is gaining access to restricted areas. Some locations within a hospital require Wi-Fi coverage but they are the hardest locations to get access into. Operating rooms, labs, and record-keeping areas seem to be impossible at times to enter. When planning for the design, always discuss access and escorts for restricted areas. If you do not, it could burn days waiting for access to the areas.

Coverage cell overlap is a complicated issue in many healthcare facilities. Many hospitals will not allow APs to be mounted in patient rooms, leaving mostly hallways for positioning. This becomes a nightmare in coverage when you need to provide service inside the patient rooms. Look for closets and alcoves along the hallways to help you offset the APs. Placing APs in the center of nurses' stations also helps to keep multiple APs from overlapping each other too much.

Additionally, specially designed rooms, like x-ray rooms with lead in the walls, can make coverage a challenge. These walls lead to heavy attenuation and may not allow needed signal strength into the room from APs outside the room. For this reason, if Wi-Fi is required in the room, an AP installed within the space may be required.

Another challenge you will always face in healthcare is multiple areas overrun with interfering frequencies. There are paging systems, MRIs, Bluetooth devices, microwave ovens, and other unknown RF sources in hospitals. These make a design difficult at times, and you should always remember to perform a thorough spectrum analysis of the entire facility to find specific locations that house RF interferers.

Finally, because of dust and contamination threats, you will find it difficult (sometimes impossible) to raise ceiling tiles to find cable paths or existing wireless systems. In the past, it has been a popular practice to place wireless APs in the ceiling instead of below it. When you are looking for the interferers, your wireless tools are key in locating these items. Learn the tools of the trade well. They will save you hours, even days of frustration. Think creatively and take your time designing. Slower is the fastest way when compared to having to go back and redesign.

Education

In educational facilities such as high schools and colleges, both capacity and CCI tend to become big issues. Too many school systems attempt to hand you a pre-meditated design of one AP per classroom. If this is sufficient, then a CWDP isn't needed. Do not fall for these popular illusions. A single AP per classroom has been known to cause CCI while not providing for capacity requirements. Take the time to go through the steps of determining what is needed for application support before you start placing APs on a floor plan. Treat a school system just as you would a large corporation full of congested office spaces and figure out what devices will be used, how many devices, what applications, and the throughput requirements of these applications. You may very well find the need for several APs in a single room in some cases. It's not uncommon to have 20 or more students in a room using streaming media during class hours. In addition, video monitoring systems may be installed using Wi-Fi, other applications such as e-mail and Google documents will also be in use. With this much use, in comes cases, multiple APs will be required for adequate capacity. Don't assume coverage based on square footage alone.

Also, provide a solution that is easily scalable. Schools expand their programs, they add applications, they grow in number, and they require more bandwidth for all the above. Be sure to account for these unseen additions before designing.

Retail

Not many challenges exist in a retail environment. Most of them require coverage for a few of the devices that employees use to price-check, and sometimes take payment and check customers out no matter where they are in the store. Coverage and capacity are usually not the issues with these situations. Payment Card Industry Data Security Standard (PCI-DSS) compliance for credit card processing is the critical issue to be concerned about. Several large retail department stores have been hacked in the past few years, making thousands of customer credit cards accessible to the hackers.

All Wi-Fi networks in retail environments should meet the PCI-DSS requirements. This standard is an industry regulation requiring organizations that handle cardholder data to meet a minimum set of security standards to minimize the risk of cardholder data theft or loss. There are stiff penalties and fines associated with noncompliance. Be sure you know the most recent security requirements for PCI compliance before designing a WLAN in a retail environment.

Hospitality

Hospitality markets are similar to those of retail in relation to PCI-DSS. A travel agency promotes a vacation package to you in hopes that you buy it. Just like the retail store, the travel agency will most likely utilize a credit card processing service on a device connected through a WLAN. Hotels may also use wireless networks for payment processing. PCI compliance is equally important in this market.

In addition to this, in hotel environments, you can think of them like a hybrid space. First, you have the business side where the employees need wireless access. This may include push-to-talk devices for communications among maintenance and cleaning staff employees. It may also include the previously mentioned payment processing as well as VoIP implementations.

Second, you must consider the "residential side." The hotel rooms can be considered similar to a large apartment complex. Each room needs proper Wi-Fi coverage and the entirety of the rooms need appropriate capacity. Placing APs in hallways is an old-school method for deploying Wi-Fi in hotels and it just doesn't work best anymore. Back then, one in every 5-10 guests actually used Wi-Fi. Today, it is more like 3 for every 1 guest because each guest has a mobile phone they want to connect, a laptop, a tablet, mobile gaming devices, and more. Hotel residential spaces should be considered medium density deployments in most cases.

Third, you must consider the conference facilities. The number of client devices within these areas varies significantly depending on events occurring within the space. At times, it will require high density performance and at others it will require medium density or even low density. The WLAN must be able to provide for peak usage times and the design must accommodate for this.

High-Density Scenarios

The biggest challenge in high-density scenarios is capacity. The most common complaint with these facilities is slow network speeds, and this is caused by more users rather than throughput. Wireless coverage is too often mistakenly thought adequate when the signal is strong throughout the area. If this were the case, then we could simply install a single AP with high-gain omni antennas and turn it up to full power. That would be the solution to every coverage issue. As a CWDP, you understand the many different factors that are needed to allow specific client devices to function properly on the WLAN. That being said, any high-density scenario is going to need attention given to more than just the size of the area requiring coverage. Go back to the device and find out its requirements, what applications are involved in the use of the device or devices, and then get an idea of how many of these devices will be using the WLAN on a regular basis. Calculate the throughput needed for average everyday use and then add onto that for expansion and growth. Once you have that information available, it can be compared to the capabilities of APs being considered for deployment.

Almost always, a high-density area will require APs set to lower power and multiple APs positioned in the area for adequate bandwidth to meet the client and application demands of users in the area.

Nearly without exception, WLAN engineers agree that the Aruba Networks high density design guides are among the best in the industry. They can be found at: cwnp.link/arubavhd

Public Hotspots

The first thing anyone should think of at the mention of a public hot spot is security. The majority of these deployments are done by the store manager or an employee who ran to the local electronics store and picked up a wireless router and turned it on so the business could post a sign saying, "Free Wi-Fi."

Design is usually not even needed when implementing a public hotspot. The CWDP should focus on how the "Free Wi-Fi" can best be secured for its users. Unfortunately, even if there were a pre-shared key for customers to use, it doesn't mean the connection is secured. A very popular attack in a situation like this is a hacker using a laptop with AP software installed on it and passing it off as the hot spot. Users join the laptop, the hacker joins the "Free Wi-Fi", and all of the users' data flows through the hacker's laptop for the keeping.

Any public hot spot that is not open should be installed with a strong pre-shared key that is changed regularly, and once a user connects, they should be redirected to a captive portal that explains the risk of information transmitted on the wireless connection being compromised. All users should then be asked to click on an agreement to use the hot spot knowing the risks.

Many in the Wi-Fi engineering field are not fond of captive portals and, here at CWNP, we are not either, when they are implemented poorly. For example, when the captive portal doesn't work at all, when it comes up on every connection rather than remembering the user, and when it is just too cumbersome to use and intrusive. However, when implemented properly, it serves a true business purpose and can be beneficial to the organization.

It is always advised to have a VPN installed on any device that connects to a public hot spot for Internet use. If the device is an enterprise-controlled device, such solutions can be installed on the device by the IT staff and configured to be used anytime the device is connected to any Wi-Fi network other than the enterprise network.

Outdoor Networks

Free space is an outdoor network's friend and enemy at the same time. When specific areas need to be covered, outdoor APs and/or mesh APs can be installed to cover the requirements. Many challenges come up when dealing with outdoor WLANs, and the first one is signal propagation outside the desired coverage.

Outdoor wireless coverage is always a target for war drivers and hackers. When a hacker wants to connect to an outdoor network, they do not have to be near the actual network. A Yagi antenna or high-gain directional antenna will allow a hacker to work at compromising wireless security measures from a great distance away, and if proper intrusion prevention mechanisms are not in place, you will not even know the hacker is working on the network.

Most outdoor-grade APs are required to be registered in a Mac filtering table on the WLC before they can join. Even though this seems like a great step in security, Mac addresses can be discovered with scanning tools and then spoofed on laptops posing as APs, quickly providing a portal into the corporate network.

Any deployments of outdoor wireless networks should have layers of security in place beyond the Mac filtering table and a pre-shared key. If possible, 802.1X EAP authentication should be available for outdoor wireless deployments to prevent hackers from using it as an easy back door into the corporation.

Another challenge facing outdoor wireless designs is the division of bandwidth at each mesh hop. Mesh hops can greatly increase latency while decreasing available bandwidth. This can prove to be a problem with real-time applications such as VoIP. To avoid VoIP issues on mesh networks, make every effort to reduce the number of hops in the network and provide redundant paths back to a root connection.

One more challenge worth mentioning is finding adequate power sources for outdoor wireless networks. In most indoor WLANs, PoE from a switchport provides sufficient power for the APs. This is also true in most cases concerning outdoor wireless APs providing root connections for mesh deployments. Many outdoor locations requiring wireless coverage are in odd locations that do not have any power sources available. Some APs can be deployed on existing light poles with the use of light pole taps. The challenge to this comes into play when the light poles are all on a common timer and/or solar eye so that all outdoor lighting shuts down from a single control. At this point, other power sources will have to utilized. A couple of options are available:

- Solar-powered units
- Calculated battery banks

The use of solar is extremely flexible and the power supply can be designed to provide any length of backup power desired. These options can be expensive, but they are quickly installed, saving labor hours and costs.

Calculated battery banks are a hybrid option that utilize the same backup power in the solar unit, but the source providing the power is the light pole while powered. The AP runs, and the batteries charge all night from a tap into the light pole's power source, and then the battery bank keeps the AP powered all day until the light pole's power source comes back on.

Mesh networks are discussed in more detail in Chapter 6.

3.9: Chapter Review

In this chapter, you learned about the processes and challenges of designing a WLAN based on necessary requirements. You learned about quality requirements, which are directly related to the demands of requirement areas. You considered the various components, including security solutions, that must be planned for to provide an enterprise WLAN solution that meets and exceeds requirements.

3.10: Facts to Remember

Be sure to remember the following facts as you prepare for the CWDP certification, and be capable of explaining the details related to them:

- When selecting WLAN architectures, the basic choices are among controller-based, cloud-based, and autonomous architectures.
- Roaming is a critical function for voice related applications.
- RRM can be used for dynamic channel assignment for APs.
- Manual channel assignment may be necessary in outdoor deployments.
- The primary goal in channel planning is to avoid/reduce CCI and ACI.
- RF profiles consist of customized configurations applied to groups of APs.
- 802.11ac does not operate in 2.4GHz, though dual-band 802.11ac devices are sold as they use 802.11n in the 2.4 GHz band.
- A PKI consists of CAs that issue certificates used for authentication and encryption procedures.
- The PTK is derived out of the four-way handshake after 802.1x authentication.
- Planning for capacity requires an understanding of the applications to be used and the number of clients to be associated.

3.11: Review Questions

1. Designing a WLAN for capacity means that ____________, ______________, and ______________ should be factored into the design.

   a. Client devices, applications, throughput

   b. Applications, switch models, client devices

   c. Throughput, WAN connection, client devices

   d. Applications, Bandwidth, category cabling

2. The Pairwise Transient Key is derived during what frame exchange?

   a. Open authentication

   b. 802.1X authentication

   c. AP join request

   d. Four-way handshake

3. ____________ are used on network routers and WLCs in order to control and restrict traffic and different traffic types that traverse the network.

   a. PKIs

   b. PMKSAs

   c. ACLs

   d. WPA/WPA2/WPA3

4. A(n)________________ is used to manage the behavior of APs by assigning channel width, power settings, data rates and other configurations by assigning it to groups of APs at a time.

   a. WLC

   b. Autonomous configuration

   c. RF Profile

   d. RRM

5. Wi-Fi networks used to connect payment card processing equipment to the internet have to comply with a minimum set of security standards known as ______________________.

   a. Cardholder Data Environment Security Standard (CDESS)

   b. Payment Card Processing Standard (PCPS)

   c. Payment Card Industry Data Security Standard (PCI DSS)

   d. PCI Audit Compliance Standard (PCI ACS)

6. An intrusion prevention system that utilizes existing APs on the network as spectrum scanners while not servicing clients is known as___________________.

   a. An overlay WIPS

   b. A part-time WIPS

   c. A connected WIPS

   d. An integrated WIPS

7. Implementing QoS on wireless networks that support real-time applications is a necessity for reducing ________________ in critical packet delivery.

   a. Collisions

   b. Latency

   c. Overhead

   d. Packet size

8. A minimum of three, preferred four APs should see a device at a minimum signal level of –75dBm or higher for best results in ________________________.

   a. Minimizing packet loss

   b. Location services

   c. Data throughput

   d. Fast roaming

9. When a client device roams from one BSS to another BSS without moving to another subnet, it is known as ____________________.

   a. Layer 3 roaming

   b. Layer 2 roaming

   c. Fast transition

   d. Fast, secure roaming

10. An example of an application-specific requirement would be __________________________.

    a. VoIP phones work best when fitted with diversity antennas.

    b. To secure the wireless handsets, the new WLAN must support WPA2- PSK or WPA2 – Enterprise.

    c. The VoIP Phones we are going to use only function on the 2.4 GHz band.

    d. Both B and C.

3.12: Review Answers

1. The correct answer is A. Designing for capacity means that client count and type are considered along with the specific applications to be utilized and how much throughput is required for the client devices and applications. Capacity can be calculated and designed for once these factors are determined.

2. The correct answer is D. Open authentication frame exchanges only establish a connection between the Authenticator (AP) and the Supplicant (client). An AP join request takes place when an AP is trying to join a specific WLC. After the 802.1X authentication process is complete, the four-way handshake takes place and derives the Pairwise Transient Key during those frame exchanges.

3. The correct answer is C. Of the options listed, an ACL is the only security option mentioned that is applied to routers and WLCs to control and restrict traffic. The other options deal with authentication.

4. The correct answer is C. RRM manages all APs on a WLC but mainly for power levels and channel settings when dealing with CCI and coverage holes from failing APs. RRM does not manage data rates and channel widths, although its adjustments to avoid CCI will affect data rates that connected clients experience. However, through the use of RF profiles, all AP behavior can be adjusted and managed, including channel widths and data rates. The RF profiles are configured and administered on the WLC along with the help of RRM on some of the parameters within the RF profile.

5. The correct answer is C. The minimum set of security standards that deals with processing credit cards is the Payment Card Industry Data Security Standard (PCI DSS). This standard lays out the security requirements for Internet connections, the requirements for securing stored customer data, and the penalties for non-compliance.

6. The correct answer is D. The WIPS option that is deployed as separate APs is called an overlay WIPS. This option allows client servicing APs to be dedicated to clients full time while the WIPS APs are dedicated to full time scanning. The integrated WIPS utilizes the WLAN APs as part-time scanners as long as the AP is not servicing clients.

7. The correct answer is B. The main objective of QoS is to ensure that time-sensitive applications get their packets delivered quickly to avoid dropped packets, lost connections, and excessive buffering. The way that QoS accomplishes this is by

prioritizing packet types as they arrive at routers along with other application packets. The time-sensitive packets get priority and are sent out first even if they arrived last. This process drastically reduces latency in critical packet delivery.

8. The correct answer is B, The only wireless application that would utilize three to four APs at the same time would be location tracking Services. The other three answers mentioned are not affected by how many APs a client device actually sees.

9. The correct answer is B. Layer 2 roaming is when a client device moves between APs that are on the same subnet. This roaming never requires a change of IP address or a tunneled IP address. The client device keeps its original IP address with no special routing or tunneling required as it roams. (The client device remains inside the same broadcast domain.)

10. The correct answer is D. VoIP device security requirements are specific to the application to be used on the WLAN. When a company has VoIP phones they wish to use, the phone's frequency band is a specific requirement for the VoIP application. So, when designing the WLAN for VoIP, two application-specific requirements are made known: Security must be WPA2-PSK or WPA2 Enterprise, and VoIP has to function on the 2.4 GHz band only.

Chapter 4: Site Surveys

Objectives Covered:

2.2.1 Select and use appropriate design tools

2.2.2 Select and use the appropriate design methodologies

The WLAN site survey[47] is commonly seen as an unnecessary step in the design process as many believe that automatic RF (auto-RF) tools built into controllers and management software will "control" any issues that develop from not performing the survey. This simply isn't true. As stated later in the chapter, the designer simply cannot design a wireless network for a site without some form of site survey having been performed. In fact, the auto-RF tools can and will cause more issues than they will solve if the survey isn't performed properly. As the site survey is performed, a consistency is found across the coverage area in things such as power levels of APs and BSSID overlap.

To be clear, much of the confusion comes from legacy use of the phrase site survey. In the early days of Wi-Fi the site survey and the design were effectively synonymous as we lacked quality design tools with propagation modeling and other needed features. We would place an AP and test the signal from it. If it was good, we would mark that as the location for that AP and then move to the next location and repeat until the design was complete. We weren't just surveying, we were designing on-the-fly and that was OK then, but we have better tools today and the survey can be limited to what a survey should be[48].

In this chapter, and the modern recommendation of CWNP, we address site surveys for what they should have always been: a survey of the site used to discover the characteristics of that site for optimal WLAN design. For this reason, this chapter will be

---

[47] The term *survey* indicates the careful observance of a thing or an area to construct a map, plan, or description for some purpose. A *site survey* is simply the inspection of a location, either in a building or in a larger area such as a campus, to construct a plan for a wireless network at that site. The survey will result in information used to design the wireless network and this information will include a map of the site or a floorplan, environmental conditions that will impact wireless signals, and existing wireless networks or signals that must be factored into the design of the new wireless network. Surveying the site is about information gathering and the use of appropriate tools that gather accurate information is essential.

[48] It is understood that small networks can be designed quickly (those with 2-3 APs in a small business) using the old methods. But modern large-scale WLANs are better designed based on a true site survey and the use of proper design tools. This book and certification are focused on enterprise and large-scale WLAN design. Anyone who can do this kind of design can easily design for small and medium businesses as well.

somewhat shorter than most other chapters in this book and legacy concepts like manual site surveys and AP-on-a-Stick (APoS) will be only lightly addressed.

4.1: Selection and Use of Appropriate Tools

Performing a site survey can vary from site to site, depending on the environment in which you are surveying. It can be as simple as a small office with a few desks and then it can be as complicated as a multi-level industrial plant. From one extreme to the other, it is essential to understand the constants in the design, and those are the tools used for the survey and the education in how to use them correctly.

Site surveys are as old or older than those performed by the Greeks for military purposes, described in 300 BCE by Polybius. Torches were placed on high walls in prearranged positions to correspond to letters of the Greek alphabet, and by a suitable manipulation of the torches, messages were thus transmitted to a distance[49]. The Greek engineers would have to ensure that the walls were high enough for the torches to be seen from the desired distance. They would have to ensure that they burned large and bright enough to be seen as well. These actions would be part of their site survey[50]. In today's wireless networks, the site survey is just as critical and must always be performed. Whether it is performed by proxy (receiving documentation and information

---

[49] Maver, William Jr. *Wireless Telegraphy: Its Past and Present Status and Prospects,* Cassier's Magazine. Volume 21 Number 3. January 1902.

[50] In the eastern United States, another wireless communication system, based on audio waves and musical instruments items, was in use during the early part of the 20th century. Then, as now, the production of alcoholic beverage for sale or distribution was illegal without proper certification or licensure. As stated in an old news piece, "a species of wireless telegraphy we may be pleased to term it, is in use today by the peasantry in one of the Southern mountain sections of the United States to warn the proprietors of illicit stills when the revenue officers are on the warpath. When they come in sight of a cabin, the tenant emerges with a tin horn and sounds a blast which is taken up at the next cabin, and so on along the trail to the heart of the mountains. There is nothing to seize when the raiders arrive at the suspected place, and when they retire, business is resumed." You see, there truly is, and has been for a very long time, the desire in the human spirit to find creative ways to communicate over long distances without wires. And these "peasantry" as they were derogatively called at that time, were intelligent and creative. They would survey the area and determine the distances between cabins wherein the signals could be heard and ensure that a proper chain of such cabin inhabitants had the tools and training to get the message out.

from the remote site location) or in person (visiting the site and gathering the information), the designer simply cannot design a wireless network for a site without some form of site survey having been performed.

The proper toolkit begins with the right protection for the safety of the surveyer, which is particularly important in harsh environments. It then includes the technical tools used in the process as explained below.

Personal Protective Equipment (PPE)

It is always important to recognize the need for your safety when performing any work, no matter how simple you think that work may be. Preparing for your safety first means identifying the environment the work is to be performed in and selecting proper personal protective equipment for that environment.

Example: If you are going to perform a WLAN site survey in a busy warehouse where multiple forklifts are moving around loading and unloading trucks, it is a safe practice to wear steel-toed boots and a bright reflective vest (sometimes a protective strobe light may be necessary). In doing so, you have provided a visible way for forklift operators to recognize your presence and prevention of serious injury in the event one of these operators come too close to you and run over a toe. Another safe practice in this type of environment is to wear a whistle for alerting a forklift operator of your presence. (On a personal note, this has prevented me from getting hit countless times).

Other PPE includes the proper use of a hard hat. If you are performing a site survey in a warehouse to where materials are stacked on shelving above your head height, wearing a hard hat is mandatory in many places. While you are focusing on your computer screen performing the survey, items above you could fall and strike you without notice. Forklifts tend to bump other items on shelving while placing new items nearby. Always place your safety above the work itself and prepare for the worst.

Safety glasses tend to be overlooked often. Most workers feel there is hardly ever a need for them. Eye injuries are very common, even in an office environment. Something as familiar as a dropped pencil has found a way to bounce up and strike a person's eye. Not that we should be wearing safety glasses in our everyday activities but being wise about the environment you are working in can protect you from injury. It is not uncommon in a warehouse to have small metal shavings in the millions of dust particles lying around on all the shelving throughout the area. These particles get disturbed by forklift activity and seem always to find their way into an eye. Use your eye protection where needed.

Hearing protection is something else that gets overlooked on a regular basis. Going back to a warehouse scenario, the consistent roar of multiple forklifts in the area can develop a high decibel level of noise, and this can eventually damage your hearing. Any noise producing machinery you may be around for short periods can also cause hearing damage over time. A common mistake made by most is not using hearing protection when just making a quick walkthrough. The thought is that there will not be enough exposure to the noise to cause damage, but it does over time. Protect your hearing. It is something you cannot restore once it's gone.

Lastly, knowing when gloves and proper clothing are necessary for an environment is also crucial to your safety. Damaged or rusty handrails can cut your hands as you ascend and descend stairs. Dust particles containing metal shavings can find their way into your palm or fingers making for a painful day at work. Depending on the areas you will need to be working in, there is the possibility of objects striking you on the legs and arms. Other hazards such as arch flashes from electrical equipment or hot metallic objects from welding or cutting going on nearby. You do not have to be in the immediate area of the work to get an injury from it. Particles can become airborne and travel several feet to find a resting place on you. Keep your arms and legs covered when surveying in industrial areas.

I feel the need to share a personal injury I (Brett) sustained in an industrial area just for the sake of supporting the statements above:

I was simply waiting nearby an area where pipe cutting was going on, and I was completely dressed out in all the right PPE, never getting close to the cutting area until the work had stopped. Later that night, sitting on my sofa at home, I brushed my shoulder across my right eye to wipe it from staring at my computer screen. An extremely small metal shaving from my shirt buried up into the center of my eye, and the agony began. I had to go have the metal removed from my eye and then spend about four days placing antibiotics into the injured eye before sight was restored.

The lesson to learn here is that when you work in an environment with industrial machinery in operation, you should make a habit of changing out of the clothing you were wearing to prevent hitchhiking particles from causing injuries after you have left the workplace. Wearing protective covering (overalls) that are discarded on exit of the area can prevent injuries such as mine.

Your safety provides the means for work to continue. Don't take a chance on it. Always stop and review safety procedures in the environment, ask about any known hazards, and always wear the proper PPE.

Hardware and Survey Tools

The RF site survey is a process of examining the physical space where the WLAN will operate. In doing so, there are several tools that may become necessary while performing the site survey. These tools vary according to personal preference but remain needed often:

- Various APs – Since selecting proper positioning of new APs is the primary reason for a WLAN design, having several models of commonly used APs on hand is a must. APs for indoor, outdoor, office, warehouse, and mesh installations are good to have in your toolkit. They can be mounted on a tripod and used to test propagation within the space and to measure wall attenuation as well. The APs can provide the signal used to survey the RF behaviors in the environment.
- In addition to AP selection, you will want to include at least two of the three primary antenna types, which can be used to test propagation with different kinds of directionality[51], including:

    - Omnidirectional
    - Semi-directional
    - Highly Directional

    Highly directional antennas are more likely to be needed for outdoor site surveys, but you may wish to include this antenna type in any site survey kit just

[51] Remember, you are using these tools to survey the area and determine how RF behaves in the area based on attenuation, reflection, and so forth. The APs and antennas are not used during a modern site survey to determine every single mounting location; however, you may choose to determine the best physical mounting location in some select trouble areas that might present challenges to the propagation modeling in your design software. Experience will help you to determine candidate areas for this effort over time.

to be sure you are prepared. Multiple omnidirectional antennas with different levels of gain may also be beneficial.

By providing different antenna types, you can test an AP installation with patch or panel antennas against omnidirectional antennas. This can help you test methods for reaching challenging areas within a facility.

- **Laptop or Tablet** – While performing a site survey, you will need to operate survey software on some device with a screen that can display and save the information you are studying. Many preferences exist here, and the main concern is to be sure your choice of the device has sufficient resources adequate for operating the survey software you are using. It is a good practice to select this device with an option for interchangeable batteries and not built-in batteries. Some devices only come with built-in batteries making your survey time dependent on the battery life.
- **Appropriate Wi-Fi Adapter(s) –** To perform a WLAN survey, your laptop or tablet will need a way to communicate with the survey AP. This can be accomplished several ways: Internal, external, and USB adapters are most common. Just be sure the adapter you are using is correct for the required survey. In other words, don't use an 802.11b/g adapter if you are required to provide coverage for 802.11ac. At the time of writing, only internal adapters were available for 802.11ax in 6 GHz, but USB 3 and other external adapters will come to market within a year or so of publication. In fact, only two USB adapters were available for 802.11ax in any band at the time of writing.
- **Other Client Devices –** Sometimes it is necessary to use a survey tool that comes on a client device such as a Wi-Fi phone or scanner gun. You should find out what client devices are going to be utilized on the new WLAN and see if the customer will provide you with one of each for the survey.
- **Measuring tool** – being able to measure off distances is essential and having a tool that can do this quickly saves you time and battery life with your laptop or tablet. A laser range finder is much more efficient than other tools. You can pick these up in the sporting goods section at most department stores. This tool can help you with floor plan calibration and finding cable pull distances.
- **Camera** – Photographing AP locations is crucial to your documentation. You should always create your documents as if someone else is going to perform the installation. They will not know the exact spot you want the AP unless you have

a good visual to go by. Other uses of the camera are for documenting switch locations, and port counts, existing cable trays and penetrations, data rack layouts and available rack space, etc. The camera can prove to be extremely helpful when doing documentation.

- **Flashlight** – Some areas you need to access may be completely dark or very low lighting making it difficult to see your footing or other obstacles that you could run into or trip over.
- **Flagging or marking tape** – While a simple picture can be enough at times, marking locations with a unique identifier can prove to be helpful. Ceiling tile is ceiling tile, and a picture of a ceiling tile doesn't always get you to the right one. When allowed, write the name of the surveyed AP on a marker tape, lift the ceiling tile and place the tape under the edge so that it will display the location name while hanging. Then take a picture far enough back that a room number, fire extinguisher, exit sign or office name shows up in the photo to help identify the location.
- **Mobile stand (tripod)** – This item has been called many things, but it is used to raise and lower your survey AP at each location used during the survey. Many have designed a stand with wheels to roll it around, while others utilize a tripod option. Whatever you choose, your stand should be able to vary up to at least a twelve feet high ceiling. If the environment has twelve feet high ceilings and you perform your site survey at eight feet because that is as high as your stand will raise, then the RF propagation from the survey will differ from the final installation resulting in a larger footprint per AP and a substantial increase in cell edge overlap.
- **Battery Pack** – When performing a site survey, the most convenient way to power a survey AP is with a portable battery. Many have used a power injector and extension cord, and these are great for a backup but should not be used for the entire survey. The battery pack allows you to keep the AP associated with your survey software while moving from place to place. This saves a lot of time.
- **Survey Cart** – This is another excellent tool to use if the environment allows for it. It is helpful having all your documentation on a portable surface so that you can take notes, move through drawings, set your laptop down and move the survey AP around without hunting a place to sit down every time you need to do something.

- **Power Injector and Extension cord –** As mentioned above, these items are great to have around as a backup in the event your battery pack goes dead one or two APs shy of completing the survey. You may also need to use the extension cord to charge batteries depending on the location.
- **Laser Temperature Sensor –** Many industrial environments have varying temperatures throughout the facility. It is common to go from one room to the next and see a huge increase in temperature. This is important because it may be necessary to use an external antenna model AP in higher temperature areas. The laser temperature sensor can be pointed at a twenty feet high ceiling and give you the temperature up there where it will most likely be significantly higher than the floor level.
- **Clipboard, Pens, and Highlighters –** While taking notes, you may want to mark switch locations in one color while marking AP locations in another. It is always good to develop a color code that you understand so that interferers, existing wireless devices, switches, and new AP locations can all be identified in your notes.
- **Long Ethernet Cable –** You will need an Ethernet cable to power your survey AP with. The length needs to be longer than the height of your survey pole so that you can be flexible with the placement of the battery pack.
- **Communication device –** When performing a site survey, it is common to need assistance from time to time. This may include getting a secure door opened or even asking a question about the environment. Having direct communication with key personnel will help in time management of the survey.
- **Global Positioning System (GPS)** units can be very handy when performing either outdoor site surveys for bridge links or site surveys for outdoor WLANs. The GPS unit can be used to document fairly exact locations for device placement or signal strength measurements. Today, most smartphones have exceptionally accurate GPS systems in them for use in such scenarios.
- **Zip Ties/Duct Tape –** you will find that having some temporary means of securing cables and survey APs will often be necessary. Zip ties and duct tape are great to have around when this need comes up.

You may find that other tools become necessary as your experience with different environments increases. This list serves as a good starting point for you.

Diagnostic and Analysis Tools

Beyond the need for the above-mentioned list of items, you will also find it is important to have a combination of software tools that detect, diagnose and analyze existing radio frequencies in the facility. In this section, we will cover many of these common tools.

- Spectrum Analyzer - One important task necessary for a successful site survey is the spectrum analyzer. In fact, this task is one of the most important when you plan to design a very large network based on an abbreviated modern survey that simple verified materials attenuation factors and locates trouble spots[52]. Spectrum analyzers come in many form factors and can range in price as well. Before starting the WLAN design, you should use a spectrum analyzer to locate existing RF sources in and around all required coverage areas. Spectrum analyzers see the spectrum, or the energy emitted within a given frequency range and not the data being carried on the waveform. Because of this, spectrum analyzers can detect items or energy that protocol analyzers cannot, such as:
    - Wireless phones
    - Bluetooth devices
    - Microwave ovens
    - Electrical energy from motors and generators
    - Cameras
    - Baby monitors
    - RF generators

Most spectrum analyzers provide RF signature samples to help you identify the source of any signals you detect. Another resource in a spectrum analyzer is the display of the duty cycle of a signal source. This will allow you to determine if a signal is intermitting or a constant. The spectrum analyzer will help you determine if a transmitting source can be removed or will need to be designed around.

[52] This statement is true because you cannot design around interference that you do not know is there. This reality is one of the primary reasons that onsite visits are nearly always necessary for effective design. Even if the acquirer tells you there are no other wireless signals in the space, it doesn't mean they are correct. A spectrum sweep of the environment will reveal the true issues that must be considered as constraints in your design (or that must be removed before implementation, if possible).

- Protocol Analyzer - Protocol analyzers are used to capture the actual WLAN frames that traverse the medium through a specific channel. If you plan on purchasing a protocol analyzer, be sure that it supports the latest 802.11 protocols.

A WLAN protocol analyzer is used to capture data packets as well as 802.11 MAC layer frames and decode their contents. They will also display information from the radio tap header, such as signal strength and noise floor information (though this data is often less reliable than that of a spectrum analyzer). The information is usually presented in a very readable format that is easier to understand than simply looking at a series of ones and zeros. For example, many protocol analyzers break the frame headers down in such a way that each bit is explained in real world terms and concepts. This process is often called frame or packet decoding.

Protocol analyzers tend to provide features that make your life easier. For example, they can be configured to filter for specific frame types or upper later protocols. They can display performance statistics for overall throughput and frame rates on the network. They can even be configured to work with remote sensors so that you can analyze and configure alerts for the entire wireless network in your organization and not just the small portion of the network near your analysis laptop.

For site surveys, use a protocol analyzer to perform the following tasks:

- Identify WLANs in the area (internal and external[53])
- Discover areas with high frame retries in the post-validation survey
- Determine the packet types traversing the current network for an upgrade site survey
- Find APs that are improperly configured in the post-validation survey

---

[53] It is very important to identify both internal and external WLANs. Many buildings are shared spaces and other organizations may have WLANs that will impact your design. You do not want to cause problems for their WLANs, and you do not want them to cause problems for yours. However, an important issue must be addressed with the acquirer in such situations. If you identify neighbor WLANs that use multiple APs and you have identified them (through MAC addresses) as APs from a vendor that you know supports automatic channel planning, the acquirer should be informed that the neighbors network configuration may change and that the change could impact the network you are designing. Methods of resolution include using automatic channel planning on the network you design, avoiding bands used by the neighbor network when possible, and configuring alerts in the monitoring solution to identify trouble areas for repair.

- Track the number of clients connecting to a given AP
- Locate the most active clients on the WLAN

Various Protocol analyzers are available to use with remote APs. The Cisco WAP371 AP is an excellent choice for building a WLAN analysis lab or for use with site surveys where you desire to perform 802.11ac analysis and have a roaming power source to power the AP with PoE. The Cisco WAP371 AP is depicted in Figure 4.1.

Figure 4.1: Cisco WAP371 APs[54]

- Cable Testers – There are many cable testers available ranging from a simple wire map option to a complete diagnostic of the cable. The phrase "You get what you pay for" applies here. With modern wireless APs and PoE solutions, it is not recommended that you use a wire map option just to see if the pinout is correct.

The Klein VDV Scout Pro 2 Tester in Figure 4.2 is a good option for testing on a budget.

The Fluke Networks CIQ-100 tester in Figure 4.3 performs more complex diagnostics of cables to determine capable bandwidth on the cable, locate crosstalk and tell you how far

[54] These Cisco small business APs are now end of life; however, they can still be acquired from used sales for very little and work will for most 802.11ac capture tasks in 2.4 GHz and 5 GHz. An alternative is the WAP125. To find 802.11ax APs that support packet capture, explore the documentation provided from the vendor. If they do not provide instructions on performing remote packet capture or capturing packets and then downloading the PCAP file, it's best to assume they do not offer it unless you have a good source that proves otherwise. Generally, you can find small business class APs for under $300 that support packet capture.

down the cable it is, discover unused switchports, etc. This unit is useful for many more functions but is significantly more expensive.

Figure 4.2: Klein VDV Scout Pro 2

Figure 4.3: Fluke Networks CIQ-100

Systems Analysis

Systems analysis is something you will need to do after initial conversations with the customer so that you have a good grasp on what is expected out of the new WLAN you are there to design. When determining the needs in your WLAN, performing systems analysis of the existing networking infrastructure is a valuable place to start to determine if it is compatible with the incoming WLAN environment. One compatibility issue, or at least a potential performance issue, is the lack of up-to-date Ethernet cabling. Another would be out of date switches such as 10/100 ports and lack of PoE or sufficient PoE. You need to evaluate the network as it currently exists while considering its future growth.

While evaluating the existing network, look at all applications running on the network and be sure to answer the following questions about them:

- How many clients use the application simultaneously?
- What is the average size of a data transmission for this application?
- What is the average number of transmissions per minute?
- What is the maximum transmission size?
- What is the maximum transmitted data seen in a given minute?
- Does the application involve bursts of large data transfers, moderate data transfers sustained over time, or both?

With the answers to these questions, you can determine the maximum throughput needed at peak times, the average throughput needed over time and any specific Quality of Service (QoS) requirements that the WLAN may need to accommodate. It is also important to check with the network administrator to see if future projects are already being planned for the network that can increase network usage and/or throughput. A WLAN can introduce hundreds of client devices and many new applications to a network's traffic load, so throughput can become a factor.

Systems analysis will take some time up front to perform correctly, but it will also save time and money on the backend due to having a well-designed plan to follow based on accurate knowledge of the existing infrastructure. Remember, the site survey is the opportunity to "see what's there" and make sure it can support the intended WLAN system. If you cannot go onsite, receiving documentation of the current network devices, cabling, and services can assist in this process.

Throughput Testers

Another software that can be used in site surveys is a throughput tester. Simple throughput testing can be implemented using returned metrics from common client tools, like FTP download clients. Figure 4.4 shows the Windows FTP client reporting the speed of a download.

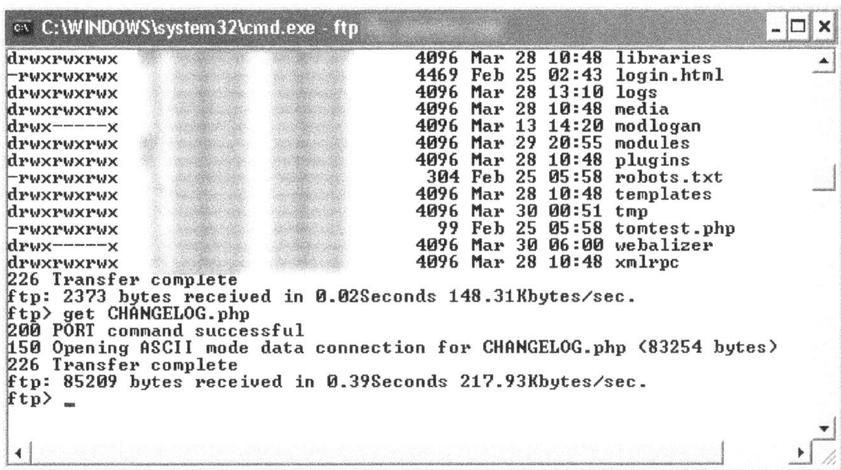

Figure 4.4: FTP Used to Report Throughput Metrics

Command line (CLI) tools or graphical engines, add capabilities like statistics for UDP or TCP, and simulating VoIP or Video traffic types.

TamoSoft Throughput Test is a free throughput testing software and can be downloaded from Tamos.com. Figure 4.5 shows the single screen of the Throughput Test software. It is comprised of a client and a server. The server runs on one computer and the client on another. Each computer is connected on either end of the link you desire to test. The test can be against Wi-Fi only (both computers connected to the same AP) or include the network infrastructure (one computer connected to the AP and the other connected at the desired point on the wired network).

A throughput Test sends data in both directions during a TamoSoft test, providing measurements for both uplink and downlink traffic. Figure 4.6 shows the server view of the Throughput Test utility.

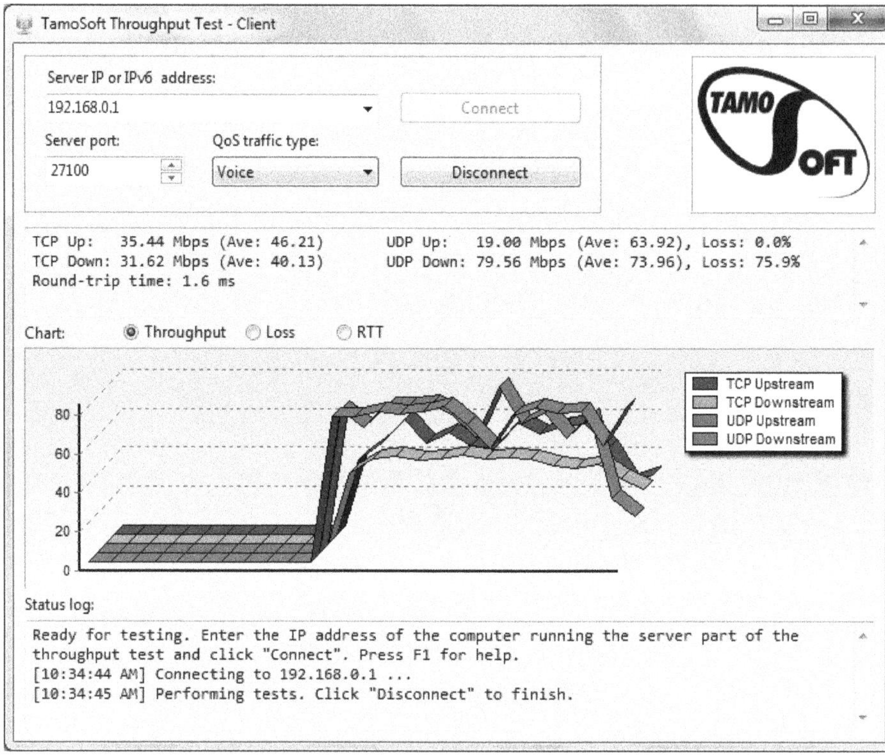

Figure 4.5: TamoSoft Throughput Test Client View

iPerf is a throughput testing tool for TCP or UDP communications and is used to measure bandwidth, report on the MTU size and supports TCP windows size adjustments. With UDP, it can also measure packet loss and jitter, which is important for VoIP testing. Currently, iPerf is available on Windows, Linux, Android and Mac OS X.

Figure 4.7 shows a basic iPerf test running on Windows. The iPerf client sends data to the server, so to test download data, you must make the wireless client the server and use the iPerf client software on a wired client. Alternatively, with supporting versions of iPerf, you can use the -d option to perform bi-directional testing. This mode is not always accurate.

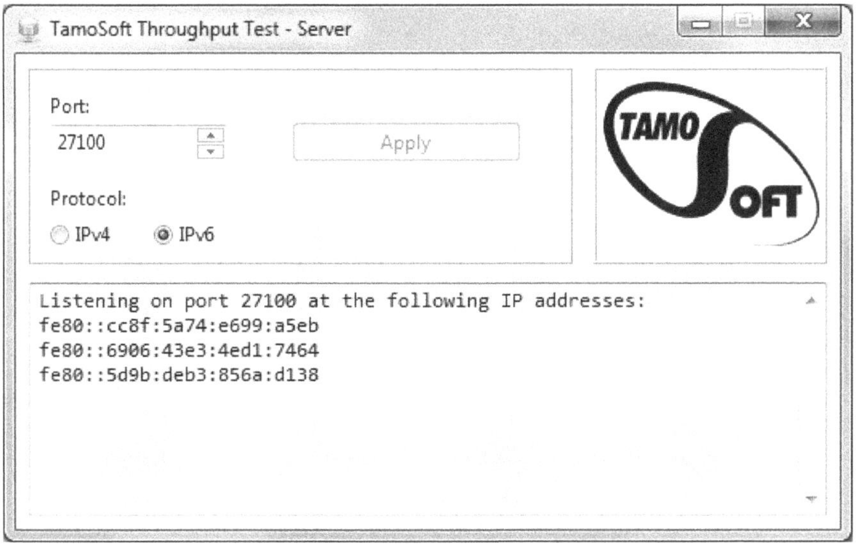

Figure 4.6: TamoSoft Throughput Test Server View

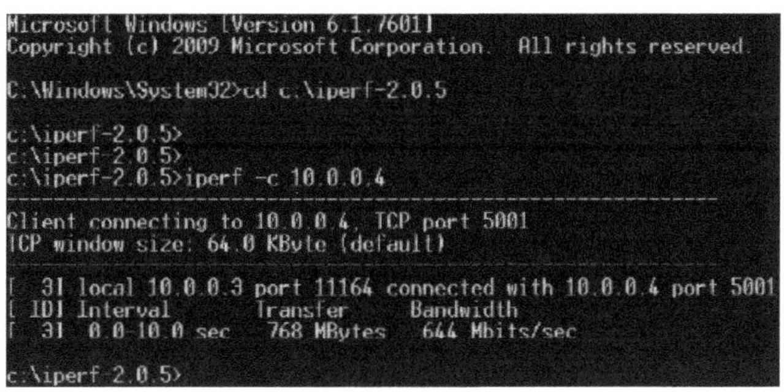

Figure 4.7: iPerf Testing

Another useful performance tool is the IxChariot software suite. IxChariot can be used to perform network diagnostics and provide detailed statistics about network performance. It also supports application emulation for testing against various popular applications. For example, the following can be emulated through the software:

- Facebook
- Google search
- Google mail
- SQL Server

- Gnutella
- VoIP calls

Some of the most useful metrics gathered from IxChariot include bitrate, loss, jitter, and delay.

Gathering Attenuation Measurements

There are times that you will need to know the decibel loss of RF transmissions from one side of a wall or object to another and you cannot assume certain materials are present in a wall just because of the way the wall looks. It is also important to note here that 5 GHz is not 5 GHz across the board when concerned with building material penetration, so you should determine which channel of a band you wish to measure with for this task. An example to explain the difference in signal physics would be to perform an active site survey in a single area where you can test beyond three normal office walls. See Figure 4.8 and 4.9 for dB loss due to channel differences.

In the example shown in Figure 4.8 you see a combination of signal loss due to free space and building materials. This result could be seen in an office area.

> This is only an example to display a potential signal performance and you should not expect to see the exact same dBm signal strengths during your tests. Building materials vary in many environments, therefore providing a difference in dB loss as compared to other environments.

As seen in Figure 4.9 you see we still have the 5 GHz band in use although there is a significant increase in the channel selection. When moving from channel 36 to channel 149 the behavior of the wave is different because the wave becomes shorter as the frequency increases. The different physical characteristics of an RF signal cause a difference in the way the RF signal behaves when striking and or penetrating various building materials.

This reality will hold even truer for 6 GHz WLAN designs. The range of spectrum is significantly greater than 5 GHz in many regulatory domains that have opened the entire band. In fact, there is 1200 MHz of space in 6 GHz when the entire range is available and there is 500 MHz available in 5 GHz. Given that there is more than double the bandwidth available in 6 GHz, the variation in attenuation between the lowest channels and the highest channels will be far more significant than in the 5 GHz band.

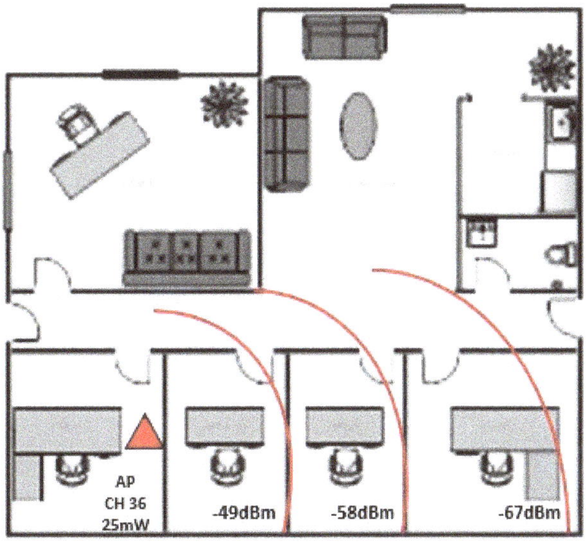

Figure 4.8: Signal Attenuation on Channel 36

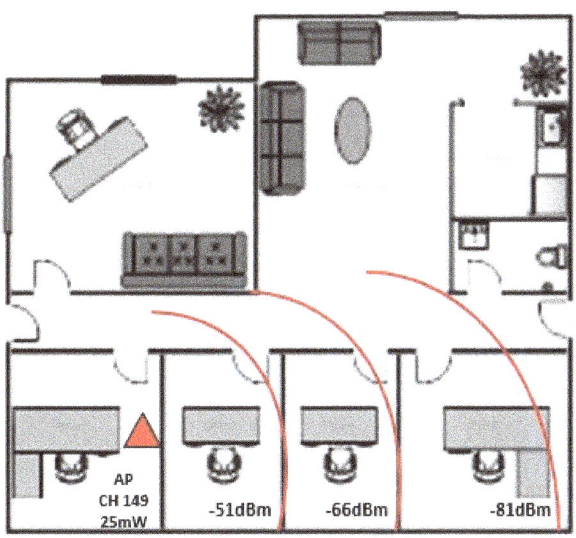

Figure 4.9: Signal Attenuation on Channel 149

As shown in Figure 4.10, we see a short waveform for higher frequencies as compared to lower frequencies. This is important to understand when performing a site survey and not only when deciding between 2.4 GHz and 5 GHz. You must understand that the physics are going to be different at every increase in channel resulting in noticeable differences in an AP's footprint as the channel numbers go ever higher and the bands increase from 2.4 GHz to 5 GHz to 6 GHz. In other words, if you are going to survey an entire office complex with an active AP set to channel 36, then you should consider limiting the channels made available to your WLC for dynamic channel selection to maintain a more accurate survey.

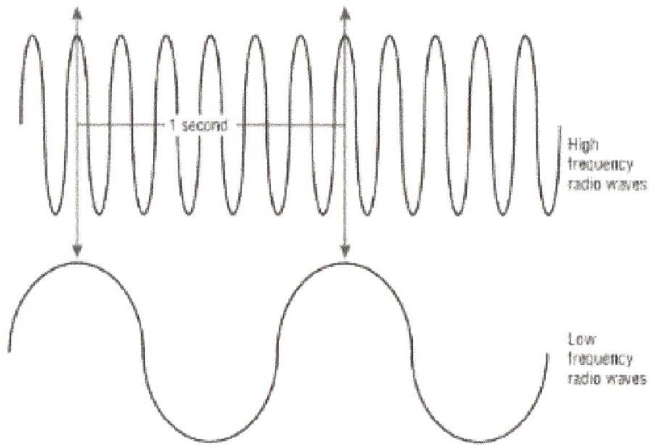

Figure 4.10: High Frequency Compared to Low Frequency

> The following scenarios are designed with 20 MHz channel usage. Many corporate WLANs still utilize 20 MHz channels with dynamic options for 40 MHz channel bonding when more throughput is needed.

Some corporations choose to utilize multiple controllers to divide up management of APs from floor to floor. An option here would be to divide the UNII-2 band channels among controllers and utilize the UNII-1 band for floors one, three, five, and seven, while utilizing the UNII-3 band on all even numbered floors. One WLC provides UNII-1 channels; another provides UNII-3 channels while half of the UNII-2 band channels are available on each. Of course, if 80 MHz channel bonding is being utilized, this option cannot be used efficiently. Below is a list of the available non-overlapping channels per UNII band[55]:

- UNII-1/Lower Band (5.150 to 5.250 GHz) Non-overlapping channels 36, 40, 44, 48
- UNII-2/Middle Band (5.250 to 5.350 GHz) Non-overlapping channels 52, 56, 60, 64
- UNII-2 Extended (5.470 to 5.725 GHz) Non-overlapping channels 100, 104, 108, 112, 120, 124, 128, 136, 140
- UNII-3/Upper Band (5.725 to 5.825 GHz) on-overlapping channels 149, 153, 157, 161, 165

It is important to note that when surveying areas of similarity, it will take more APs utilizing UNII-3 channels than it will if you are utilizing UNII-1 channels if you are surveying similar floors with the same square footage so expect to see a slightly higher AP count.

A best practice for surveying would be to choose a channel in the center of the band you are using for your survey AP. In other words, choose channel 157 of the UNII-3 band to survey with, so there are two channels below it and two channels above it. This option will allow for much more accurate design.

As a CWDP, you should always seek to provide the best design possible for your clients. This starts with a good understanding of RF physics and signal propagation across the spectrum(s) you are using in your design.

Now that we have covered differences in frequencies based on channel selection let's look at how to determine attenuation from building materials.

Survey software providers have many pre-set calculations built into their predictive modeling options. These calculations are close in most cases but should never be assumed correct in all cases. It has been found that building contractors sometimes double up the layers of drywall for noise reduction. If this happens, the pre-set

---

[55] You will never be tested in relation to the term UNII in CWNP exams. The term is not a global term, though vendor literature uses it for global documentation. When discussing the 6 GHz channels, CWNP has opted to use the terms lower channels, middle channels, and upper channels. It is up to the professional WLAN designer to know the channels available in the target regulatory domain and know the terms used for the various sections of bandwidth if required, but such information is not tested on the CWNP exams.

calculations in your survey software will not be accurate. Other scenarios have been found to have metal pipes, large amounts of wiring and various insulations. All these materials change the signal propagation behavior and the only way to determine how much is to test it.

As we see in Figure 4.11, once you have determined the channel with which you wish to test the attenuation, you place the AP next to the wall in question (typically 2-3 meters from the wall) at a height you can reach with your client device containing the survey software. Hold the laptop (client device), at the same height as the AP and record the signal level next to the wall on the AP side. As we see in Figure 4.11, the signal level is at -38 dBm in the room. Next, you move the laptop to the opposite side of the wall and record the signal level after it passes through the wall at a distance that would be as close to twice the distance from the AP as compared to the last measurement location. In Figure 4.11 we see that signal reading is -47 dBm. After performing this simple test, we now know this particular wall presents approximately a 3 dB signal loss[56]. This is because you lose 6 dB for every doubling of the distance. By intentionally factoring this in, you can achieve more accurate attenuation measurements. However, see footnote 56 for more important information.

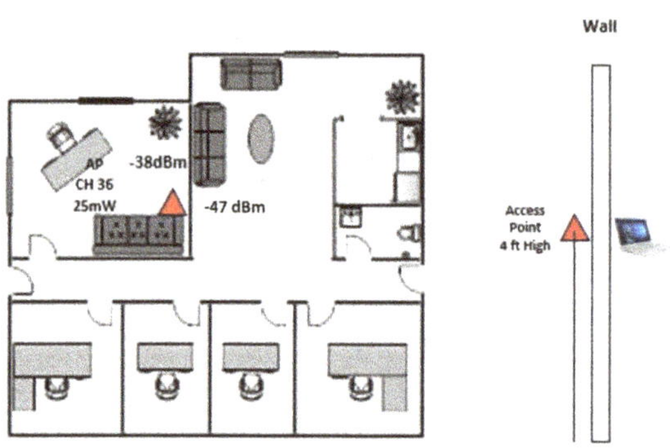

Figure 4.11: Measuring Attenuation

[56] Realize that reflections in the space can impact such measurements. That is, the signal may be stronger in the room with the AP at two meters away due to reflections. Additionally, due to reflective paths, the signal may be stronger on the other side of the wall than the wall alone would allow. However, the difference in signal strength on the other side of the wall at twice the distance from the AP than the laptop was positioned in the room with the AP, will give you a rough calculation of the attenuation factor.

This method can be used to gather attenuation for walls, windows, doors, and objects that RF may encounter. Once you have gathered attenuation values, they can be applied to a facility floor plan in the WLAN design software to help determine facility coverage and AP placements. This is accomplished differently from performing a traditional AP-on-a-stick (APoS) survey.

4.2: Performing a Pre-Design Site Survey

Some organizations do not budget for a pre-design site survey at all. They simply start with the predictive design and either use best guess for wall and ceiling attenuation factors or provide proper blueprints and accurate materials that have information that can be used to determine attenuation factors. This method is covered first in this section. However, it is recommended that the CWDP perform a spectrum scan of the target environment and perform attenuation measurements at a minimum in an onsite survey whenever possible. An entire active or passive survey is not always required or preferred, but when it is, this section provides an overview of the processes as well.

Predictive Design

In predictive design, a floorplan is loaded into the survey/design software and then calibrated or scaled for accurate building dimensions and square footage. After this step, walls are drawn (if not available for import) and attenuation values are assigned to walls, windows, doors, file cabinets, etc. that exist across the entire floorplan. Once you have assigned the attenuation values, you can then simulate APs and RF propagation. This allows you to choose a brand, model, antenna type, and power settings of the AP before assigning it to the predictive model design. Now, the configured AP can be placed anywhere on the floor plan and the software will provide you with a visible RF footprint based on the assigned attenuation values of the building materials. You can move the simulated AP around the floorplan to find the desired placement before adding the next AP. Once coverage is simulated throughout the facility, APs can be moved slightly, as needed, to adjust for best coverage and proper cell overlap.

An additional use of a predictive design is building a Bill of Materials (BoM) and a budget for a project. A predictive model can get you close to the number of APs needed before visiting the site and performing a site survey. This gives you and/or the project manager a great idea for setting a budget[57] for equipment, planning for cabling, and how many switch locations will need to exist or be added. Predictive designs can also provide you with a good starting point for your site survey. By building a predictive model first, you can better determine how far an AP will propagate and from where when starting a site survey for a large facility.

Performing a Spectrum Analysis

When an onsite survey is performed, a spectrum analysis should always be performed to avoid any RF surprises that might arise after design and implementation.

When performing spectrum analysis, it is best served to walk slowly through the facility, one floor at a time while documenting any unusual signal patterns, existing Wi-Fi signals, and/or signal spikes intermittently happening as you walk. Most spectrum analyzers have a recording feature that will allow you to capture all activity on the monitor and play it back later for references. You should name captures in a manner that will allow you to separate floor to floor findings and then plot the findings on a floor plan so that the spectrum analysis data becomes part of the design documentation.

Figure 4.12 shows a screen from Metageek's Chanalyzer Dual-Band spectrum analyzer. This product allows you to see multiple screens simultaneously so that you can view signal strength, signatures, saturated channels and more. It also has a recording feature for capturing data and playing it back later. One other nice feature about such a product is the ability to build reports from the data gathered. When choosing a spectrum analyzer, you should be sure the needed features are all available and compatible with the client device you intend to use it with.

In Figure 4.13 you can see other information provided by Chanalyzer such as the SSID, channels used, signal strength, security type, max data rate, vendor name, and 802.11 protocols used on the AP. When performing analysis of a facility, it is really helpful to choose a tool that will provide as much information about spectrum utilization as possible.

---

[57] At the early stage, this is likely to be a rough order of magnitude (ROM) budget estimate. These estimates are typically considered at ±50% accuracy, but can be helpful in the early stages to determine if a project is feasible.

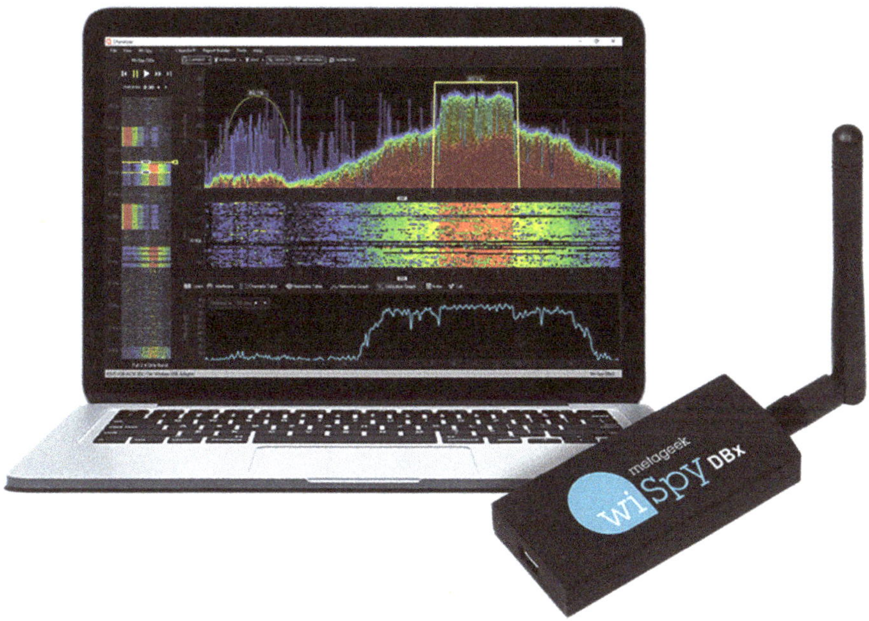

Figure 4.12: Metageek wiSpy DBx

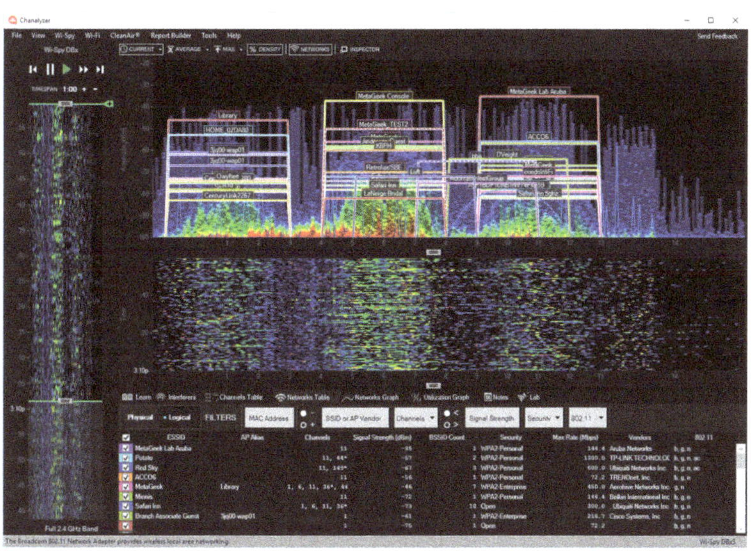

Figure 4.13: Metageek Chanalyzer

In Figure 4.14 Air Magnet Spectrum XT is shown displaying four separate screens. It displays the real-time FFT and the FFT duty cycle of signals found. Below those screens are the sweep spectrogram views of both FFT screens. This program will also provide channel usage and a list of devices found within the range of the Spectrum XT sensor. The bottom line is that whichever spectrum analyzer option you choose, it is going to be extremely beneficial to you and your design process.

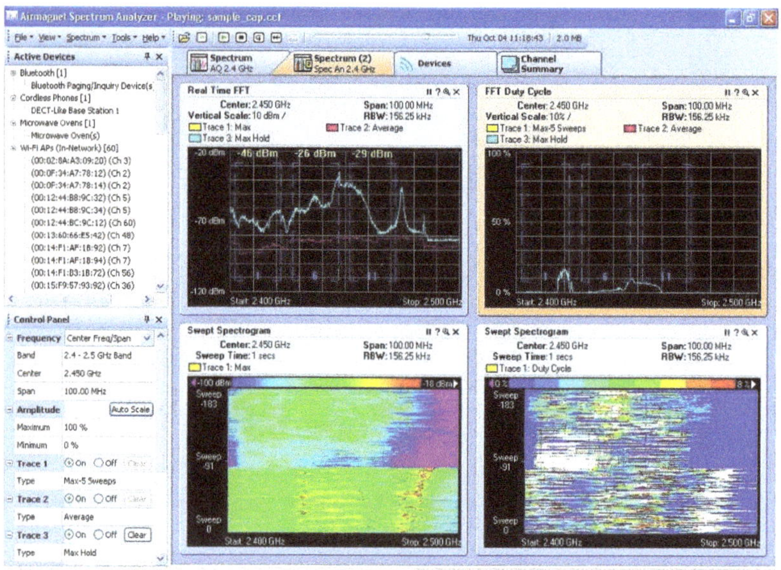

Figure 4.14: Spectrum XT

Survey Software

Survey software is designed for performing WLAN site surveys by utilizing several options. Site survey software is available from iBwave (iBwave Wi-Fi), Ekahau (Ekahau Site Survey (ESS)), NETSCOUT (AirMagnet Survey Pro), and TamoSoft (TamoGraph Site Survey) as well as several other software providers although many others are not capable of the accuracy needed for commercial WLAN design. The software typically provides the following features and capabilities:

- Capability to import floor plans[58] for documenting AP locations and coverage area metrics
    - Signal strength
    - Data rates
    - Throughput rates
    - AP coverage
- Active and passive survey options for performing manual surveys
- Provides pre-set attenuation metrics for performing predictive designs
- Capability of surveying multiple floors
- Floorplan calibration for RF footprint accuracy
- Simulation of different APs (within reasonable non-calibrated boundaries)
- Simulation of different clients (within reasonable non-calibrated boundaries)
- Configuration of supported PHYs

It is important to calibrate all floorplans after importing them. This process involved selecting a long-distance range on the floor plan that has a known distance in the real world and entering that distance into the software. This step allows the software to ensure proper calibration for RF signal coverage.

Figure 4.15 shows the use of site survey software for the planning of AP placement and coverage within an area. Capacity is essential, but capacity is impossible without coverage. Coverage analysis is an essential part of site survey software.

Figure 4.16 displays the capacity planning feature of ESS. If providing coverage only, users may not be able to perform the required functions of their jobs.

[58] As much as possible, request CAD-based floor plans that include pre-drawn walls and doors, etc. This will make the initial steps much simpler because the walls and other architectural items can be imported into the design software. Otherwise, you will have to spend, in some cases several hours or days, drawing in all the details, depending on the size of the area and the number of floors and variations across floors. However, if no such files are available, the WLAN design tools allow you to import standard graphics files with line drawn floor plans and then manually draw the needed objects.

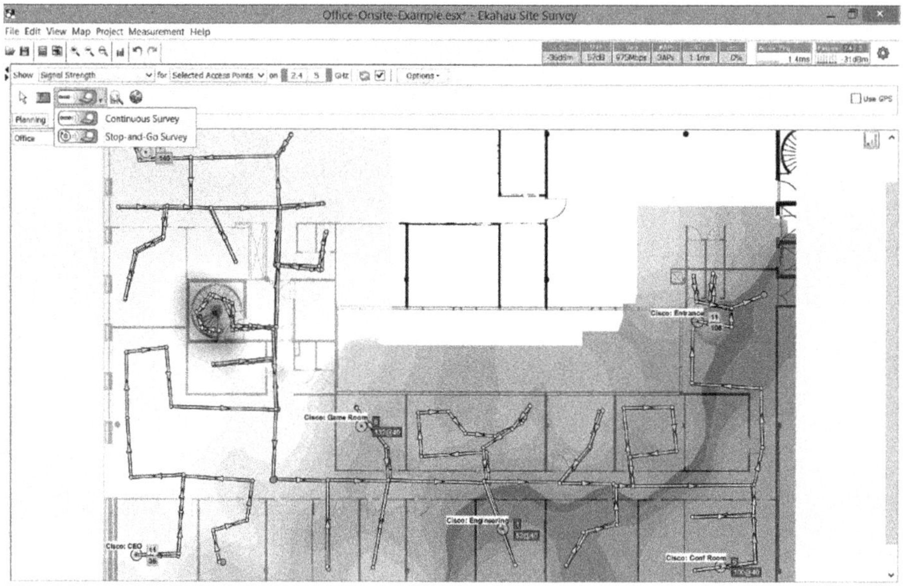

Figure 4.15: ESS Showing Coverage with a Heat Map

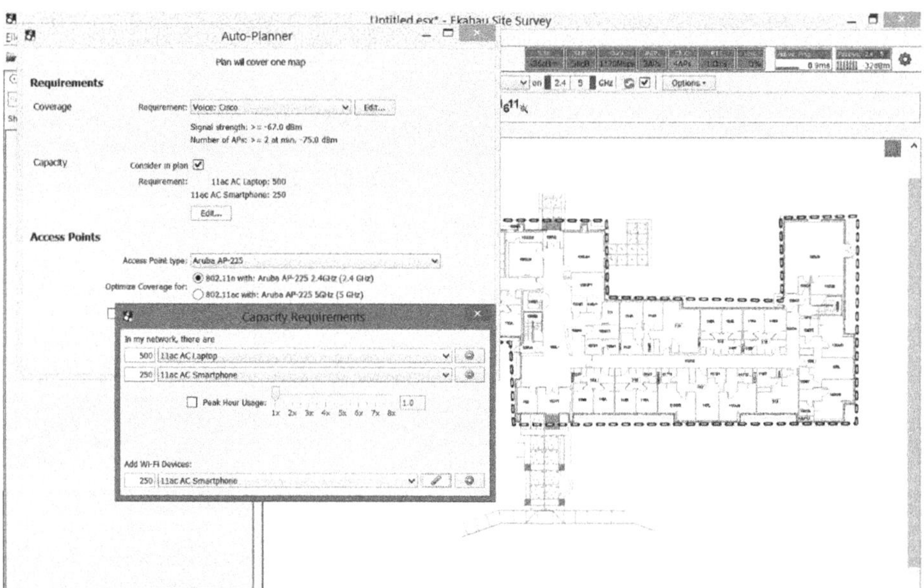

Figure 4.16: Defining Capacity Requirements in ESS

The capacity planning feature of ESS allows you to specify requirements, such as voice requirements specified by a given vendor. These requirements result in two important metrics:

1. Signal strength
2. Number of visible APs

For voice capability, the required signal strength is at least -67 dB at the edge of each AP's coverage cell with a 20 – 25 percent overlap between individual coverage cells. This ensures adequate time for roaming voice clients to perform proper steps needed to associate to the next AP without interrupting the call.

When possible, you should optimize AP placement so that less used areas have fewer APs and client congested areas have more APs.

Part of capacity planning is throughput analysis. ESS offers a special view showing throughput when an active survey is performed. This feature is useful in both site surveys and validation surveys. Figure 4.17 shows the throughput heat map.

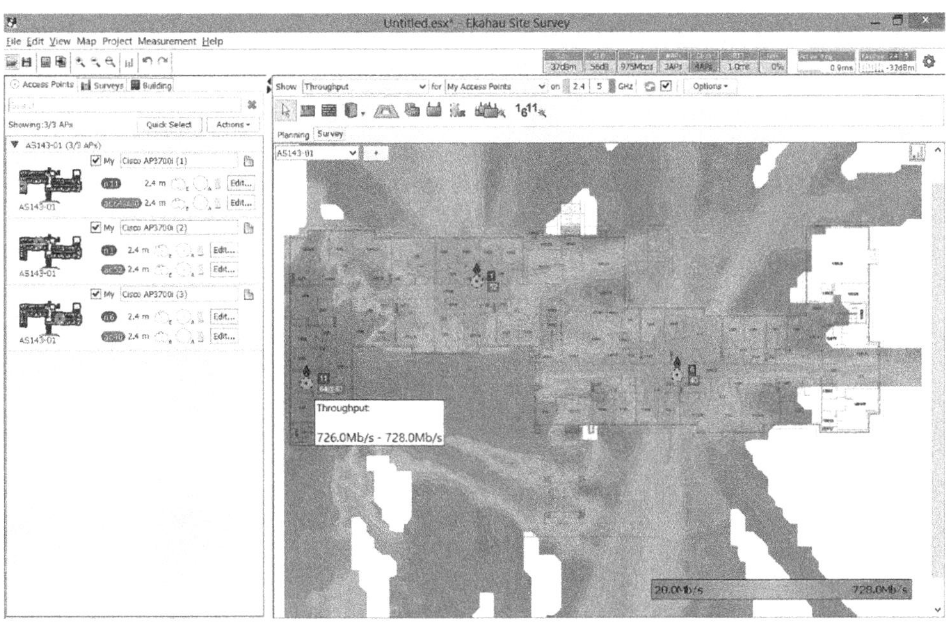

Figure 4.17: ESS Showing a Throughput Heatmap

An additional feature of ESS related to capacity planning is the ability to view metrics based on the number of clients per AP (see Figure 4.18). With the estimated number and types of clients per AP, the software can better represent capacity demands expected in areas and allow you to provision the appropriate APs and number of APs to all areas. An example would be an area with a high-density client count. You would not want to leave an area of fifty or sixty client devices to be covered by only a single AP especially if many of these devices are VoIP clients. In reference to the appropriate AP, this scenario would be less than ideal for an 802.11bgn AP with only one or two spatial streams. Maximum performance would be reached rather quickly, and users would be lining up at the network administrator's door to complain. Plan for capacity based on existing client devices and their usage, future growth of the area in question, and appropriate APs with more than enough capability to achieve these requirements.

Note that the view in Figure 4.18 also shows 2.4 GHz vs. 5 GHz clients. This comparison is very important as the designer's ability to implement multiple APs in 5 GHz within a coverage area is much greater than the implementation of multiple APs in 2.4 GHz. For example, deploying an AP on channel 48 and another AP on channel 36 can be performed easily without requiring much separation between the APs. In the 2.4 GHz band, no such options exist.

An additional feature of survey software is the ability to select different antennas, which is particularly useful for predictive modeling. Figure 4.19 shows the antenna selector on the ESS software[59]. This shows the antenna model along with the elevation and azimuth charts, which are covered in the CWNA Study Guide.

Survey software may also provide detailed information on channel overlap between surveyed APs. This is essential for proper roaming. If users need to be mobile while connected with real-time applications, such as VoIP, multiple APs need to be seen by the

---

[59] While this section has focused heavily on ESS, it should not be taken as a sponsorship of that product. CWNP is vendor neutral, and you can use the design solution that works best for you. In fact, you will see screenshots of other tools throughout the book. ESS is an excellent design software solution, but so are iBwave Wi-Fi, Airmagnet, and TamoGraph. They all have their own pros and cons. Always evaluate several options before deciding, and do not take recommendations from just one or two industry experts, we all have our preferences. The best solution for you may be different from the best solution for them.

client device while roaming so it can choose which AP it will roam to next, in sufficient time, to avoid signal interruption. Figure 4.20 shows an example of channel overlap.

Figure 4.18: ESS Showing Clients per AP

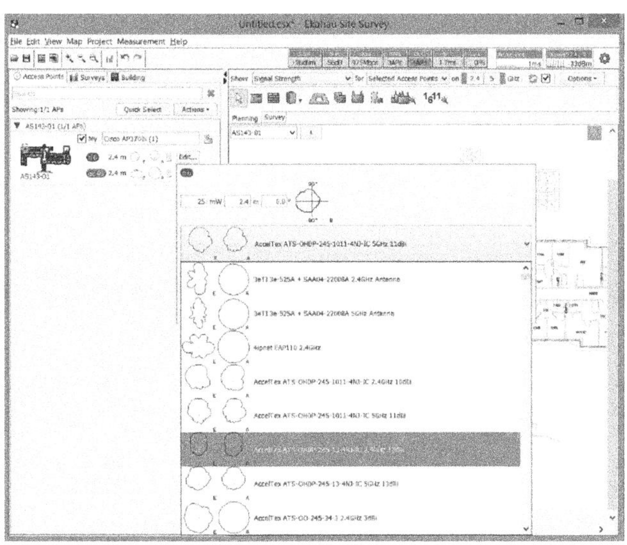

Figure 4.19: ESS Showing the Antenna Selector Interface

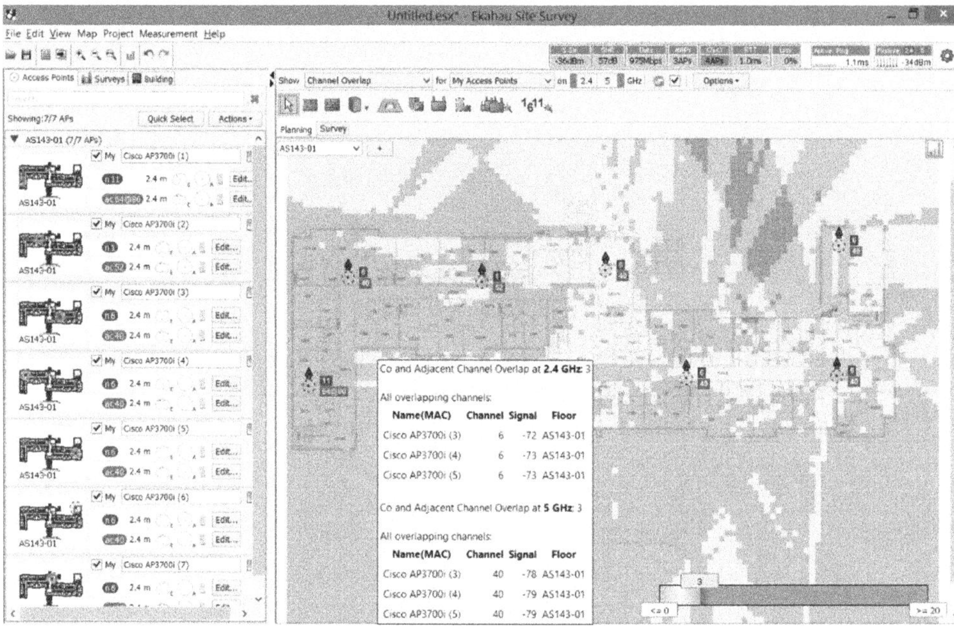

Figure 4.20: ESS Revealing Overlapping APs within an Area

Some site survey software applications will allow you to simulate a network adapter other than the one you actually use to capture WLAN metrics. It is important to know that, while this feature may show the results in a manner that is closer to the simulated card, the only way to ensure you are capturing data as that card would see it is to capture with that card.

Traditional Design while Surveying Methods

As previously stated, in the early days of Wi-Fi, most designs were performed during the site survey. The engineer performed an APoS survey and documented the locations for AP installation as the AP was moved to various locations and signals were tested throughout the coverage area. Focus was primarily on coverage because client densities didn't really exist.

This is not to say that capacity was not an early concern. An article in IEEE Communications from 2001, *Large-Scale Wireless LAN Design, Alex Hills,* stated that, "one would like to carry out a design that is coverage-oriented in most (low-density) areas, minimizing the number of APs, but capacity-oriented in some (high-density) areas, assuring adequate capacity to serve all users in these areas.[60]" The author goes on to describe what we eventually called APoS surveys, stating that, "radio propagation inside a building is frequently anomalous and seldom completely predictable, the design of an indoor wireless installation must be an iterative one." He then proceeds to provide the following five-step procedure:

1. Initial selection of AP locations.
2. Test and redesign, which is adjusting the AP locations based on signal strength measurements.
3. Creation of a coverage map.
4. Assignment of frequencies (channels) to APs.
5. Audit, which is documenting the AP locations and a final set of signal strength measurements at the frequencies (channels) selected[61].

At this stage in WLAN design, the author literally had to create a design tool that allowed them to "measure AP coverage areas by merely walking around the target space." If you've used a modern WLAN design tool to perform an active or passive site survey, this probably sounds quite familiar to you, but you didn't have to design the tool as we, thankfully, have several to choose from today.

In addition to getting better tools, we have learned and documented much in the intervening years. For example, early design plans often called for the placement of APs in hallways (because the primary focus was coverage). This decision was driven by the existence of cable drops and the simplicity of causing signals to travel down hallways and into rooms with sufficient coverage for 1, 2, 5.5, and 11 Mbps in 2.4 GHz. As time went by and more 802.11 client were introduced to our networks, we realized the need to take APs out of the hallways and into rooms so that the walls helped reduce signal propagation and provide for smaller cells, which allowed for more APs to be used in

---

[60] This is also one of the earlier references to the concept of requirement areas for WLANs, without calling them that. The author clearly indicates that different areas have different requirements and the same is still, and even more, true today.

[61] Hills, Alex. *Large-Scale Wireless LAN Design.* IEEE Communications. 2001.

higher capacity designs. In one popular book from these early days, which was an excellent book in its time, *Designing a Wireless Network: Understand How Wireless Communication Works, Wheat, Hiser, et. al. 2001,* there was very little about measuring signal strength within the facility and the primary focus was just on theoretical best practices for mounting locations in specific cases. At the time, this book provided valuable insight because very few WLANs were provided for every location and every person in the building. Today, this is no longer reality.

In this section, we will explore manual site surveys, the traditional APoS model and hybrid site surveys. Both implementing portions of design-as-you-go concepts. Both methods are still applicable today in some cases and similar actions may be performed to truly survey the environment, but most design is performed in well-structured tools[62] with propagation modeling accurate enough to have very high success rates in actual implementations.

Manual Site Surveys with Design

The traditional site survey is the manual site survey or the APoS survey. In this type of survey, the wireless engineer enters the physical area that is to be provided with WLAN coverage with an RF site survey kit. The engineer will place one or more APs at strategic locations and walk throughout the intended service area to test signal strengths and coverage. The APs may then be moved to other locations to start finding further AP placements for site coverage. The use of more than one AP will allow you to discover more exact roaming points and determine the best AP placements resulting in the most

---

[62] One of the earliest tools designed specifically for Wi-Fi surveys was InterpretAir (IA) by Fluke Corporation in 2005. It allowed you to visualize both recorded (from surveys) and predicted coverage areas with data rates and other WLAN characteristics for 802.11.a/b/g radios. It could also be used to locate APs and find the best locations for new AP installs. These were the nostalgic days (for us older Wi-Fi folks) when we used Netstumbler on a regular basis and actually performed site surveys using client utilities (that had survey features built-in in many cases). We even had a little trick early on: place an AP in a far edge corner and move out from it until you find the location where the signal strength is the minimum you desire, then move the AP to that location and move out from it (further away from that corner) in a couple of directions to find the location for the next AP, and then repeat, repeat, repeat. It was fun! However, I am thankful that proper design tools limit the need for this in modern WLAN design. For historical purposes, early design tools available from around 2004 or 2005 forward included SpecraGuard Planner, Wireless Valley LANPlanner, AirDefense Survey, Trapeze RingMaster, and Bluesocket Wireless LAN Planner. The propagation modeling was better in some and worse in others, but they offered varying degrees of accuracy.

effective coverage of the service area. As you are performing these takes on a repeated and sometimes iterative bases, you document the best locations for APs (the beginnings of your design).

Manual site surveys are very accurate for determining RF behavior and signal attenuations in an environment, assuming the engineer understands the process and results and uses the same APs that will be used in deployment. Because the engineer is making decisions based on actual RF information in the intended coverage area, guesswork is not part of the equation.

However, a complete manual site survey does take more time to perform, but the accuracy may be worth the effort, when compared to that of a predictive design, for some locations (though such locations are become fewer and fewer with better tools). If you expect to deploy more than a few dozen APs, and time is not available for a single engineer to perform the entire survey, you may choose to divide the coverage area up between two or more engineers to speed up the process. This is especially effective when the area can be divided up in floors of a multi-story building with little to no floor-to-floor bleed through. In some cases, larger site surveys can be performed utilizing a combination of techniques which will lead us to our next topic of hybrid site surveys and predictive modeling.

It is important to note that when selecting APs for performing a manual site survey, you should always survey with the same APs you intend to implement. You should also survey with the same power level configuration as you intend the WLAN to function at, even if you plan to utilize RRM on your controller(s). For example, for a standard density deployment, you may set power levels to 20 or 25 mW. For a high-density deployment, you may choose to set power at 10 or 15 mW. When performing a site survey, you are creating an environment on paper that must behave the way you design it. Utilizing different equipment or power settings during the survey, other than what is going to be implemented, will most certainly change the coverage and behavior of the actual WLAN and possibly lack coverage in areas as a result.

The APoS method will continue to be a valid tool in the designer's toolkit because it can be used to gather measurement data useful in the actual design of the WLAN. The following key measurements can be taken with an APoS method:

- Wall Attenuation: measuring the attenuation factor of walls and other blockages allows you to enter more accurate metrics into the WLAN design software.
- Challenging Area Measurements: areas with lots of metal shelving, large metal equipment, and other oddities are challenging to model correctly in predictive design tools. Performing APoS surveys in these areas can often be faster than modeling, implementing, testing, adjusting, modeling, implementing, testing, adjusting, well, you get the point.
- Spot Checks for Accuracy: If you have modeled the WLAN in design software, you can place a few APs where the modeled design specifies and verify the accuracy of the modeling before attempting an installation of hundreds or thousands of APs.

Hybrid Site Surveys and Predictive Modeling Design

Hybrid site surveys with predictive modeling and design include elements of manual surveys and predictive designs. When utilizing the attenuation discovery method shown earlier in Figure 4.11, a more accurate attenuation level can be used to provide data for some areas on which you wish to perform predictive modeling. Other options of hybrid site surveys include the discovery of existing Wi-Fi coverage and then simply building a design around that coverage, so it can be utilized instead of removing the AP(s) and re-surveying the area. Lastly, many facilities consist of large complex areas that certainly need to have a manual survey performed to provide the best accurate coverage for it, while outlying areas are isolated buildings just a few meters from the main structure. These isolated areas can be surveyed with predictive options to save on time while larger critical areas are designed utilizing a manual survey method. The beauty of modern tools is that you have options to optimize the process that simply were not available in the early days. The WLAN designer should take advantage of these options rather than remaining stuck in traditional methods[63].

---

[63] The methods chosen will vary by project, but the modern tendency is toward significantly more predictive design with highly accurate propagation modeling and partial manual surveys to gather useful information for the WLAN design tool to use. iBwave offers multiple propagation modeling algorithms to choose from and we will see more of this in other tools as time goes by as well. Each tool has its strengths and weaknesses, so you should evaluate several of them to select the best tool for your needs.

When engineering, you are not guessing. Instead, you are providing factual data. Therefore, actual scientific findings from utilizing real RF in the environment provide an excellent input to accuracy. Perform the manual survey as needed to gather factual data every chance you are given.

Site Survey/Design Type	Pros	Cons	Applications
Manual	Very accurate; real readings in a real environment. Less time utilized on adjustments and tweaking after implementation.	More time utilized on the front end; high level of RF knowledge required.	Small networks that can be measured and designed quickly. Challenging areas in larger networks.
Predictive	No or little physical activity required; RF map generated automatically to assist in AP placement.	Costly software. More adjusting and tweaking may be required after implementation.	Most modern WLAN deployments. Primary design tool with inputs from manual surveys.
Hybrid	Time and budget saving	Requires both skill sets: manual and predictive.	WLAN environments with challenging areas. Networks with high quality requirements.

Table 4.1: Site Survey Types: Pros, Cons, and Applications

Surveying for Applications and Architectures

Surveying for specific applications and architectures is about understanding unique requirements they impose on the network. This section provides an overview of common applications and architectures, and their requirements.

Voice over IP (VoIP)

Voice applications are the most critical when it comes to Wi-Fi design and providing enough throughput for the application to work alongside other applications is crucial. The standard for VoIP wireless design is to provide a minimum signal strength of -67 dBm throughout the facility, a 20 – 25% overlap of individual coverage cells[64], and power settings on the AP to be at or below the power level of the actual client device performing the VoIP functions. Other important restrictions in a VoIP design is the SNR should be 25 dB or higher for the required throughput, and packet loss should be less than 1%. These numbers are important to note because a signal strength of -67 dBm doesn't automatically place coverage into the qualifying category. If the noise floor increases, then the signal strength should increase to compensate the dB loss due to noise. Using the Active Survey option in your survey software, you can associate to an SSID on an existing network or the SSID of the survey AP and see the noise floor, signal strength, SNR, and packet loss displays running in real time. With these measurements, you can monitor the area in question to see if it meets the necessary criteria. If the noise floor moves up 3 dBm, then the required signal strength should be 3 dBm stronger to ensure the 25 dB SNR. In other words, as long as the noise floor averages -92 dBm a signal strength of -67 dBm is sufficient for a 25 dB SNR. If the noise floor becomes a constant -89 dBm, then the signal strength of your AP needs to be -64 dBm or greater to keep the SNR 25 or above.

Many client types are available to use for VoIP services, and that can present a challenge for your design. Before surveying for VoIP coverage, be sure to know the clients that will be using the VoIP services. Select the VoIP client that operates at the lowest transmit power of all VoIP clients to be used, and then use that client device to design your WLAN around.

Video

Video is a high-bandwidth application. Video comes in two primary forms: live and streaming. Live video has real-time requirements and often requires multicast support as well. Streaming video typically uses buffering to reduce the latency demands on the

---

[64] Most engineers to not "measure" this percentage of overlap. The math is more challenging than the benefit. Some design so that an AP can be seen at all times that meets the signal requirement and one or more other APs will meet the requirement as you near the cell edge. Others design so that at least two APs (in 5 GHz and 6 GHz) can be seen at all times that meet the requirement. If you use these two methods or another, you are unlikely to actually measure percent overlap.

application. For example, the player may buffer anywhere from five seconds to one minute of video so that network delays do not prevent smooth play of the video. When designing for video, be sure to consider the kind of video being implemented.

Data

Standard data networks are much easier to design than real-time data networks. However, unless you are implementing a WLAN for real-time applications and another for standard data applications, you will need to accommodate for QoS configuration in all modern enterprise WLANs. This is because applications such as Skype, FaceTime, and additional real-time applications will be used even if VoWLAN is not implemented.

When planning for a standard data WLAN, you must ensure that the required throughput demands (part of capacity) are met. For example, if the average user will require 2 Mbps and 25 users will be connected to each AP, the APs must provide 50 Mbps data throughput. Given that throughput is often 50 percent of the data rate, the users will require a data rate higher than 100 Mbps to ensure throughput demands are met.

Location Services

Location services are implemented, in WLANs, in two ways: 802.11-based or non-802.11. With non-802.11 systems, RFID or some other RF mechanism is used to track the device, and the systems may use Wi-Fi to communicate with central servers, but the actual location tracking is performed using non-Wi-Fi tags or devices. 802.11-based uses various triangulation techniques to locate devices. When designing for location services, it is important to ensure that multiple APs can see each possible location. This design calls for a minimum of three and preferably four APs seeing client locations at a minimum of -75dBm or greater to obtain the best results. The other critical criteria for location services is the requirement of APs being placed in the corners and along the outside walls of the facility requiring location services. Sometimes these APs can be set to a "Listen Only" mode and not be used to service clients. This helps to avoid Co-Channel Interference issues caused by too many APs near of each other.

Single-Channel Architecture (SCA)

SCA is implemented based on density requirements. That is because the APs are configured by the controller and are all operating on the same channel, SCA surveys are usually about determining the number of clients and types of applications to be used.

Once this is known, an appropriate number of APs may be selected for the deployment within a given area.

Outdoor Surveys

When designing outdoor wireless bridge links, you will be implementing one of two primary types: point-to-point or point-to-multipoint. In a point-to-point implementation, two wireless bridges are configured to communicate with each other and provide the link across which network traffic is routed as seen in Figure 4.21.

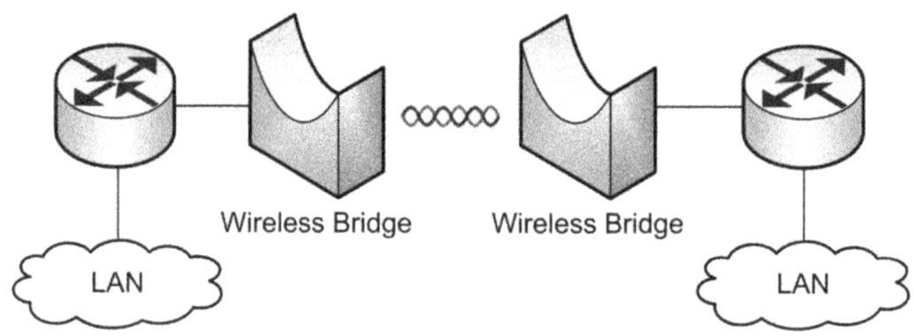

Figure 4.21: Point-to-Point Bridge Link

When creating a point-to-multipoint bridge link, you will configure one bridge to act as the centralized bridge (root) and the other bridges to act as remote bridges (non-root). The remote bridges will communicate with each other through the centralized bridge and users from all LANs may be allowed to communicate with all other connected LANs via the wireless bridge. Figure 4.22 shows a point-to-multipoint bridge link.

When creating these bridge links, you must ensure that proper radio line of site exists and that the Fresnel Zone is at least 60% clear. This may mean gathering terrain maps from online and local government sources. It may also require that you photograph the surrounding areas to document buildings, trees, and other structures that may factor into your link configuration. You may also benefit from local weather information that can be used to predict the amount of fog, smog, and other weather factors common to the area in which the link is expected to operate.

With this information gathered, you can begin to plan the links. You will need to determine the optimum height in order to overcome any possible obstructions that were discovered during the information gathering phase. Next, the link budget can be calculated as well as the EIRP. The FCC regulates the output of outdoor transmitting signals so be sure to adhere to these regulations when designing the bridge.

The FCC max transmit power for bridges can be seen at this link:
cwnp.link/fccbridgerules

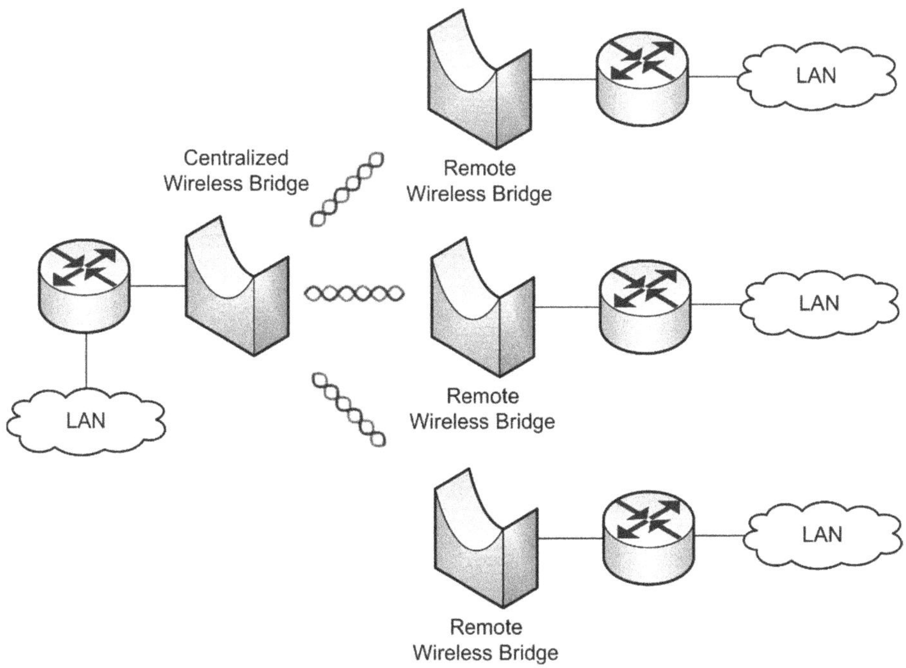

Figure 4.22: Point-to-Multipoint Bridge Link

Gaining Access and Clearance

It is important that you have access to all required areas to perform the site survey for the sake of time conservation. Many organizations have secured areas within the facility that require special access privileges before you can enter. You may be given a temporary access card, key code or a designated employee to escort you while accessing these areas. Other locations require training and/or background checks to gain access even with an escort. This is something that should be requested as early as possible to have a plan in place when the actual survey begins. In some organizations, access to areas could take days or even weeks to get it approved. Inquire about this before arriving on site to start requirements analysis so that access issues do not delay the design process. You should also plan access for the implementation staff at this point to arrange for any training or background checks they may be required to go through so that access does not delay the implementation.

Documenting the Survey

Once you've completed the work of the site survey, you must document your findings. Site survey software will usually generate a set of documents that can prove to be beneficial to the documentation process. The main objective of the documentation process is to provide enough pertinent information about the design so that someone else can come in behind you and implement the design without you being present to guide them through it. The deliverables will vary depending on the purpose of the survey. For a WLAN survey used in the purest sense of the term (to gather information for the architecting (design) of the WLAN)[65]:

- Floor plan containing locations of surveyed APs, the model of APs to be used, configuration used with the APs, and existing (previous networks) AP locations (if any).
- A floor plan showing spectrum analysis results such as where and what an interferer is and notes explaining the need to remove it or work around it and the frequencies on which it was detected.
- A floor plan showing existing switch location, model of switch(s), type of PoE available and the remaining PoE budget from each switch, and the number of switchports available to be used for the WLAN. If new switch locations are needed, they can be added to this document as well.
- Photographs – It is always a good practice to photograph everything and provide a marker in the photograph as to where the survey APs were positions as well as photos of any interferers and/or any other oddities around which the design must work.

For a design-as-you-go survey, or an APoS survey with plans to determine the location and configuration of every AP for the design:

- Everything from the preceding list where appropriate.
- An AP placement design including the proper placement of APs and the configuration settings for those APs.

---

[65] Business and stakeholder requirements should exist before performing any kind of survey. Without them, you do not know the target for which you are surveying. Technical requirements or functional requirements may evolve during and after the survey based on information discovered.

- A Bill of Materials (BoM) – This item provides the project manager with a way to manage/set a budget. It also shows the installers the number of materials being installed so they can determine a time frame for their work. It should include every item that will need to be purchased for the new WLAN including an idea of cabling unless it is included in an installer's bid. Sometimes cable contractors include the cabling items such as cable, RJ45s, jumper cables, termination blocks and patch panels.
- Implementation Plan – Sometimes this document is included as part of the survey documentation and other times it is simply a separate document altogether[66]. This should be decided during the initial planning stages of the project.

The above list of documentation deliverables is a starting point, and you should always discuss with the customer ahead of time to determine what is desired in this documentation. Figure 4.23 provides a sample page of what may be required in a survey document.

In Figure 4.23, you see floor plans containing AP locations, both new and existing. The model of each AP is listed along with a number that coincides with numbers on each provided picture. In the pictures, you see exact mounting locations and mounting instructions along with which switch the AP will be powered from. This example is by no means a standard. It is only to provide an idea of information needed. Many reports are created differently, but the actual information in the report is the critical necessity. It is always a good practice to show your customer a few examples of design documents you have created so that they can help decide how their documentation will look. You will find that most of the time customers will like and accept a version you already have as long as the needed information is included in it.

---

[66] It is not practical to perform 100 percent of the design during a design-as-you-go survey. Instead, you will design the AP placement plan and configuration settings related to RF during the survey. Then, you can design other requirements related to non-RF AP configuration settings, network switching plan, network services plan, etc., after the survey is complete. This practice of design-as-you-go still works well. It's not an issue of effectiveness, it's an issue of efficiency. I (Tom), personally, would not use it for a design requiring more than a couple dozen APs, unless the client specifically requested it and is willing to pay for the extra hours.

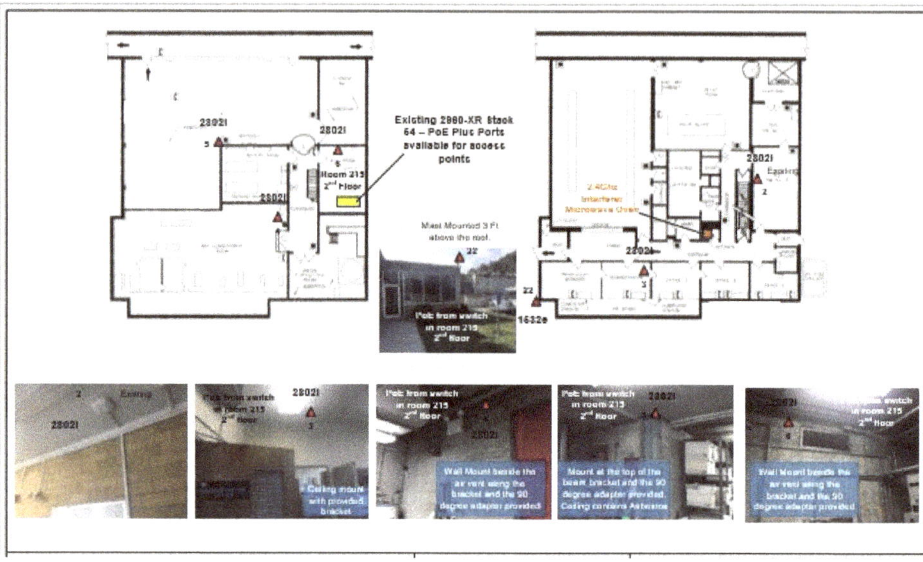

Figure 4.23: Floor Plans with AP Locations

Creating an Implementation Plan

After performing the site survey and documenting the results, the next step is to create a plan for implementing the new WLAN design. Creating the implementation plan is not part of the survey process, it is the end of the design process. This plan can vary significantly depending on the outcome of the survey.

Example: The new WLAN is designed, and all parties have agreed to implement it. Several problems exist in the current network that will hinder implementation of the new WLAN, and those are:

- Not enough switch ports available
- Not enough switch locations to support the new WLAN coverage area
- Not enough IP addresses in the VLANs to support the project
- Not enough rack space to mount the WLC and the HA-WLC

Now, with these issues comes a decision on how to implement the WLAN knowing there will be these obstacles to overcome. From a project manager's point of view, one might deal with all the issues in parallel and have multiple tasks being executed at any given time. Even by doing this, time waiting on products, time utilized to redesign subnetting and reconfiguring switches with the expanded IP pools, and time to install a taller rack or place one beside the existing rack for the WLCs must all be considered. In these types of situations, having some good project management experience will pay off.

It will be necessary to create the implementation plan in the same way a project manager would manage a project and its resources. First, sit down with all parties involved and discuss the timeframe within which the project needs to be completed and then build a WBS so that subsets of work can be divided up among staff and contractors. Once you have a good handle on the work that needs to be done, discussions can be held to determine which work will be started first, what tasks must wait on other tasks to be complete before they can start, and which tasks can run in parallel to each other and independent from the progress of other tasks. Once all this information is made clear, a plan of implementation is already near completion. The exception may be in certain industry-specific environments. With some regions, the implementation plan may need to adhere to specific industry requirements as well as mandatory training for all who perform labor on site. Below are a couple of those know requirements.

Some industries have industry-specific requirements mandating union workers are to assist with certain tasks, or you can't perform the work. In this case, you may be required to factor in the use of union labor to meet these requirements. If so, design the Work Breakdown Structure so that union workers can pull cable, mount APs, hang conduit, as well as perform other non-technical tasks.

When operating within healthcare environments, it is essential that you honor patient privacy. This is both an issue of respect and legal compliance. Additional industry-specific requirements include OSHA regulations, PCI DSS security guidelines, SOX guidelines, and more. Before building an implementation plan, be sure of any industry-specific requirements that could delay or stop work if not complied with.

Training requirements are also important. A specific requirement is training in safety and operations. Safety training may be required for installation personnel. This includes OSHA-type compliance training (or additional local safety regulations). Many companies have site-specific training that is mandatory for everyone performing work at that location. This training will take time to administer, and it should be included in the implementation plan and project timeline.

No matter which route you take on designing your implementation plan, time and resource management is a key factor in the project success so be sure to provide a plan that can take advantage of every opportunity to save time and resources without sacrificing quality.

Chapters 5 through 7 provide more information on the creation of an implementation plan.

Validation Surveys

After implementing the WLAN, you worked so hard to design, there is still the need to validate the design to ensure it functions as intended. A validation survey, or simply a validation or an acceptance test, is performed to ensure that requirements have been met. If there were specific throughput requirements given to you by the customer, then you should test the throughput for compliance with these requirements. Any other specific requirements such as VoIP quality coverage throughout the facility, or capacity specifics, should be validated as well.

In testing coverage for the new WLAN, you can utilize your survey software in active survey mode and connect to an SSID on the WLAN. This will allow you to generate a heat map that will display RF coverage from this SSID in colors that represent signal strength (same as the site survey). After walking the entire facility and documenting the required coverage, you now have validated the required coverage and recorded the results in a heat map. Throughput can be tested with one of the tools mentioned earlier on in this chapter. Just understand that whatever requirements you were given upfront may be required to be validated and documented before the project is considered complete.

Chapters 8 and 9 are dedicated to this process.

Defining and Documenting Metrics

Metrics are the data values gathered during the survey that represent effective or ineffective operations of a potential WLAN. They are used to establish minimum expectations for particular data rates and other WLAN expectations. The following metrics should be defined and documented after a survey is performed:

- **RSSI**—Received Signal Strength Indicator (RSSI) is a measurement of the signal strength at a given location. Devices have receive sensitivity ratings and must have a signal at a particular level to receive transmissions at a particular data rate. RSSI is a relative number as each vendor calculates it using a different range. But higher numbers mean a better signal up to some maximum and lower numbers mean a worse signal down to some minimum.
- **Noise Floor**—Noise is effectively any signal or RF energy other than the signal being monitored. The noise floor is the level of energy in the environment without the introduction of local intentional signals, typically measured in dBm. For example, -94 dBm is a common noise floor reading.
- **SNR**—The signal-to-noise (SNR) ratio is not so much a ratio as simply the difference between the noise floor and the desired signal's strength. For example, if the noise floor is -94 dBm and the signal is -53 dBm, the SNR is 41 dB. The ability to properly interpret (demodulate) signals is directly related to the SNR. When SNR values are higher, more complex modulations may be used that result in higher data rates. The reality is that SNR is the key factor in getting a data rate, assuming CCI is not a problem. RSSI is relative, SNR is actual and real.
- **Interference Levels**—Interference levels can be higher in some channels than in others. For example, video cameras and other devices may be operating on channel 6 in an area, but not on channel 11. The result is that the interference levels are higher on channel 6. Effectively, the interference level can be thought of as the noise floor on that channel as it is a measurement of signals on the channel and the strength and duty cycle of those signals.
- **Cell Coverage**—Cell coverage is a reference to the size of a single AP cell. That is, how much physical space should an AP cover? This is typically defined as a signal strength metric, and then the actual size is simply accepted given some output power level determined for the APs, such as 10 mW for high density or 25 mW for standard density or even 50 mW for low density.
- **Cell Overlap**—Cell overlap is the measurement of overlap among cells, which allows for effective client roaming. Vendors sometimes encourage 25 percent or more overlap, but of course, this is not possible to measure. Instead, the goal should be two or more APs (depending on density) at each measurement location. This suggesting is particularly true for 5 GHz WLANs, though a bit harder to achieve for 2.4 GHz WLANs, where accepting that two APs should be

visible to a client for an acceptable distance to allow roaming while mobile, for example, for 20–40 feet, is more realistic.

- **Data Rates**—The speed at which bits can be sent across the RF medium is the data rate. The data rate is determined based on several factors, including, signal strength, channel width, guard intervals, modulation, and coding methods.
- **Throughput**—A measurement of the usable data passed through the network. For example, the speed at which a file is transferred as opposed to the data rate. On WLANs, the data rate is always significantly higher than the throughput. To ensure application functionality, throughput, and not simply data rates, should be used to test for achieved capacity[67].
- **Latency**—Latency, also called delay, is a measurement of the time it takes to move data from one point to another. That is to say, how long does it take for data to get from a VoWLAN phone to the other VoWLAN phone in the conversation. This value should typically be less than 150 ms unidirectionally for VoIP implementations.
- **Jitter**—Jitter is the variance in delay (latency). If one packet takes 130 ms to get from point A to point B and the next packet takes 80 ms, it can cause problems in VoIP communications. Jitter buffers are typically used to circumvent this problem.
- **Loss**—Loss is a generic term used to reference packets that do not reach their destination for any reason. Packet loss is typically measured in percentages.
- **Retries**—With connection-oriented protocols and the 802.11 MAC, when frames are lost (or packets with TCP), they are resent. This resending is called a retry. High retry rates indicate a problem in a WLAN[68].

---

[67] For each requirement area, consider defining throughput in three ways: minimum required, normal load requirement, and VID (very important device) requirement. The minimum requirement is the throughput that must be achieved when the requirement area is at maximum capacity. For example, if normal load is 130 client devices in the area, maximum capacity (based on the design) might be 250 client devices. Minimum required throughput might be 12 Mbps and normal load might be 28 Mbps. Using special network configurations, you can optimize throughput for VIDs, which might be specified at 40 Mbps. This model is just one way to be more precise in your requirements and to achieve customer satisfaction.

[68] What is a high retry rate? It depends on the scenario. For VoIP, three or four percent may be a high retry rate. For standard data (checking email, browsing the web, etc.) seven or eight percent might be a high retry rate. You should learn the demands of your application and design to minimize the retry rate based on application requirements. (continued on next page)

With all these metrics, clear definitions of expectations should be set in requirements analysis so that the site survey can be performed such that the resulting network design accommodates the expectations. What is an acceptable loss rate? What is the acceptable delay in communications? What SNR must exist (or RSSI)? How many APs should cover each area? Clearly defining these metrics gives you something to measure in the site survey. Without their definition, you are simply designing for coverage and, sometimes, not even that.

Always remember that some TCP retransmissions will occur, which do not show up as 802.11 retries. Therefore the 802.11 retry rate must be kept low if there are TCP retransmission rate requirements as well. For example, if an application requires fewer than 10 percent retransmissions when operating on a wired network (due to lack of a TCP acknowledgement), then the 802.11 retry rate should be kept even lower (remembering that a single frame may not be a complete TCP segment if fragmentation is occurring).

4.3: Chapter Review

In this chapter, you learned about site survey tools, various surveying procedures, software tools for assisting you in your designs, and applicable scenarios. Also covered were topics such as proper PPE use and how to ensure you are safe while performing your work. This information will not only help prepare you for the CWDP exam but will also assist you in preparing to perform site surveys in real-world scenarios.

4.4: Facts to Remember

Be sure to remember the following facts as you prepare for the CWDP certification and be sure that you can explain the details related to them:

- Manual site surveys are also called AP-on-a-stick and involve placing APs in the physical environment and then testing the signal propagation throughout that environment.
- APs used in site surveys should be the same APs being deployed.
- Predictive modeling surveys use software to simulate RF behavior on a floor plan of the environment.
- Calibration of floor plans before performing surveys is a must.
- Proper material definition is key to accurate predictive modeling surveys.
- Protocol analyzers can locate WLANs and their configuration parameters which are revealed through the beacon frames.
- Spectrum analyzers can locate both Wi-Fi and non-Wi-Fi devices that may cause issues on specific channels.
- Co-channel interference (CCI), also called co-channel contention (CCC) is caused by two or more APs existing on the same channel in close proximity (from a signal strength perspective) to each other.
- Adjacent channel interference (ACI) is caused when a near cell is using an adjacent channel; for example, when a cell is on channel 1 and another near cell is on channels 2, 3, 4, or 5.
- VoIP requires low latency and low jitter with low packet loss of less than 1 percent.
- Throughput on a WLAN is often 50 percent or less than the data rate.

4.5: Review Questions

1. A wireless design tool kit should include______________________.

    a. A power injector, Range finder, Network cables

    b. APs, Marking tape, Battery pack

    c. Omni-antenna, Duct Tape, Patch antenna

    d. A, B, and C

2. A spectrum analyzer should be utilized in WLAN design for the purpose of ____________________________.

    a. Locating employee cell phones

    b. Finding existing 802.11 devices

    c. B and D

    d. Determining the duty cycle of interferers

3. When designing a WLAN for VoIP communications, _________________, _________________, and _________________ are essential to proper device performance.

    a. Minimum of -67 dBm signal, less than 5% packet loss, >= 25 dB SNR.

    b. 3 APs seen at -75 dBm, a 25 dB SNR or better, optimized QoS settings

    c. Less than 1% packet loss, minimum of -67 dBm signal, >= 25 dB SNR

    d. 3 to 4 APs seen by the device, <5% packet loss, -92 dBm noise floor

4. If concerned with how many client devices are being utilized in a specific area while designing a WLAN, what design parameter is in question?

    a. Location Services

    b. Capacity planning

    c. QoS traffic monitoring

    d. Priority Throughput

5. Once you arrive on site to begin implementation of the new WLAN the first thing you should do is ask if there are any facility access requirements you need to meet before starting work.

    a. True

    b. False

6. Site survey software can simulate APs, antenna types, and AP coverage on an accurate floorplan without knowing anything about the facility or its building materials by performing a Predictive design.

    a. True

    b. False

7. A document(s) that can be used to help form a budget for the WLAN implementation is a ______________________.

    a. Project Charter and SOW

    b. Spectrum Analysis data report

    c. Resource Management Plan

    d. Wireless Design Report and BOM

8. Two friends are playing video games across the WLAN while utilizing VoIP headsets on the game console to communicate with each other. Multiple VoIP packets are traversing the WLAN one right after the other. The difference in time that it takes one packet from another to get from the send station to the receive station is best described as_____________.

   a. Latency

   b. Time of Arrival

   c. Jitter

   d. Delay of Game

9. A wireless bridge can be implemented to connect one LAN to another four city blocks away without the need for overhead wires or underground fiber. During configuration, both APs will be set to bridge mode, but a decision will have to be made as to which one will be configured as the _________________ and which one will be configured as the____________________.

   a. Primary, Secondary

   b. Master, Slave

   c. Root, Non-root

   d. Server, Client

10. You set up an AP on a stick next to an exterior wall of a building. Next you record the signal strength of the AP while standing next to the AP, then walk to the opposite side of the wall. While standing with the wall between you and the AP you record the signal strength again to see the signal loss caused by the wall. What best describes what you are looking for?

    a. dBm loss

    b. signal refraction

    c. Quality of service

    d. Attenuation

4.6: Review Answers

1. The correct answer is D. All the above-mentioned items are good to have on hand when performing a manual site survey.

2. The correct answer is C. A spectrum analyzer is a tool used to determine if the area to be surveyed is free and clear from existing 802.11 signals and other interferers that might cause functional problems with a new WLAN implementation. One of the ways of doing so is providing a duty cycle display for the signal in question.

3. The correct answer is C. The criteria for a VoIP quality WLAN include having less than 1% packet loss, a minimum of -67dBm signal strength from the AP, and a 25dB SNR or greater. The main requirement of signal strength is the 25dB SNR. The -67dBm signal strength is a minimum requirement and partially because the average noise floor level is -92dBm or lower. In the event the noise floor average remains higher, the minimum signal strength from the AP will have to increase to compensate for it.

4. The correct answer is B. Average voice, and data coverage is provided through basic survey practices. When the concern is raised about the number of client devices in an area, then the issue is with capacity and not signal strength or other parameters. Too many client devices functioning in the same BSSID can divide the AP's bandwidth enough that none of the client devices receive adequate throughput. This is corrected through capacity planning, and it starts with finding the number of client devices that will be operating in a single area and planning coverage from multiple APs to ensure proper throughput is available for all client devices.

5. The correct answer is B. This concern should be address early on at the beginning of the project. Required training and/or background checks could take several days to complete and waiting until the day work starts is not the time to address this issue. This could cause unnecessary delays in the work schedule.

6. The correct answer is B. For a predictive design to be performed you first need to have adequate knowledge of the building materials and objects inside of the building. Without this information, proper attenuation levels cannot be assigned to the floorplan to provide simulated coverage from APs.

7. The correct answer is D. By looking at the wireless design report, an estimate of the number of cabling hours can be provided. The bill of materials provides a list of all equipment that will be installed so this document can provide equipment costs and an idea on equipment configuration time. The two documents together can provide a lot of information that will help in forming a budget for the implementation

8. The correct answer is C. Latency is the time of delay in packet delivery. Time of arrival is used in location services, and jitter is the difference in delay time between packets going from point A to Point B. One packet might take 120ms to arrive while the next one takes 170ms. 120ms and 170ms is the latency of the two packets. The difference in latency between the two packets is 50ms which is called jitter. Delay of Game is a penalty given in a football game.

9. The correct answer is C. When configuring APs to be a bridge, you must also select one of two options for each of the APs to exist as in the bridge link. Only one AP in a bridge can be a Root Bridge while the other must be configured as a Non-Root Bridge.

10. The correct answer is D. By placing an AP near a wall then taking signal strength measurements on both sides of the wall; you are utilizing the best practical form of finding out the attenuation of the given wall.

Chapter 5: Designing the WLAN

Objectives Covered:

2.1 Define WLAN architectures and select the appropriate architecture for a design

2.2 Produce a design to meet requirements

2.3 Create, distribute, and communicate design documentation

In the previous chapters, you learned about the details of requirements engineering and the information you need to design a WLAN. This chapter will bring together this knowledge to help you design typical wireless networks based on the 802.11 protocol. By typical, we simply mean common office type deployments. Special deployment cases are addressed in Chapter 6. You will learn about WLAN architectures, RF considerations, design software, and documentation. The journey begins with WLAN architecture selection.

5.1: WLAN Architecture Selection

The WLAN architecture defines the way in which the WLAN operates and is managed. Does it send client data directly onto the network from the AP routed to the destination, or does it pass all data through a controller within a tunnel? Is it managed onsite or from the cloud? Will you deploy an autonomous solution? These questions can be answered based on the information in this section. As a CWDP, it is your responsibility to understand WLAN architectures and select the appropriate solution for the installation. Many vendors offer multiple architectures, so, even when the organization has already chosen the vendor, you may be required to select the best architecture for their installation. This section provides the information you need to make this decision.

Management Solutions

Management solutions typically fall into three categories:

- controller-based
- WNMS-based
- cloud-based
- autonomous

Controller-based solutions include virtual controller solutions where one of the APs acts as the controller for the other APs. Aruba Network's Instant AP solution is an example of this, and other vendors offer it as well.

Controller-based solutions were the first popular centralized management solutions (except for some wireless network management systems (WNMS), which have now become embedded in what some vendors call controller-based solutions but are really

WNMS solutions). They allow administrators to connect an AP to the network, have it receive updates and configuration settings appropriate for it, and go live on the network automatically. Of course, someone had to set up the controller, but the AP configuration was made much simpler.

Wireless Network Management System (WNMS) solutions are focused on automatically configuring APs and monitoring them. They do not typically provide for centralized data forwarding and do not typically process any 802.11 frames directly.

Cloud-based solutions, from a configuration standpoint, work much like controller-based solutions. The APs are connected to the network, then they connect to the cloud management software, and finally, they are either configured automatically or manually within the cloud then brought online. A key difference is that cloud-based solutions do not use centralized data forwarding in most cases so that the data can pass directly through the AP onto the local network. Some cloud-based providers are now offering on location centralized data forwarding through appliances or localized servers for organizations that desire to centralize filtering of WLAN traffic.

> Remember that cloud-based architectures may offer localized servers or appliances for centralized data forwarding and that it is often used for filtering or traffic management in relation to the WLAN.

Autonomous APs are directly configured by the WLAN engineer during installation. This model is still very popular in small businesses that require only a few APs. An engineer can typically configure four or five APs in the amount of time it would take to install a controller and configure the profiles and then connect the APs. Ongoing management may take more time, but the reality is that enterprise-grade APs often run for months without requiring maintenance resulting in the autonomous model continuing to thrive in very small WLANs.

The following sections provide more details about each of these three management solutions. Additionally, data forwarding models are explained so that the WLAN designer can make effective selections.

Controller-Based Architectures

Controller-based architectures are sometimes called split MAC[69] models because a portion of the 802.11 MAC is managed in the AP and a portion is managed in the controller. The APs used in this model are often called thin APs or lightweight APs because they do not perform as many functions as the autonomous APs (fat APs or think APs). The split MAC model is very popular in large networks today and is becoming more popular in smaller networks, as well. The primary benefit of controller-based architecture is simplified, centralized management on a large scale. Again, the costs and benefits associated with the split MAC model must be considered. On smaller networks with a dozen or a couple dozen APs, the extra cost of controllers may not be beneficial[70].

The controller-based model is not an end-all solution despite its excellent capabilities. For example, the controller itself could be a central point of failure for the WLAN depending

---

[69] Split MAC indicates that a portion of the 802.11 MAC is handled in the controller and a portion is handled in the APs. Local MAC indicates that the entire MAC is in the APs. This is a key differentiation between the traditional controller-based architecture model (split MAC) and a WNMS (local MAC), even if the WNMS is called a controller by the vendor. Again, terminology is blurred here, so check with your vendor to see what is truly meant by the term controller. Is it controlling the configuration of the APs or both the configuration of the APs and a portion of the 802.11 MAC.

[70] The cost/benefit analysis should include the cost of ongoing management of the network. If you have available administrators who can manage the autonomous network and it will take no more than 20-30 hours of their time per quarter, the cost of a controller-based solution may not be justified. However, when it requires 40-60 hours of the administrator's time per quarter to management the WLAN, a controller-based solution is likely to cut this administration time in half or reduce the time even more. If you calculate the cost of the administrator at $80 US per hour, then 40 hours is $3200.00. Half of that is $1600 times four quarters over three years (assuming a minimum of a three-year life cycle), the total savings could be nearing $20,000 by having a controller-based architecture. If the network requires 40 hours of the administrator's time for management tasks per quarter, it likely has more than one or two dozen APs and a controller or a WNMS (whether cloud-based or on premises) is likely in order.
Consider that a single AP, the WAP371 802.11ax AP, had nine firmware updates from 2014 to 2019. That's almost two per year. If you must manage 12 APs manually, and each firmware update is applied and each update requires one hour, that is about 22 hours per year just applying firmware updates. Add in the additional administration tasks (periodically changing channels due to new interferers, troubleshooting problems, adding/removing SSIDs, etc.) and you will easily see that just 12 APs can consume above 40-60 hours per year easily. Therefore one or two dozen APs becomes the practical cost/benefit limit that makes centralized management appealing.

on the configuration. While a vendor can suggest that redundant controllers in an N+1 configuration resolves this issue, it does so at an increased cost as well. Additionally, increased wired traffic is required to manage the wireless stations because typical actions like authentication and roaming now involved the controller. This is a good thing from a centralized management perspective, but it does add extra traffic to the wired network that must be accounted for. If the network is using centralized data forwarding, this can result in increased latency as well.

None of these issues should prevent an organization from using a controller-based WLAN, but they should be addressed. Ensure the wired network can handle the extra capacity for wireless AP and client management. Ensure the central point of failure issue is addressed. Make sure any latency issues are addressed, which can be as simple as configuring the network do use distributed forwarding.

On the positive side, controller-based architectures give you centralized administration, potential increases in stations per AP, and centralized filtering of traffic as well (when centralized forwarding is used).

Controller-based architectures use Lightweight Access Point Protocol (LWAPP) or Control and Provisioning Wireless Access Point (CAPWAP)[71] tunnels between the APs and the controllers. Traditionally, all data is forwarded to the controller, and then the controller passes the data onto the wired network. LWAPP uses UDP ports 12222 and 12223 by default. CAPWAP uses UDP ports 5246 and 5247 by default. So LWAPP or CAPWAP is the communication protocol between the APs and the controllers. The 802.11 frame is tunneled to the controller so that the controller can see all the 802.11 details and make appropriate decisions.

---

[71] CAPWAP is the more popular tunneling protocol today. Both are defined in IETF RFCs. CAPWAP was introduced in 2009 and is based on LWAPP. Cisco uses these, but other vendors may use other protocols. For example, Aruba Networks APs and controllers use PAPI (Process Application Programming Interface) for control tunnels and GRE (Generic Routing Encapsulation) for user data tunnels. Extreme APs and controllers use the WASSP (Wireless Access Station Session Protocol) for tunnels. These examples represent just a few vendors but illustrate the reality that various options are used. Additionally, some vendors support implementing an IPSec (IP Security) tunnel under the tunneling protocol used. Therefore, for example, the WASSP tunnel would be inside of an IPSec tunnel, both existing between the APs and the controllers.

To understand what is typically performed in the AP, when using a controller-based model, and what is typically performed in the controller, consider the following AP responsibilities (based on Cisco documentation):

- Frame exchange handshake between a client and AP
- Transmission of beacon frames
- Buffering and transmission of frames for clients in power save mode
- Response to probe request frames from clients; the probe requests are also sent to the WLC for processing
- Forwarding notification of received probe requests to the WLC
- Provision of real-time signal quality information to the switch with every received frame
- Monitoring each of the radio channels for noise, interference, and other WLANs
- Monitoring for the presence of other APs
- Encryption and decryption of 802.11 frames

Now consider the following wireless controller responsibilities (based on Cisco documentation):

- 802.11 authentication
- 802.11 association and reassociation (mobility)
- 802.11 frame translation and bridging
- 802.1X/EAP/RADIUS processing
- Termination of 802.11 traffic on a wired interface, except in the case of REAP and H-REAP configured APs

Figure 5.1 provides a high-level comparative difference between autonomous APS and controller-based APs.

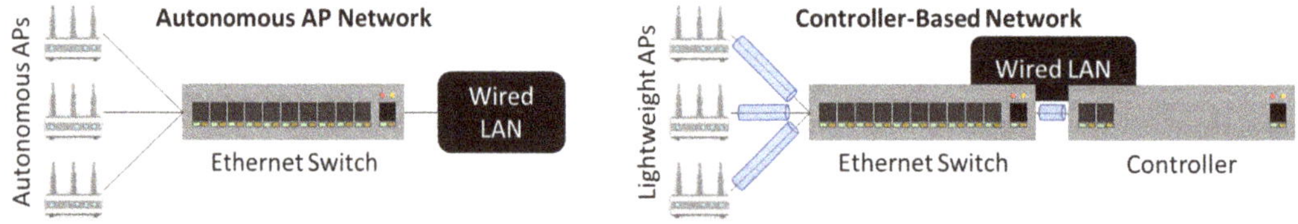

Figure 5.2: Autonomous AP Network vs. Controller-Based Network

In summary, as shown in Figure 5.2, a controller-based architecture depends on a wired network device called a controller (in the past they were called WLAN switches). The lightweight APs connect to the controller to update the firmware, be configured, and then provision the radio for the WLAN. The radio provisioning occurs using WLAN profiles (wireless network profiles) that stipulate requirements such as:

- SSID
- Security configuration
- Channel
- Output power
- Channel width
- Fragmentation settings
- Filtering options
- QoS settings
- VLAN IDs
- Data forwarding mode
- Band steering

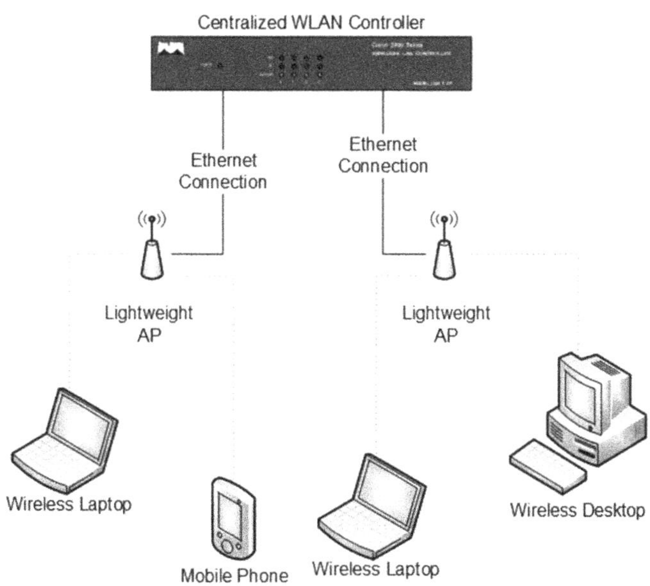

Figure 5.3: Controller-Based Architecture

In some controller-based solutions, settings are configured at a high level for all WLAN profiles, and then specific settings are configured in each profile. In other solutions, all settings are configured in each individual profile.

WNMS-Based Solutions

Some vendors call their centralized management device a controller, but they do not fit into the controller model just described[72]. Instead, the device acts as a wireless network management system (WNMS). To be clear, a controller can be configured in such a way that it acts much closer to a WNMS, but the primary distinction is that a WNMS manages the APs while the APs perform all 802.11 operations. All frames, control, management, and data are processed by the AP and the AP forwards data frames directly to the appropriate network destinations.

Early WNMS solutions were server-based. You installed the software on a server and the APs were configured and managed by the server. Newer WNMS solutions use an appliance that looks very similar to a controller, or they run the WNMS in a switch that is added to the network (you can't use just any switch, it has to be the switch the vendor provides).

An example of this modern WNMS solution was the EnGenius line of APs and WNMS solutions. Newer EnGenius solutions use a cloud-based management architecture or allow the installation of a server on-premises to manage the network. In either case, the APs are configured, monitored, and managed by the (now ezMaster) solution, but no 802.11-level processing actually occurs in the management system. That is, no 802.11 frames from the clients are tunneled to the management system as 802.11 frames for processing. However, it does allow for use of the AP radios (through the firmware in the APs) to scan the RF environment periodically to learn about the characteristics for automatic channel configuration, transmit power adjustment, and rogue AP detection.

---

[72] Honestly, if these vendors want to call their WNMS a controller, I don't blame them. The term controller is so generic that it can apply to literally anything. However, in the history of WLANs, the term controller has traditionally had a different meaning than many modern systems apply to it. When CWNP refers to a "controller-based" WLAN, we are specifically referencing that kind of system that at least offers a split-MAC concept. When we speak of a particular vendor, the term controller should always be taken to mean what that vendor means.

Another example of this is the LigoWave Infinity Controller[73]. Either the first AP acts as the controller or a server plays the role. The controller simply provides configuration management for the other APs. Monitoring statistics are sent back to the managing AP or server for monitoring.

The key different that quickly indicates a WNMS instead of a controller is that tunnels are not established between the APs and the WNMS for data forwarding or processing of any 802.11 frames directly. Controllers use a tunnel for processing of some 802.11 frames directly and may require that data pass through them for data forwarding (though this latter option can be disabled to allow for distributed data forwarding in nearly all controller-based environments today).

Cloud-Based Solutions

Cloud-based solutions work across the Internet for device management. Modern cloud WLAN vendors provision APs, monitor APs, reconfigure APs, and provide other maintenance operations from the cloud; however, most APs work in a distributed edge manner, and they fully function as MAC stations without requiring access to the cloud. This provides for faster data communications and allows for continued local operations if the Internet connection fails.

The advantages of cloud-based management include "anywhere access" to monitoring and configuration options and the lack of need for on-premises gear (WLAN controllers) to run the WLAN. The disadvantages of cloud-based management include consumption of Internet bandwidth, inability to manage the WLAN when the Internet is down (with the common exception of some basic console commands), and the monthly or annual service contract fees required to use the system.

Some cloud-based providers offer on-premises solutions as well. This solution allows for their APs to be used without requiring cloud management. However, it is not the common way in which their solutions are used as customers typically choose them because of the cloud platform.

[73] www.ligowave.com/public/downloads/Infinity_controller_guide.pdf

Autonomous Solutions

Autonomous APs work in either a standalone deployment model where each AP is individually configured and managed or in a centralized configuration management model where they are configured from a centralized software application or server console, but other than monitoring and configuration they perform completely independent as WLAN APs. They include all the logic and ability within them to run one or more WLANs.

Figure 5.3 shows the typical autonomous AP deployment. Such a deployment is still very common for smaller networks.

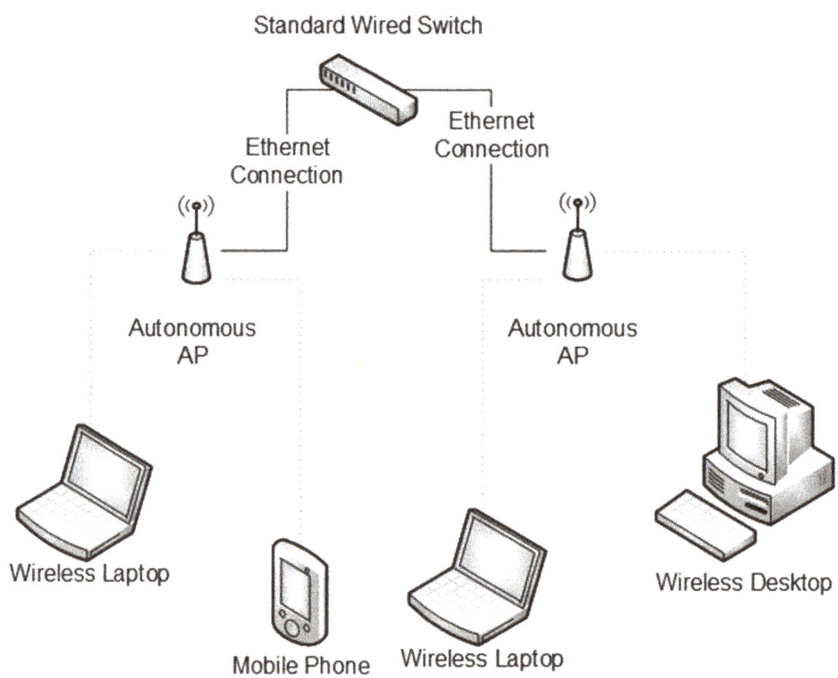

Figure 5.4: Autonomous AP Deployment

In all architecture models, RF Profiles may be independent of the WLAN profile, or they may be included in it. An RF profile effectively defines how the radio will be used, such as output power, channel selection, and channel width.

Data Forwarding Models

Data forwarding models are either centralized or distributed. Early WLAN controllers (switches) all used centralized data forwarding. And, Autonomous APs all use distributed data forwarding. Over the years hybrid models have arisen.

Cloud-managed WLANS mostly use distributed data forwarding, which means that the actual user data goes from the AP directly to the destination and need not go first to the cloud. This makes sense for cloud-managed WLANs because data destined for an internal server would not perform well when first going to the cloud provider network and then back into the local network to the server. Distributed data forwarding then indicates that the AP determines where to send the data next along the journey to its destination rather than that WLAN controller.

Centralized data forwarding indicates that all data is tunneled to the controller, with varying tunneling protocols depending on the vendor, and it is then forwarded from the controller to the actual destination. While most networks and most controllers are well able to keep up with such demands, it may not be a very efficient route. In many cases, the data is destined in the data center where the controller may well reside anyway, in many scenarios this results in great redundancy. For example, if the wired network is designed such that many services (printers, DNS, Internet proxy servers, etc.) are located near the access layer and not near the data center, it simply results in data traveling the path twice to reach the destination. The outcome is a greater delay in communications. If the network has sufficient extra bandwidth to handle this, it will be no problem. If the network is already heavily loaded (for example, utilizing many 100 Mbps links instead of 1 or 10 Gbps links), the extra communications could cause performance issues for other time-sensitive communications.

Modern controller-based solutions allow for the enabling of distributed data forwarding. However, it is important to read the vendor documentation closely. While they may support distributed data forwarding instead of centralized data forwarding, some features may be lost in such scenarios. For example, certain monitoring and alerting features may be lost if the data is no longer passing through the controller.

Choosing among the architectural models will be based on client needs, budgets, and technical abilities to implement and manage the solution.

5.2: RF Management

RF management involves controlling the RF activity in an environment. Specific to WLANs, it involves choosing an MCA or SCA solution, understanding RRM and addressing interference. These topics are covered in this section.

MCA and SCA RF Management

RF management, in MCA, is about planning channel assignments. This will be discussed more in the next section, Channel Plans. With SCA solutions, RF management is automatic within the channel selected, and more density or WLANs dedicated to additional purposes can be added by adding new channel blankets[74]. For example, if you first implement SCA on channel 6, you can later add a network on channel 11 to use for VoIP only. The remainder of this section will address areas with a high density of clients and mobile and non-mobile clients and their needs.

WLAN clients can be either mobile or non-mobile. Non-mobile clients generally have consistent results because the AP does not move and neither does the client. Unless a significant change occurs in the environment, such as the addition of new walls, doors, or filing cabinets, the non-mobile client is likely to continue connecting to the same AP day after day.

Mobile clients will utilize many APs throughout their life. Mobile clients include laptops, tablets, barcode scanners, mobile phones, and VoWLAN handsets among others. These clients will have varying results depending on the signal strength and CCI and ACI in the area where they are used. The WLAN should be designed to provide a minimum level of service, and the users should be trained to understand that varying levels of performance can be expected above that minimum level. Service Level Agreements (SLA) can be written to document the minimum level required of the WLAN.

Whether using MCA or SCA architectures, the WLAN designer will encounter areas with a high density of clients. Such high density (HD) and very high density (VHD) WLANs must be carefully engineered to provide the required coverage and capacity. In the later section, Selecting Access Points, you will discover a useful tool freely available that

[74] A modern example of SCA is Allied Telesis. They refer to it as Channel Blanket. Fortinet also continues to offer SCA, which they reference as a Virtual Cell deployment model. In both vendors, either MCA or SCA can be used as desired given the best fit for a scenario.

assists in capacity planning. But the general guideline for high-density networks is more APs with lower output power settings. This allows for a greater density of APs in a physical space. In many cases, these APs will be configured at 5 or 10 mW of output power to accomplish the small cell size required. More information will be provided on cell size planning in the later discussions of co-channel and adjacent channel interference.

Radio Resource Management (RRM)

RRM is often defined as a dynamic channel assignment and dynamic RF output power control, and this definition is true. However, modern implementations of RRM evaluate additional factors and provide additional benefits because of this extra evaluation. RRM is ultimately the MCA response to SCA. Rather than using a single channel model to allow for simpler configuration deployment, the idea is to use algorithms for channel selection based on radio metrics provided to the controller or WNMS from all of the APs deployed. The APs can see each other as well as clients from their and other BSSs and this information can be provided to the centralized system. The centralized system can then make configuration changes dynamically with better knowledge of the environment, including transmit power adjustments, channel widths, channel selection, and, in the newest of systems, possibly even BSS color[75] adjustments.

The basic features of all RRM solutions are dynamic channel selection or assignment and transmit power level management. Most RRM solutions use neighbor communications that provide information to APs about their neighbors and then this information is passed up the chain to a WLAN controller that uses the data to make decisions about channels and power levels. The basic process is as follows:

1. APs send neighbor communications.
2. APs listen on all channels for neighbor communications.
3. The information gathered is sent to the controller.
4. The controller adjusts AP configurations based on the information

---

[75] BSS color is a new capability in 802.11ax that allows each BSS to be assigned a color and then devices can treat frames differently if they are intra-BSS frames (same BSS color) or inter-BSS frames (different BSS color). Under the hood, the "BSS color" is just a number. Therefore, devices are comparing the numeric ID of their BSS with that in the frame. If it is a different BSS color, they can ignore the frame, in relation to updating their NAV, at a stronger signal level than if it is the same BSS color.

However, there are additional features, such as coverage hole detection and interference detection. Coverage hole detection is focused on clients. When an AP hears from an associated client below a configurable dB threshold, it is assumed that the client has no better AP as a roaming option. The controller can either increase the output power of the AP (while hoping not to create a link mismatch) or alert an administrator of the potential problem.

Interference detection occurs by scanning all channels, including adjacent channels, and listening for communications. When detected, the controller is notified. When the frame is not recognized by the controller, it is identified as interference and the controller may adjust appropriate AP channels to avoid the interference and it may provide an alert to the administrator.

> Some WLAN engineers are completely opposed to RRM, and others feel it must be carefully managed. The CWNP position is a neutral position in that many WLANs benefit greatly from RRM and others can experience severe problems. It is a technology that provides exceptional value in the right environment.

Software Defined Radios/Software Configurable Radios

Many vendors now offer APs with radios that are referenced as Software Defined Radios (SDRs) or Software Configurable Radios (SCRs). In most cases, the AP will ship with one fixed radio that is designed for 5 GHz operations and a second configurable radio that can either operate as a 2.4 GHz radio or a 5 GHz radio. In some cases, both radios could be software configurable. Additionally, it is possible to have three radios with either one or two of them operating as configurable radios.

The benefit to the designer of the configurable radios is that some APs can be implemented with one 2.4 GHz radio and one 5 GHz radio, while others are implemented with both radios configured for 5 GHz. Given that 2.4 GHz signals can be received at greater distances in most environments, this feature can be quite useful. Additionally, with the extra channels available in 5 GHz, it allows the designer to implement three, or more times the channels in 5 GHz than would be possible if all APs were fixed with a 2.4 GHz radio and a 5 GHz radio. When configured with both radios on 5 GHz, this is sometimes called a dual-5 GHz AP. Again, APs are available with one 2.4 GHz radio and two 5 GHz radios.

> Debates have occurred over whether an AP with two 5 GHz radios performs as well as two APs separated by a few meters with one 5 GHz radio each. The reality is that, even if the dual-5 GHz APs performed at 20 percent less than the two separated APs, the reduced cost and increased capacity achieved is usually worth it.

If significant problems are to occur with dual-5 GHz APs, it would most likely happen when one radio is receiving a frame, and the other is transmitting a frame, and the two radios use side-by-side channels like channels 36 and 38. This interference would occur because the transmitting signal would be very loud and only a few inches from the received signal. The sidebands from the transmitting radio could raise the noise floor for the receiving radio such that it cannot receive at the data rate of the incoming frame.

This problem scenario can be resolved by separating the channels significantly. For example, if your regulatory domain allows, use a U-NII-1 or U-NII-2Achannel (36-64) for one radio and a U-NII 3 channel (149-165) for the other radio. With the frequency separation, such interference is less likely to occur or be as severe. When the regulatory domain does not allow it, separate the channels by at least 3, for example, channel 36 and 48. When deploying multiple dual-5 GHz APs, channel planning will be more complex, but with the number of channels in 5 GHz bands and the shorter distance at which signals are significantly strong, it is possible – particularly for 20 MHz channels.

Co-Channel Interference

Co-channel interference has been defined in previous chapters. It occurs when two cells are close enough to each other and are operating on the same channel that some clients and often even the APs will be required to contend with each other to gain access to the medium. As was previously noted, for this reason, many are now calling it co-channel contention (CCC) as it is a more precise reference to what it is. However, the industry at large still uses the term CCI.

The mitigating solution to CCI is cell sizing. By adjusting cell sizes through transmit power adjustment (and shapes with directional antennas), you can reduce CCI in 2.4

GHz and all but remove it in many 5 and 6 GHz 20 MHz channel deployments. To understand how this works, you should understand two boundaries of a cell: the association boundary and the CCI boundary (Figure 5.4). Additionally, you should be aware that every client associated with a cell (an AP) will have its own CCI boundaries, as well.

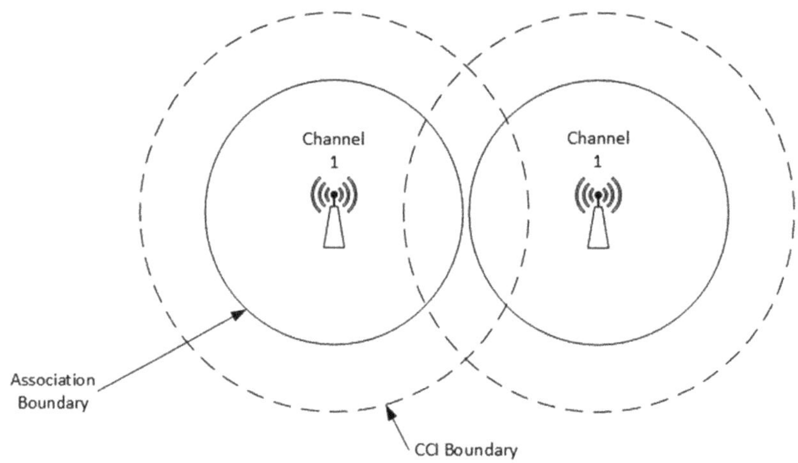

Figure 5.5: Association Boundaries for APs

The association boundary is the edge of the cell where the minimum supported data rate can be achieved. If the lowest data rates of 1 and 2 Mbps have been disabled, for example, this will result in a smaller association cell (reduced association boundary) than when they are enabled. However, just because clients can no longer associate beyond that point does not mean that CCI cannot exist beyond that point.

The CCI boundary exists where the listening stations that are not part of the cell must be silent because they have detected a signal at a minimum threshold. This behavior is part of the CSMA/CA protocol used in 802.11 WLANs. Even when lower data rates are disabled, such as 1, 2, and 5.5 Mbps, 802.11 standard devices must still process and wait for 1, 2, and 5.5 Mbps signals that are heard in the 2.4 GHz band. The same is true in 5 GHz if low data rates like 6 and 12 Mbps are disabled. Therefore, CCI can occur based on low data rate signals even though those signals are not enabled in the local cell.

Additionally, just because the signal cannot be fully processed does not mean that the signal cannot be detected (using the PHY preambles and headers) and initiate a backoff based on the duration for a STA.

When considering association boundaries (also called data rate boundaries) and CCI boundaries, the AP perspective is almost always considered, and it should be. Additionally, the client perspective must be considered. The AP perspective is the simplest and is represented in Figure 5.4. The APs in the image have low data rates of 1, 2, and 5.5 Mbps disabled. Therefore, the association boundary is smaller than the CCI boundary by a significant amount. The image shows only a 2.4 GHz representation, but 5 GHz boundaries work in a similar manner. In Figure 5.4, the CCI boundary does not reach all the way into the space of the other AP, but this does not mean it will not cause problems. The client perspective will show that problems still occur in such scenarios.

The client perspective of the association boundary, remember, is also called the data rate boundary and the CCI boundary is much different. Like APs, the clients transmit, typically, in an omnidirectional manner. Therefore, they both see the boundaries of other nearby APs and create their own. That is, they see cells, and they also create a cell of RF energy of their own. Figure 5.5 illustrates this.

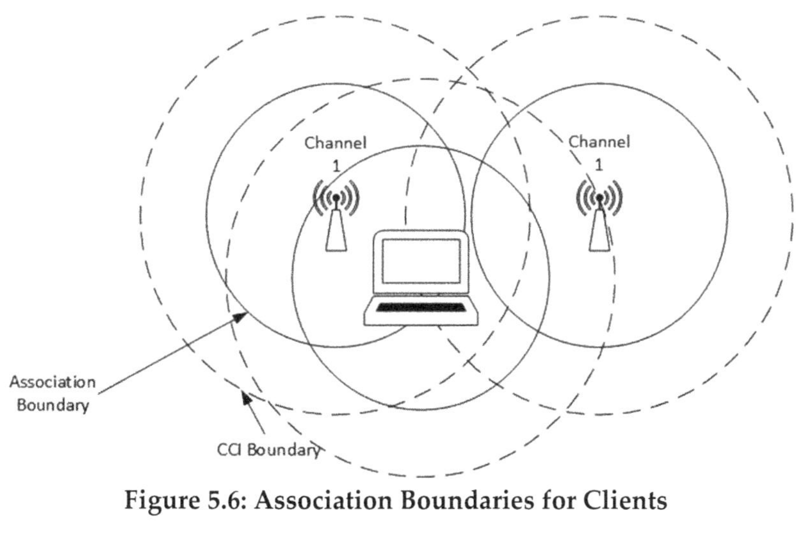

Figure 5.6: Association Boundaries for Clients

Notice in Figure 5.5 that the signals from the client station that are associated with the AP on the left are reaching the AP on the right. The client is causing CCI for the AP on the right, and the AP on the right is causing CCI for the client even though the APs are not causing significant CCI for each other directly. Therefore, the cells created by the two APs are causing CCI for each other. Again, remember, that CCI is the normal contention process defined in CSMA/CA. The client station in Figure 5.5 must be silent for the duration of any frames it sees from the AP on the right, or any stations associated with the AP on the right, that it can see. Additionally, the AP on the right must be silent for the duration of any frames it sees from the client, or any other clients it can see, that are associated with the AP on the left. Therefore, CCI is not an AP-only issue, but it is an issue impacting clients and created by clients, as well.

The issue of CCI is one of the primary motivators of the introduction of BSS color in the HE MAC and PHY (802.11ax). BSS color assigns a number to each BSS. When a station perceives an 802.11 frame on its radio interface, it can inspect the BSS color and, if it is different from that of the BSS in which the station is participating, it may ignore the frame at a different signal level than a frame having the same BSS color as its own. A frame with a different BSS color is called an inter-BSS frame and a frame with the same BSS color is called an intra-BSS frame. Based on the CSMA/CA algorithm, a station may avoid updating its NAV based on an inter-BSS frame if it is below a specific signal amplitude threshold. This inter-BSS threshold will be a stronger signal strength than the intra-BSS threshold. That is, the station must update its NAV based on weaker signals from its own BSS and may not update its NAV based on stronger signals from a different BSS. In the end, this function can theoretically minimize CCI communication delays.

Adjacent Channel Interference

Adjacent channel interference (ACI) is caused when channels overlap in their sidebands[76], such as 2.4 GHz channels 1, 6, and 11, or when non-overlapping channels cause sideband interference, such as 5 GHz and 6 GHz channels. 2.4 GHz channels are 5 MHz apart and 20 or 22 MHz wide; therefore, it is a certainty that channels will overlap

[76] The terms sideband and side-lobe are often confused among WLAN professionals and in the literature. Technically, the sideband is the energy above and below the intended signal bandwidth that is the natural result of the generation of the signal. Side-lobes are the lobes of radiation from the antenna in the unintentional direction of radiation.

and cause interference if channels 1, 2, and 3, for example, are used in the same area. For example, Figure 5.6 shows the overlap of channel 1 and channel 2 in the 2.4 GHz band.

In Figure 5.6, the spectral mask for an OFDM channel is shown with an overlay of the center frequency for channel 1 and channel 2. Given that the spectral mask would be that of channel 1, it is obvious that channel 2 would overlap with it if its spectral mask were in view. These channels overlap and cause interference with each other even though the standard does not call them adjacent channels (they are simply overlapping channels).

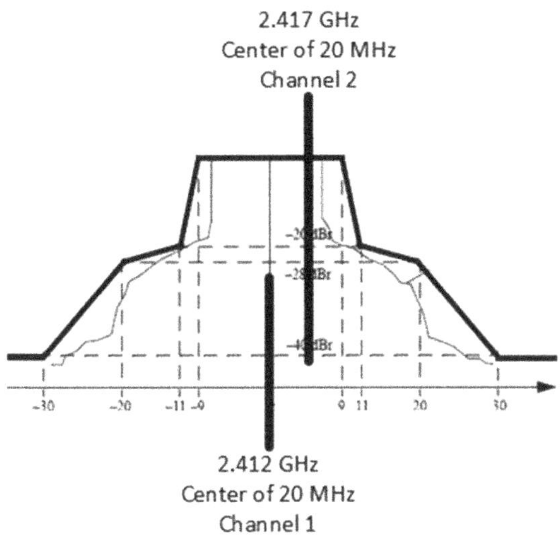

Figure 5.7: Channel Overlap in 2.4 GHz

Channels 1 and 6, however, would be considered adjacent channels and they cause ACI. When adjacent channels overlap, they cause ACI, which can manifest as corrupt frames or simply prevent communications due to energy detect in 2.4 GHz. In 2.4 GHz, ACI can be minimized within your controlled networks by using channels 1, 6, and 11 with appropriate transmit power settings. However, in areas where neighbor WLANs can be seen, if they are using channels 2-5, 7–10, or 11–14, ACI may exist regardless of your network settings. Additionally, ACI is not caused only by APs. Like CCI, it is caused by APs and clients. Therefore, if a client device is near the edge of a channel 6 2.4 GHz cell and the neighbor cell is on channel 1 or 11, the sideband can cause ACI for the neighbor cell (and vice versa). For this reason, part of ACI control is ensuring that no 300 mW client devices are implemented. They sound great (from a hacker's point of view), but they can wreak havoc on your network. This ACI can also be minimized by carefully planning channel selections nearest to these neighbor WLANs.

In 5 and 6 GHz, ACI is caused by sideband energy when the adjacent channel is implemented either very close to the AP or with high-output power or a client associated with the other channel AP is near the interfered AP/cell or uses high transmit power. 5 and 6 GHz ACI is typically manifest as corrupt frames. In 5 and 6 GHz it can be minimized far more easily than in 2.4 GHz (because you have more than three channels to work with) with proper channel plans that do not place adjacent channels near each other.

In regions of the world where the 6 GHz band is completely available, 40 MHz channels can certainly be used in many deployments that will allow for significant channel separation in neighboring cells that will reduce ACI. In fact, 80 MHz channels may even be possible in many deployments. In 5 GHz, it is far more challenging to implement 80 MHz channels with reduced ACI and 40 MHz channels can only be implemented with significantly reduced ACI if the entire band is available and Dynamic Frequency Selection (DFS) channels are used[77]. In 2.4 GHz, bonded channels should simply never be used.

In summary, ACI problems are increased with channel bonding and careful planning should be implemented to minimize its impact. Using narrower channels is one strategy and providing sufficient separation between neighboring bonded channels is another.

5.3: Channel Plans

With an understanding of CCI and ACI, channel planning is clearly essential. This section provides information to help the WLAN designer in the channel planning process. Channel planning involves the selection of channels, power levels of APs, and planning for channel reuse. This task requires a knowledge of channels available in 2.4 GHz, 5 GHz, and 6 GHz, which you learned in CWNA studies, but they are addressed here again in specific relation to the design of a WLAN channel plan. Even if your

---

[77] The preceding statements assume a larger network that will use dozens of APs at a minimum. If only a few APs are used, 40 MHz and even 80 MHz channels may be implemented properly in 5 GHz as to ACI reduction.

solution uses automatic channel selection, understanding the channels available will help you to tune the parameters used by the automatic algorithms for improved channel selection and performance.

802.11 Channels

The channels available for use in WLANs are defined in the 802.11 standard and constrained by local regulatory agencies. For example, in the United States, only channels 1–11 may be used in 2.4 GHz, but in other regions of the world channels 12–13 and even 14 may be used. The first thing you must ensure is that you follow local regulations related to channel usage. Most enterprise APs can be configured for operation in a regulatory domain, and they will automatically remove the option to use channels not allowed in that domain.

However, the regulatory domain is only part of the picture. After you have determined the channels that you can use, you must then determine how to use those channels. Later portions of this section will address channel widths, reuse patterns, and advanced channel features. Here, you will find detailed information on the channels available in frequency bands (2.4 GHz, 5 GHz, and 6 GHz), automatic channel selection, and output power level requirements (like Dynamic Frequency Selection (DFS) and Transmit Power Control (TPC).

Figure 5.7 shows the channels available in 2.4 GHz. Channels are either 22 MHz wide (802.11-prime and 802.11b) or 20 MHz wide (802.11g and 802.11n). 802.11ac does not operate in the 2.4 GHz band even though some consumer-grade products have implemented 256-QAM modulation in 2.4 GHz; it is not part of the standard[78].

Remember that 802.11ac does not operate in 2.4 GHz. Any time a scenario references 11ac, it should be assumed that 5 GHz is in view for enterprise WLANs.

[78] Some vendors have implemented 256 QAM in 2.4 GHz, but the standard does not specify this. Therefore, client devices are unlikely to support it even if it is implemented in the 2.4 GHz band. Therefore, the claims of high data rates based on 256 QAM implementation in 2.4 GHz are highly overstated and you are unlikely to benefit from them in any real-world implementations at any significant level.

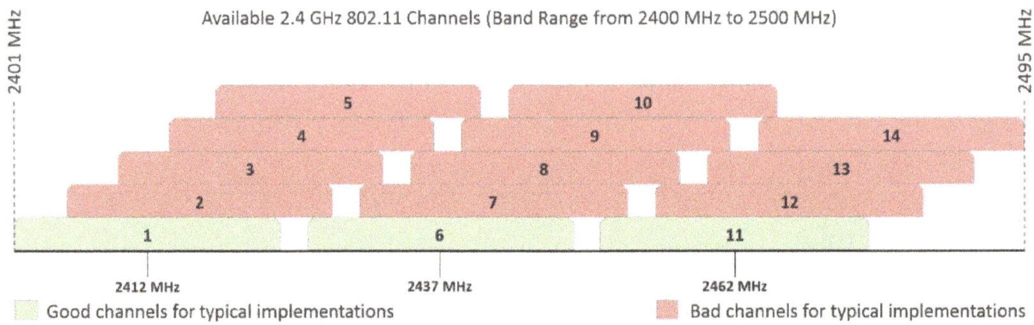

Figure 5.7: 2.4 GHz Channels

Table 5.1 shows the channels available in 5 GHz and Figure 5.7a shows them visually. Many more channels are available in the 5 GHz frequency band with a total of 25 available.

Frequency Range (GHz)	Number of Channels	Channel Number	Center Frequency (MHz)
5.170–5.230	4	36, 40, 44 48	5180, 5200, 5220 5240
5.250–5.330	4	52, 56, 60 64	5260, 5280, 5300 5320
5.490–5.730	12	100, 104, 108, 112, 116, 120, 124, 128, 132, 136, 140, 144	5500, 5520, 5540, 5560, 5580, 5600, 5620, 5640, 5660, 5680, 5700, 5720
5.735–5.835	5	149, 153, 157 161, 165	5745, 5765, 5785 5805, 5825

Table 5.2: 5 GHz Channels with Only Used Frequency Ranges Listed

Not all channels are available in all regulatory domains due to DFS requirements in some regulatory domains, and Terminal Doppler Weather Radar (TDWR) also constrains the use of some channels and additional regulatory reasons. However, even if the lower and higher ranges alone are used, this provides nine 20 MHz channels, and the channels do not intentionally overlap like 2.4 GHz channels. For this reason, channel reuse plans are much simpler in 5 GHz than in 2.4 GHz. Additionally, because there are so many channels, 5 GHz is typically far less busy (congested) than 2.4 GHz.

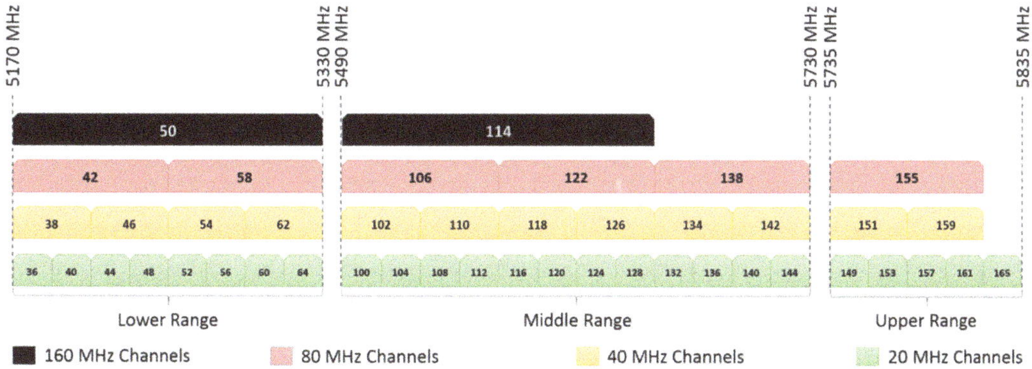

Figure 5.7a: 5 GHz Channels

The 6 GHz band varies significantly around the world. Some regulatory domains implement a very limited portion of the available bandwidth. The map in Figure 5.7b illustrates the state of 6 GHz regulatory adoption in late 2021.

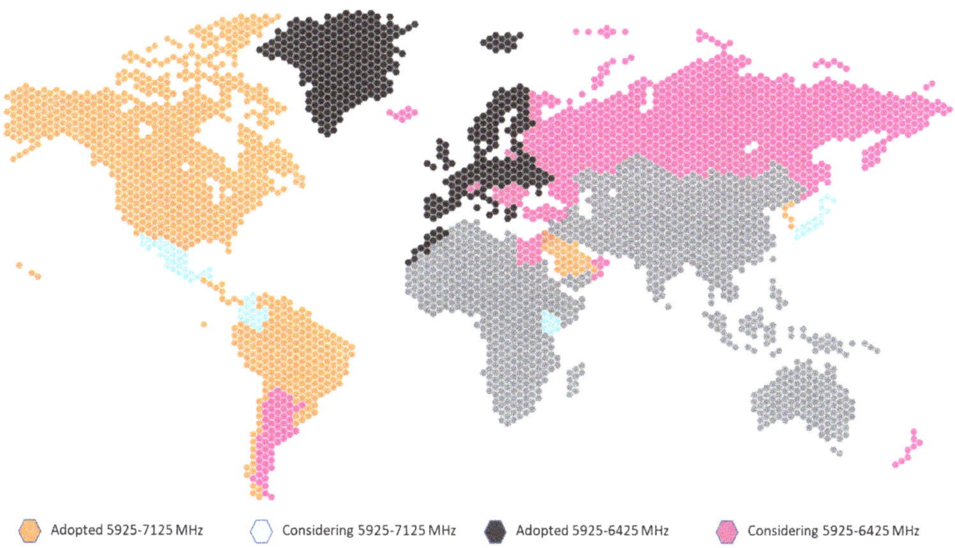

Figure 5.7b: 6 GHz Regulatory Adoption as of Late 2021

The entire 6 GHz band ranges from 5.925 GHz to 7.125 GHz. So, while we call it the 6 GHz band, it ranges from slightly below 6 GHz to slightly above 6 GHz. The channelization possibilities are shown in Figure 5.7c and 5.7d.

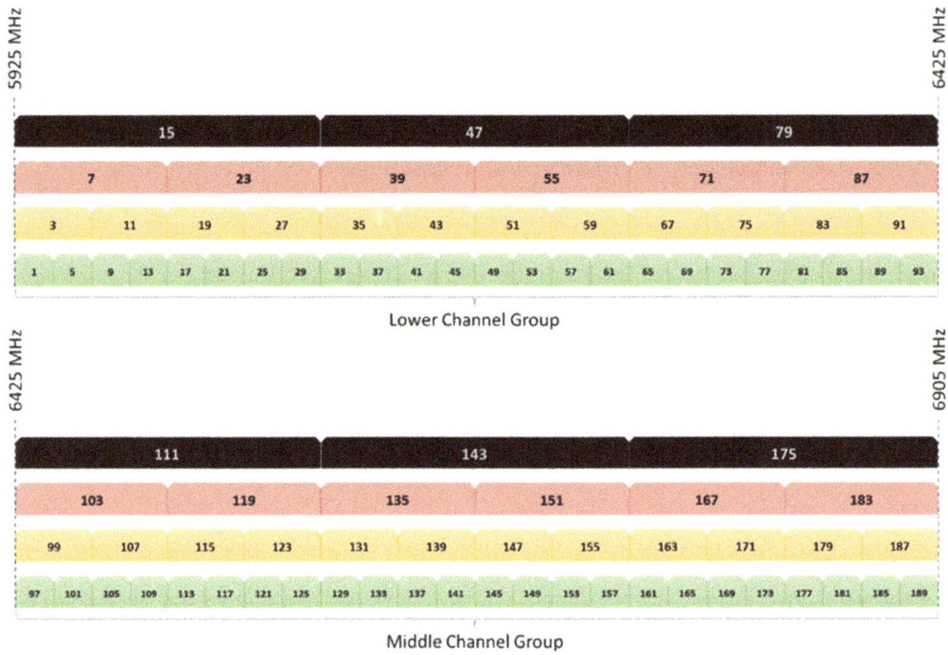

Figure 5.7c: 6 GHz Band Channel Structure (Lower and Middle Channel Group)

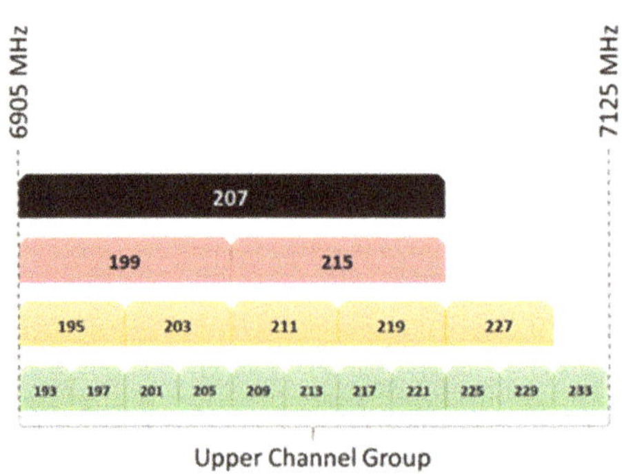

Figure 5.7d: 6 GHz Band Upper Channel Group

Note that the channels in 6 GHz restart at 1, unlike the channels among the 2.4 GHz and 5 GHz bands. It is essential that you determine the regulatory band in which the 6 GHz WLAN will operate. Several regions have already limited the use to only the lower channel group. Still, this provides an additional 480 MHz of bandwidth or 24 channels when using 20 MHz channel widths. In regions that allow only the lower channel group, channel planning will be like that in 5 GHz. In regions that allow all the channel groups, it is very likely that 40 MHz and even 80 MHz channels will be very common[79].

When selecting channels, you must also select the output power for the channel or allow RRM to configure the output power automatically. Output power selection is driven by three factors:

- Capabilities of the AP
- Desired cell size
- Regulatory requirements

APs have output power capabilities ranging from 5 mW to 1 W. As a WLAN designer you will need to select APs that have the output power capabilities you require, and then you can adjust the output power to create the desired cell size. However, at the same time, you must ensure that you are operating within regulatory requirements. Again, assuming the AP allows you to select a regulatory domain or region of the world setting, it will likely do this part for you. The software will ensure that the AP does not transmit on disallowed channels or at power levels beyond the allowance. This is good for operations, but it does not always help you during design. During design, you must still select the proper channels and power settings, and you must know the regulatory requirements so that your design can actually be implemented.

The CWDP exam does not test on specific requirements of regulatory domains; however, you should know that DFS may be required and that the regulatory domain requirements must be considered when creating channel plans.

[79] You will not be tested on the intricate details of regulatory requirements for any band. This constraint is because they vary too significantly, and it would be unrealistic to expect engineers from all over the world to know the requirements for the rest of the world. However, you should know that they do vary and that the local regulatory agencies provide guidance.

Selecting Channel Widths

After selecting the channel on which an AP will operate, you must select the channel width. 802.11n APs will support either 20 MHz or 40 MHz channel widths using the bonding of two 20 MHz channels. 802.11ac and 802.11ax APs will support 20, 40, 80, or 160 MHz channels. 160 MHz channels are not likely to be used in enterprise networks for the foreseeable future due to insufficient spectrum availability with the possible exception of the 6 GHz band when all channel groups are available. In 5 GHz, 80 MHz channels may be used in some installations, but it will likely only be used in specific sections of the WLAN based on the demand for very high throughput levels in that area. In 6 GHz, the same guidelines apply as those in 5 GHz for regulatory domains limited to 480 MHz and bandwidth. For regulatory domains with full 6 GHz bandwidth, 80 MHz channels may be used with much greater flexibility. For links to individual clients, with today's application demands, 80 MHz channels are simply not commonly warranted on a universal basis across an entire WLAN in the 5 GHz band. Time will tell how the 6 GHz band is utilized. Of note, all APs certified for Wi-Fi 6E must support 80 MHz channels, though narrower channels may be implemented for a BSS.

In the end, 20 and 40 MHz channels will be used for most 802.11n/ac deployments. The 20 MHz channels are used in 5 GHz for high-density deployments, and 40 MHz channels may be used for medium or low-density deployments. However, even in medium or low-density deployments, the throughput levels offered by a 40 MHz 802.11ac channel may not be required, and the decision must be made on a case-by-case bases. The decision is made based on the aggregate throughput requirements in a cell. In the later section on selecting APs, you will learn to use a free tool that helps in determining capacity requirements and discovering the difference between 20 and 40 MHz channels in an implementation. Finally, in 6 GHz, 40 MHz is likely to be the new 20 MHz for regulatory domains with all three channel groups available and 80 MHz is the new 40 MHz[80].

Channel Reuse Patterns

Channel reuse patterns are used to diminish or remove CCI and ACI. In 2.4 GHz reuse patterns help to mitigate the CCI though never remove it, but in 5 GHz reuse patterns

[80] Attributed to Chuck Lukaszewski of Aruba Networks, an HPE company.

properly implemented with 20 MHz can nearly remove it and, in some networks, it can be completely removed in 6 GHz with 20 MHz channels (and possibly 40 MHz channels). Diminishing ACI is a bit more complex, but the more bandwidth you have and the narrower channels you use the greater the likelihood of removal of ACI. The concept of a channel reuse pattern is simply to avoid placing APs near each other that are on the same channel or, in 5 GHz, on an adjacent channel. Figure 5.8 shows an example channel reuse pattern in 2.4 GHz.

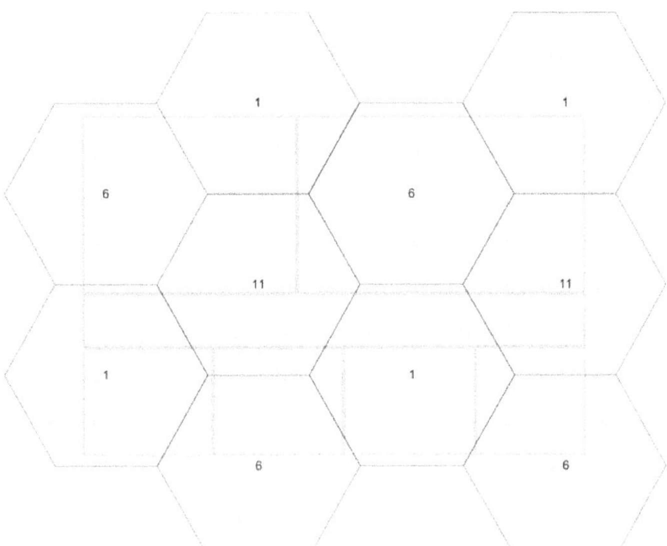

Figure 5.8: Channel Reuse Patterns

Figure 5.9 shows more realistic patterns, though still not close to what is actually seen in physical spaces. The point is that APs do not generate cells that are circular or hexagonal. The most important thing to know is that such reuse patterns are essential in avoiding CCI and ACI. Notice in both Figure 5.8 and 5.9, the plans separate all APs on the same channel from other APs on the same channel by an entire cell. This is all that can be accomplished in 2.4 GHz due to that fact that there are only three non-overlapping channels. In 5 GHz, you can often separate APs on the same channel by two cells because so many more channels are available. In 6 GHz, the same statement holds true for all regulatory domains that allow it and those that allow the full bandwidth can even support this with 40 MHz channels in most networks and 80 MHz channels in many networks.

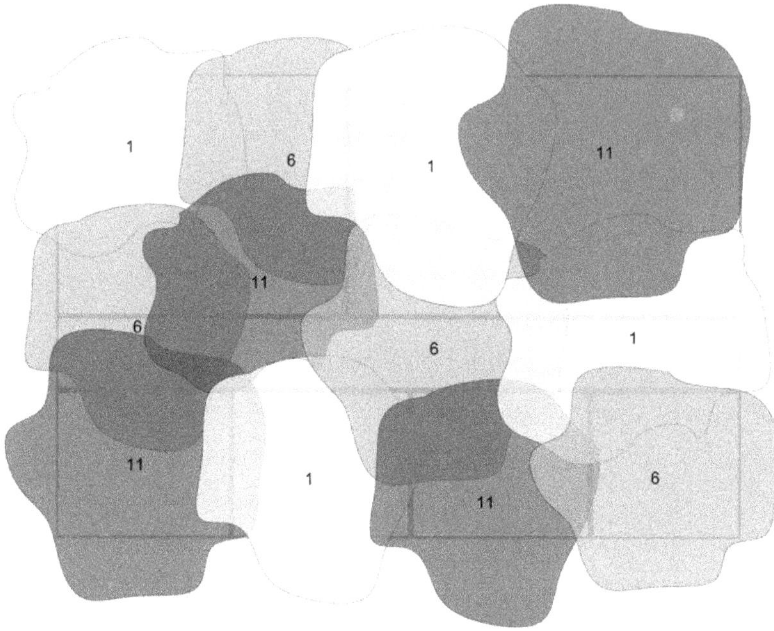

Figure 5.9: Realistic Channel Reuse Patterns

The algorithm for channel reuse in 2.4 GHz with a single floor deployment is simple[81]:

```
ChannelSet = 1, 6, 11
SC != NCᵃ, NCᵇ
```

[81] We are not concerned with channel selection in consumer spaces at this point. We are assuming complete control of most of the RF activity in an enterprise sized deployment in this text. It is not uncommon, in consumer spaces, to have three neighbor channels on 1, 6, and 11 of equal or very near equal strength. For this reason, the channel selection should be made based on channel load. That is, if all three channels have detected APs of near equal strength, the new AP should be placed on the channel with the lowest utilization of airtime. To make this decision most accurately, monitoring must be performed over a significantly long period and when the neighbor networks are most active. If channel 1 has a load of 67 percent utilization, channel 6 has a load of 35 percent utilization, and channel 11 has a load of 62 percent utilization, all at the busiest times, channel 6 is likely to be the best choice under channel load evaluation models. Finally, channels 1, 6, and 11 should be used to avoid ACI-type interference, which is destructive, as opposed to CCI-type interference, which is cooperative. Both have a negative impact on communications, but ACI tends to be worse in such spaces.

Where NC^a is the strongest neighbor channel and NC^b is the next strongest neighbor channel that is not the same channel number as NC^a and SC equals the selected channel for the AP under planning. Looking at Figure 5.9, this algorithm holds true. For example, with the upper-left-most cell that is on channel 1, the selection for channel 1 makes sense per the algorithm. Channel 6 is the strongest detected channel (NC^a) and Channel 11 is the next strongest detected neighbor channel (NC^b), therefore SC equals 1. Given that only channels 1, 6, and 11 are available, channels cannot be skipped in such a plan and so the algorithm is simple[82].

To implement a skipped channel plan in 5 GHz using only the lower channels, with 20 MHz channel widths, consider the following linear model:

```
ChannelSet[5 GHz] = ch36, ch36(+8), ch36(+16), ch36(+24),
                    ch36(+4), ch36(+12), ch36(+20),
                    ch36(+28)
```

This model results in the following channels:

```
ChannelSet[5 GHz] = 36, 44, 52, 60, 40, 48, 56, 64
```

The intention is not to "use these channels in a row" but to be sure that neighbors fit into this pattern at a minimum. Stated differently, no immediate neighbor channel is the adjacent channel among available channels. Implementing this model in an actual network looks like Figure 5.9a. Note that immediate neighbor channels, in this model, are all separated by two channels at a minimum. If you can use the upper 5 GHz band and also the DFS channels, such a channel plan is simpler still.

6 GHz is similar to 5 GHz if only the first 480 MHz is available for use. However, when the entire 6 GHz band is available, a similar channel plan that uses 40 MHz channels can be used. In fact, if you consider Figure 5.9a and use the 80 MHz channel numbers for 6 GHz, you can see that a similar model can be achieved when the entire band is available (however, with increased ACI due to the use of wider channels).

---

[82] Of course, in addition to channel selection, transmit power must also be calculated so that the cells are not excessively large causing undo CCI for the next cell using the same channel number. At the borders, some CCI will exist between same-number cells, but it is reduced through proper transmit power selection and proper antenna selection and use.

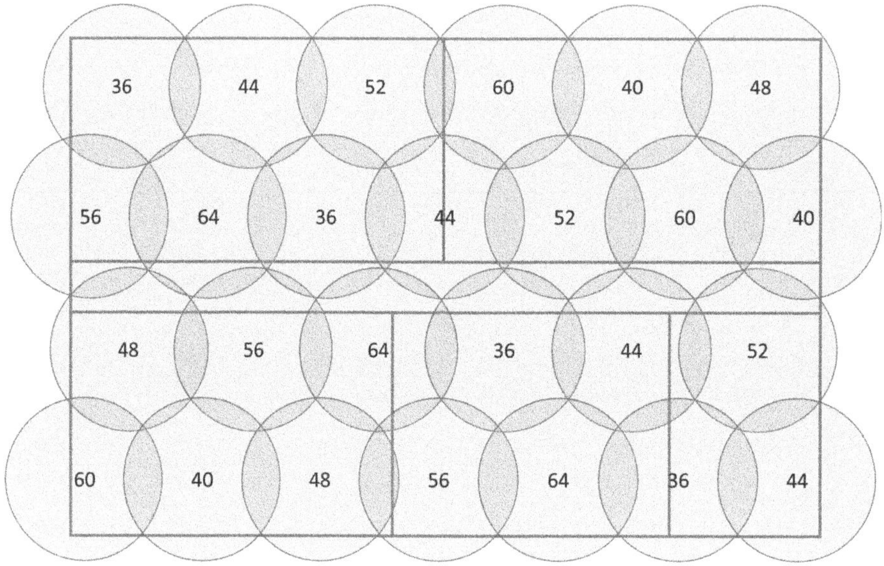

Figure 5.9a: 20 MHz Skipped Channel Plan in 5 GHz to Minimize ACI and CCI

As you can see, channel planning is a bit complicated by the constraints of regulatory domains, but several options are available for most designs in 5 GHz and 6 GHz. 2.4 GHz is certainly the most constrained.

Advanced Channel Features

802.11ac added some operational channel features that should be understood and planned for by the WLAN designer as well. These include 80+80 channels and RTS/CTS enhancements.

The 80+80 channels, available in both 5 GHz (802.11ac/ax) and 6 GHz (802.11ax) use two 80 MHz channels that are not adjacent in the frequency band. This allows for the use of a 160 MHz configuration even if two adjacent 80 MHz channels are not available. This feature is not likely to be used in enterprise 5 GHz deployment in the near future due to that fact that such high-data rate client connections are not typically essential, and currently, there is insufficient frequency space available to use 160 MHz of bandwidth for a single channel and still have enough bandwidth for additional APs in the area. However, it may find some use in the 6 GHz band with 802.11ax.

The RTS/CTS enhancements have been made to 802.11ac to allow two cells to share the same space. That is, one cell can use channels 36, 40, 44, and 48 for an 80 MHz channel, and another cell in the same area can use channels 44, 48, 52, and 56 for an 80 MHz channel. The process works such that RTS frames are sent out on all four 20 MHz channels, and the 80 MHz of bandwidth is only used to transmit a data frame if all four channels respond with a CTS frame. Therefore, at times, the cell will communicate as a 40 MHz channel and at times as an 80 MHz channel when the other sharing cell is not using the top or bottom two 20 MHz channels.

5.4: Infrastructure Services

WLANs do not stand alone. Instead, they depend on several infrastructure services to operate effectively. These services include DHCP, DNS, NTP, Firewalls, and the management of additional settings and services. This section provides guidance in implementing these settings and services to support the WLAN and ensuring the design appropriately defines them.

Dynamic Host Configuration Protocol (DHCP)

DHCP is used to provide IP configuration sets to computers on the network. It is sometimes said that DHCP provides IP addresses to computers, but this is an oversimplification. DHCP can provide the following and more:

- IP addresses
- Subnet Masks
- Default gateways (routers)
- DNS servers
- Lease durations
- Configuration options
- Service locations

When planning for DHCP support of the WLAN, at least three things should be considered:

- Will you use an existing scope or create new separate scopes for the WLAN operations?

- What options should be provided by the DHCP server for proper WLAN functionality?
- How long should leases endure?

The selection of a DHCP scope for the WLAN is actually very important. It can impact roaming capabilities and result in a large subnet if you're not careful in your planning. For this reason, vendor recommendations should be considered when answering the first question. In general, it is best to place WLAN traffic on a separate IP subnet from wired traffic so that it can be managed accordingly. IPv4 is still the most commonly used IP solution, with IPv6 very slowly gaining momentum.

When designing a WLAN, you must determine how subnetting will be performed and the networks on which WLAN clients will operate. Will the clients operate on the same subnet whether they are internal users or guest users? Will VLANs be used to separate them, separate subnets or tunnels to the DMZ? These decisions should be made and specified in the WLAN design documentation.

Lease durations or timeouts should also be considered relative to DHCP. This is particularly important in networks where users come and go, such as guest WLANs or mobile areas of the WLAN like a lobby. Setting the lease duration to a few hours instead of a few days can reduce DHCP pool (scope) exhaustion and allow the DHCP operations to continue.

DHCP options can be used to locate services as well. For example, they can be used to define the IP address of a WLAN controller so that an AP, receiving a configuration set from the DHCP server, will know exactly where the controller is located to begin synchronizing with it and receiving configuration settings from it. This setting is often specified with DHCP option 43.

Domain Name System (DNS)

DNS is used to resolve domain names to IP addresses. It is used in WLANs to service clients for hostname resolutions, but it is also used for specific WLAN functions including locating controllers. For example, when a Cisco AP receives its IP configuration from the DHCP server, if option 43 is not specified, the AP will try to resolve the DNS hostname *CISCO-CAPWAP-CONTROLLER.localdomain*, where *localdomain* is the domain configured by DHCP. This is one example. Vendors may vary the hostname used, of course, but a DNS record is often used for the location of the controller or WNMS. Of course, DNS is nearly always used with cloud-based management solutions; however, these DNS host names would be resolved through your Internet Service Provider (ISP).

When using CAPWAP, four controller discovery methods are available: 1) Layer 3 subnet broadcast looking for a controller, 2) NVRAM stored addresses from past controller connections, 3) DHCP option 43, and 4) DNS resolution as referenced here.

Network Time Protocol (NTP)

NTP is used to configure the time on a device using a network time server. Accurate time configuration is important for logging and other operations. Most WLAN controllers and APs can have their internal clock set using a network time server. Figure 5.10 shows the configuration of NTP in a Ruckus Wireless solution.

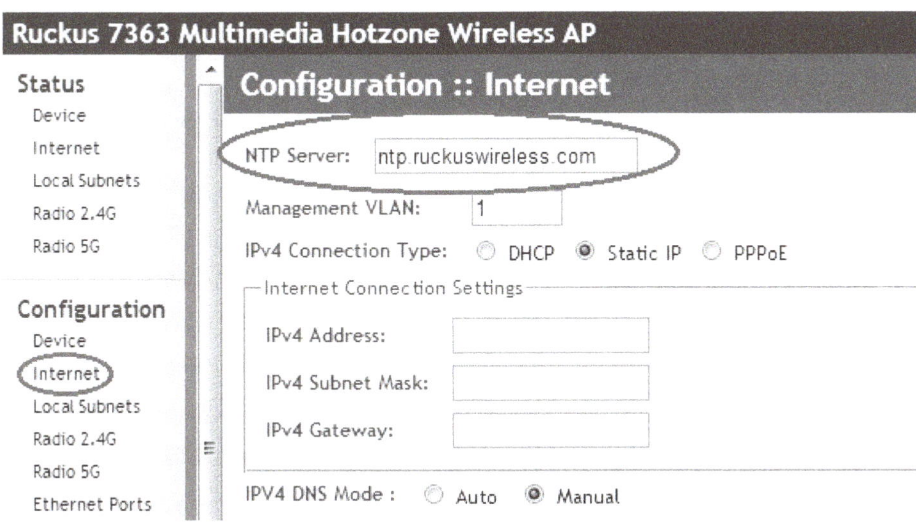

Figure 5.10: Configuring NTP Server Settings

Firewall Configuration

Firewalls may require configuration to support the WLAN. If a firewall resides between the APs and controllers, modifications may be required. The firewall should allow the

LWAPP or CAPWAP ports through. LWAPP uses UDP ports 12222 and 12223 by default. CAPWAP uses UDP ports 5246 and 5247 by default.

Additional firewall configuration settings may be required. Check the vendor literature for the ports (both UDP and TCP) used for various communications and ensure that the appropriate firewall rules are in place.

ACL Management

Access Control Lists (ACLs) are used on network routers to control the traffic that is enabled for pass-through to other networks. Like a firewall, the routers must allow the communications to go through that is required by the WLAN. In fact, ACLs on routers are more likely to cause problems than firewall rules as the controllers and APs are more frequently inside the firewall. Check the vendor literature for the ports (both UDP and TCP) used for various communications and ensure that the appropriate ACLs are configured.

VLAN Management

Virtual LANs (VLANs) are used to segregate network traffic into separate broadcast domains even when they traffic originates in the same physical switch or switch aggregate. Additionally, VLANs, through tunnels can span multiple physical subnets. In WLANs, VLANs are used to separate wireless traffic from wired traffic, and they are used to separate one wireless network from another, such as separating an enterprise WLAN from a guest WLAN using the same APs.

The VLAN can be configured in relation to the SSID within the AP configuration profile. It can also be configured through information provided by the RADIUS server at the time of authentication. Additionally, VLANs can simply be assigned based on the switchport configuration to which the AP is connected. That is, with an autonomous or cloud-managed AP, the switchport can determine the VLAN on which all traffic coming from the AP is placed.

RADIUS and LDAP

RADIUS is used for authentication, and LDAP is a protocol used to access a directory service. A directory service contains network objects like user accounts and groups. In many deployments, RADIUS servers are used for authentication, but they often depend on large-scale directories for account information such as Microsoft Active Directory. LDAP is used to make the connection from the RADIUS server to the Active Directory Domain Controller.

From a design perspective, you can recommend the use of a RADIUS server that fits well with the existing network solutions. For example, in a Microsoft Active Directory environment, using the Microsoft RADIUS server makes sense. In a Linux/Unix environment, FreeRADIUS or a commercial RADIUS product may be recommended, or the organization may choose to use a RADIUS server from the network infrastructure provider like Cisco.

Public Key Infrastructure (PKI)

Because many EAP types rely on certificates and some depend on client certificates a PKI may be required. A PKI is comprised of certificate authorities (CAs) that either authorize other CAs to distribute certificates or generate and distribute (issue) certificates themselves. As a WLAN designer, you will have to determine whether a PKI will be available or not so that you can select the appropriate EAP type. If a constraint is in place requiring an EAP type that must have client certificates, you should be sure to document the requirement of a PKI in the design plans, though the details of the PKI can be left up to the individuals who will implement that project.

RBAC Implementation

Role-Based Access Control (RBAC) is another security solution of worth. This solution allows for the implementation of authorization based on roles (or groups). Rather than manually authorizing each user, the users can be assigned roles and then perform any function the roles allow.

> As vendor implementations of RBAC vary greatly and many depend on wired-side solutions for this capability, it is not covered in detail on the CWDP exam. For the exam, be sure to remember that RBAC organizes users into groups or roles and assigns capabilities or authorizes users based on those groups or roles.

BYOD and MDM Solutions

When designing for Bring Your Own Device (BYOD), Mobile Device Management (MDM) becomes an important factor. MDM is used to onboard, manage, and decommission mobile devices (both organization-owned devices and BYOD units). An MDM solution may be cloud-based (sometimes called Software as a Service (SaaS) or on-premise. Both have pros and cons.

The cloud-based solution can be used without requiring the installation of software or hardware within your organization. It will include robust features from the start, and enhanced features may be added by simply expanding your subscription. However, the downside of cloud-based MDM is that the Internet connection must be working to onboard new devices and to manage the system.

On-premises MDM will work even if the Internet connection goes down. It is likely to have the same capabilities as a cloud-based service but may require more time and effort to add features later.

MDM solutions may integrate with Network Access Control (NAC) solutions to perform onboarding with security. For example, when a new device connects that should have antivirus but does not, it can be disallowed network access and placed into a quarantine network until it can be repaired.

As a WLAN designer, it is important to select an MDM solution that meets the needs of the organization. Consider the following questions:

- Must the MDM solution integrate with existing directory services?
- Is self-service onboarding required?
- What devices are supported and what operating systems and versions are used?
- Will containerization be required?
- Will features like remote wipe, remote lock, and encrypted storage be required?

Additional questions may be required in some installations, but the important thing is to ask the right questions and match the MDM solution to the organization's needs.

5.5: Selecting Access Points

When selecting APs, it is important to choose the right AP for the installation. You may be constrained to a specific vendor by the RFP or Project Charter, but vendors often have several models, and the appropriate model should be acquired and documented in the design. This section will provide information to help with this process.

Selecting APs

The first step in selecting an AP is determining the location of mounting. If the AP is mounted indoors or in a weather-protected space, an indoor AP should be purchased. If the AP is mounted in a weather-exposed area, an external AP should be purchased. The same basic AP in an external model is usually more expensive because costly materials are used to ruggedize the AP. Indoor APs usually use plastic covers with metal backing, and outdoor APs often use either entire metal encasements with non-reflective coating and external antennas, or they use a more rugged plastic cover with internal antennas.

The last statement reveals another important decision about APs. Should you use internal or external antennas? In general, indoor implementations, internal antennas with omni-optimized patterns are sufficient. In specific indoor scenarios, directional antennas may be used. For example, you might use a directional antenna mounted on the wall and aimed inward or an antenna aimed down a hallway in a warehouse. With these kinds of exceptions considered, most other indoor APs use internal antennas.

Outdoor APs commonly use external antennas. They will either use high-gain omni antennas or directional antennas. Just as APs that are weather-exposed must be designed with rugged enclosures, antennas intended for outdoor use should also be designed such that they can endure cold, heat, rain, and other weather factors.

In addition to choosing indoor or outdoor APs and internal or external antennas, selecting the right number of APs and the APs with the right antenna/radio options is important.

For example, an AP with one antenna using SISO can perform at a given level. But an AP with two antennas, using the same PHY and MIMO, can perform significantly better. This is not just true about data rates, but also about other features like Maximal Ratio Combining (MRC) and Transmit Beamforming (TxBF), which were introduced in your CWNA studies.

Using logarithmic math, you can calculate the maximum theoretical gains of signal reception and transmission techniques like MRC and TxBF so as to quantify their potential benefits. By adding a second antenna, the maximum theoretical gain is 3 dB (log2 x 10). By adding a third antenna, the maximum theoretical gain is 5.77 dB (log3 x10) or roughly 5 dB, and by adding a fourth antenna, the maximum theoretical gain is 6 dB (log4 x 10). In reality, MRC gains are much closer to the theoretical maximum than are TxBF gains. For that reason, MRC is generally a more useful feature. With explicit feedback standardized in 802.11ac and well-supported by infrastructure and client vendors, the real-world gains may improve as more and more networks use 802.11ac APs. Also note that some vendors advertise higher gains because the comparative reference (remember, dB is a reference against something else) is slightly misleading.

So how do you determine the number of APs you require and the number of radio chains that should be in the APs? You can use a freely available tool that can be downloaded from www.revolutionwifi.net/capacity-planner[83]. This tool, currently in the form of an Excel spreadsheet, can be used to perform predictive WLAN design, scope a project, or even create a BoM based on the number of APs required.

The first items you configure in the Revolution Wi-Fi Capacity Planner is the Network Infrastructure options shown in Figure 5.11.

Next, you will configure Client Devices and application profiles shown in Figure 5.12.

Finally, you will configure Capacity Distribution and calculate results as shown in Figure 5.13. Network details, clients, applications, and more can be modified on additional tabs. The spreadsheets come with a detailed 40+ page document explaining how to use it, and this free resource provides great value to the WLAN design industry[84].

---

[83] If a resource is no longer available on the named site, consider using archive.org to locate a previous version of the site. This is true for the revolutionwifi.net site and the URL is maintained in the text so that you can locate it via archive.org.

[84] Capacity planners such as this are an exceptional addition to your toolset; however, they should not replace proper design planning. While these tools are useful, they do not always factor in the considerations of every environment, such as non-Wi-Fi interferers and full CCI issues in a channel. Your design must factor in all such issues.

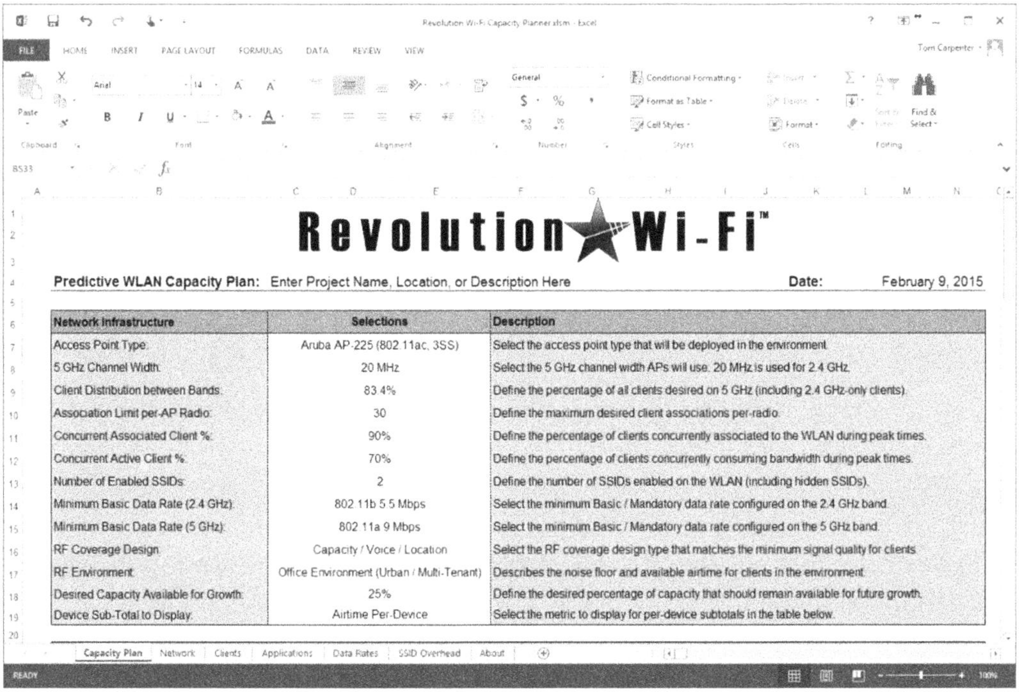

Revolution Wi-Fi™

Predictive WLAN Capacity Plan: Enter Project Name, Location, or Description Here Date: February 9, 2015

Network Infrastructure	Selections	Description
Access Point Type:	Aruba AP-225 (802.11ac, 3SS)	Select the access point type that will be deployed in the environment.
5 GHz Channel Width:	20 MHz	Select the 5 GHz channel width APs will use. 20 MHz is used for 2.4 GHz.
Client Distribution between Bands:	83.4%	Define the percentage of all clients desired on 5 GHz (including 2.4 GHz-only clients).
Association Limit per-AP Radio:	30	Define the maximum desired client associations per-radio.
Concurrent Associated Client %:	90%	Define the percentage of clients concurrently associated to the WLAN during peak times.
Concurrent Active Client %:	70%	Define the percentage of clients concurrently consuming bandwidth during peak times.
Number of Enabled SSIDs:	2	Define the number of SSIDs enabled on the WLAN (including hidden SSIDs).
Minimum Basic Data Rate (2.4 GHz):	802.11b 5.5 Mbps	Select the minimum Basic / Mandatory data rate configured on the 2.4 GHz band.
Minimum Basic Data Rate (5 GHz):	802.11a 9 Mbps	Select the minimum Basic / Mandatory data rate configured on the 5 GHz band.
RF Coverage Design:	Capacity / Voice / Location	Select the RF coverage design type that matches the minimum signal quality for clients.
RF Environment:	Office Environment (Urban / Multi-Tenant)	Describes the noise floor and available airtime for clients in the environment.
Desired Capacity Available for Growth:	25%	Define the desired percentage of capacity that should remain available for future growth.
Device Sub-Total to Display:	Airtime Per-Device	Select the metric to display for per-device subtotals in the table below.

Figure 5.11: Configuring Network Infrastructure Options

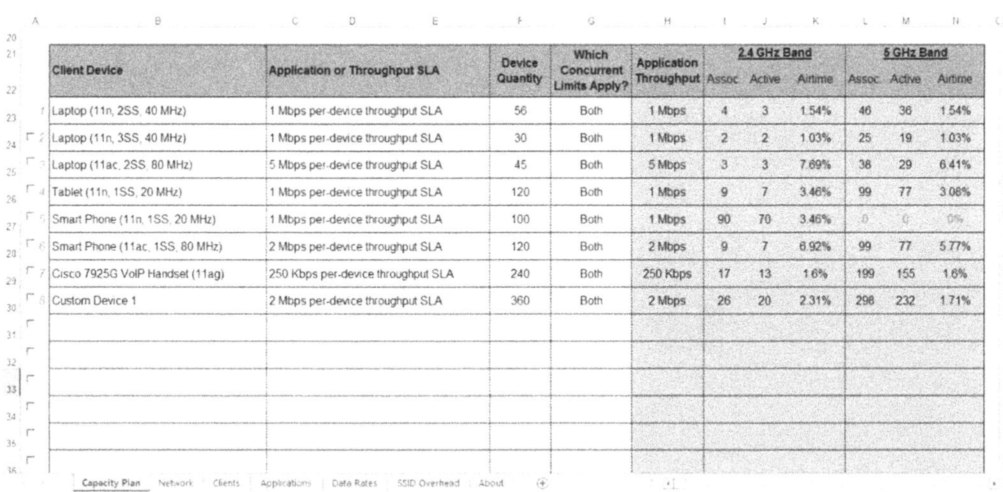

Client Device	Application or Throughput SLA	Device Quantity	Which Concurrent Limits Apply?	Application Throughput	2.4 GHz Band Assoc	2.4 GHz Band Active	2.4 GHz Band Airtime	5 GHz Band Assoc	5 GHz Band Active	5 GHz Band Airtime
Laptop (11n, 2SS, 40 MHz)	1 Mbps per-device throughput SLA	56	Both	1 Mbps	4	3	1.54%	46	36	1.54%
Laptop (11n, 3SS, 40 MHz)	1 Mbps per-device throughput SLA	30	Both	1 Mbps	2	2	1.03%	25	19	1.03%
Laptop (11ac, 2SS, 80 MHz)	5 Mbps per-device throughput SLA	45	Both	5 Mbps	3	3	7.69%	38	29	6.41%
Tablet (11n, 1SS, 20 MHz)	1 Mbps per-device throughput SLA	120	Both	1 Mbps	9	7	3.46%	99	77	3.08%
Smart Phone (11n, 1SS, 20 MHz)	1 Mbps per-device throughput SLA	100	Both	1 Mbps	90	70	3.46%	0	0	0%
Smart Phone (11ac, 1SS, 80 MHz)	2 Mbps per-device throughput SLA	120	Both	2 Mbps	9	7	6.92%	99	77	5.77%
Cisco 7925G VoIP Handset (11ag)	250 Kbps per-device throughput SLA	240	Both	250 Kbps	17	13	1.6%	199	155	1.6%
Custom Device 1	2 Mbps per-device throughput SLA	360	Both	2 Mbps	26	20	2.31%	298	232	1.71%

Figure 5.12: Configuring Client Devices and Application Profiles

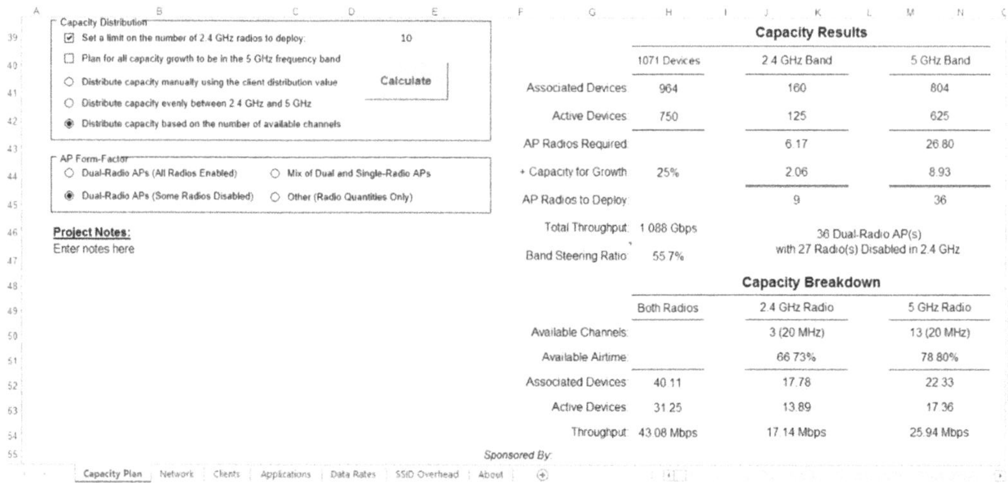

Figure 5.13: Configuring Capacity Distribution and Calculating Results

Choosing a Power Source

When selecting a power source for APs, you have two basic choices: PoE-powered APs and wall outlet-powered APs. PoE-powered APs receive their power through the same Ethernet cable as the network connection. PoE requirements come in two basic models: 802.3af and 802.3at. Both amendments are simply part of the 802.3 Ethernet standard at this point. The original difference between them was that 802.3af provided 15.4 watts of power at the power source (also called power sourcing equipment (PSE) and roughly 12.95 watts at the powered device (PD). 802.3at provided up to 30 watts at the power source and 25.5 watts at the PD. Therefore, APs requiring more power require 802.3at PoE sources, and APs requiring less power require only 802.3af PoE sources.

Taking this further, 802.3bt-2018 extends PoE to support 45-90 W of power at the PSE. With new tri-band APs supporting 2.4 GHz, 5 GHz, and 6 GHz, and some talk of housing a fourth radio for Wi-Fi and a fifth radio for BLE and/or Zigbee and other IoT protocols, the need for more power continues to rise in modern APs.

When choosing PoE PSEs, you must select between endpoint and midspan PSEs. Endpoint PSEs are switches that also provide PoE. Midspan PSEs are usually single port injectors that sit between the switch and the AP; however, multi-port injectors are also available. Figure 5.14 shows these options.

When designing a PoE solution for a WLAN, the power budget is an important consideration. For example, endpoint PoE switches may have 24 ports that can be powered with PoE, but they may only have 300 watts of power to provide. If ten devices are connected requiring 30 watts, the entire power budget is consumed, and the remaining 14 ports must be used as standard Ethernet ports without PoE or simply remain unused. Determining the power budget is not as simple as buying a PoE switch and assuming that all ports will work at the same time regardless of the PoE PD connected. One PD may require 12 watts while another requires 25 watts. Together, the two PDs have consumed 37 watts of your switch PoE budget.

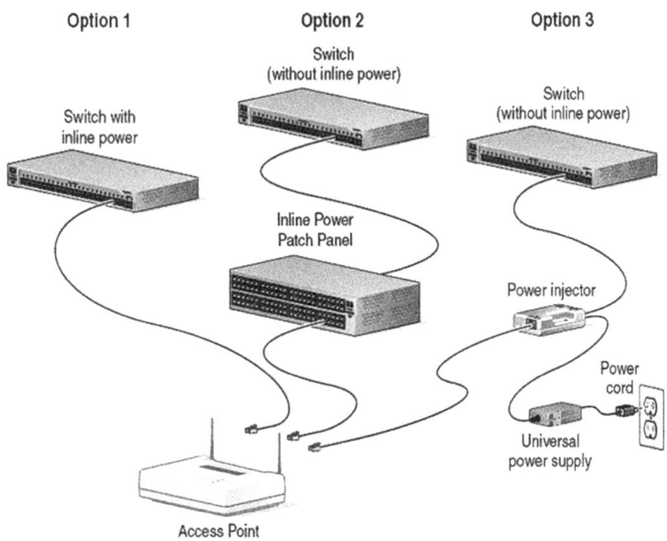

Figure 5.14: PSE Types (Option 1 is endpoint; Option 2 and 3 are midspan)

To calculate the PoE budget required:

1. Determine how many PoE devices will be deployed and what they require in power provisioning.
2. Determine the power provisioned by each switch.
3. Purchase the appropriate number of switches to provide the power.

AP Staging Procedures

After APs are selected, they must be appropriately prepared for installation. AP staging procedures should be clarified and even documented to ensure proper implementation. Staging procedures should answer the following questions at a minimum:

- Will APs be configured while connected to the LAN?
- What settings should be configured as a baseline in all APs?
- Is a particular firmware revision required?
- Will SNMP be enabled on APs during deployment?

Some of these questions may be irrelevant to your chosen WLAN management system, but they should be answered when they apply.

Mounting Solutions

The final aspect of AP selection is choosing the right mounting solution. Bad mounting decisions are frequently seen on Internet blogs ranging from APs duct taped to windows to APs hanging from drop ceilings by their Ethernet connection cables. Most vendors provide mounting options for ceiling, pole, and even desktop options in some cases. Be sure to use the appropriate mounting tools so that the AP will remain fixed. If it moves, such as with the dangling AP hanging from the ceiling, the coverage will change—sometimes significantly. Standard APs should always be fixed to a position, so you have one fewer variable in the link.

5.6: Designing for Client Devices

Chapter 3 addressed most of what you need to know about client devices in relation to design planning. In this section, we will address just a few more details to help you be prepared to design for select client device types.

Laptops

The most important things to know about laptops (or any clients for that matter) are:

- Supported PHYs
- Receive sensitivity (when available)

- Supported bands
- Supported channels
- Supported roaming methods
- Supported security options

With this information, you can ensure the designed WLAN will support the needs of the devices.

Stationary Devices

Stationary devices can be quite expensive, and so they often linger in use on the networks. Figure 5.15 shows a stationary WLAN device used in healthcare. This device monitors blood pressure, pulse rate, etc. Such a device is not replaced every 3-4 years like laptops and desktops. As the WLAN designer, you must understand the PHYs supported by these devices and ensure they are accommodated in the design.

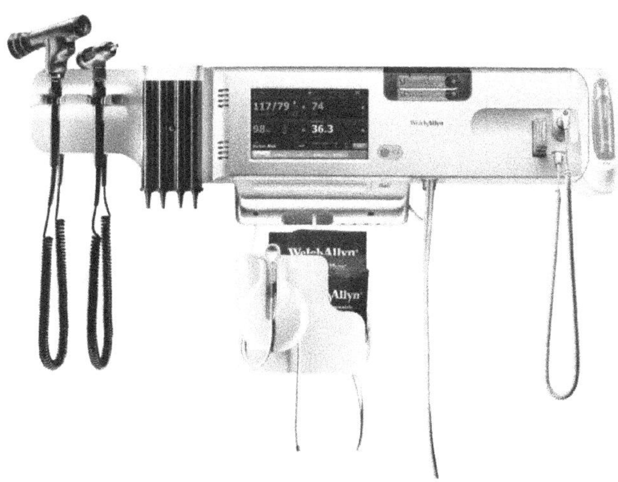

Figure 5.15: Welch Allyn Connex 8500 Series Hospital Wireless Integrated Wall System

5.7: Designing for Applications

Chapter 3 addressed many of the factors you must consider when designing for applications, including:

- Roaming requirements
- Data throughput demands
- Latency requirements
- Voice and video demands
- Location services
- Quality of Service (QoS)

These issues will not be addressed again in this chapter but should already be considered in the design based on the definition of requirement areas in the network. Additionally, security solutions were defined in Chapter 3 for requirement areas.

5.8: Band Steering

Band steering can be an effective tool in all WLANs, but it is particularly beneficial in high-density networks. It works by simply ignoring 2.4 GHz requests and responding to only 5 GHz requests once the 2.4 GHz maximum connection limit is reached. Alternatively, it will always delay response in the 2.4 GHz band to see if the client attempts to connect to the 5 GHz band. If the client does not connect on the 5 GHz band, the AP may respond to future attempts (probe requests or authentication requests) in the 2.4 GHz band. When wireless clients support dual-band operations, the MAC address is the same for the 2.4 GHz and 5 GHz radios. Therefore, an AP that sees the same MAC address on its 2.4 GHz and 5 GHz radios knows that it is the same client. From there, the AP can push the client to the 5 GHz band where many more channels are available, and higher densities can be achieved.

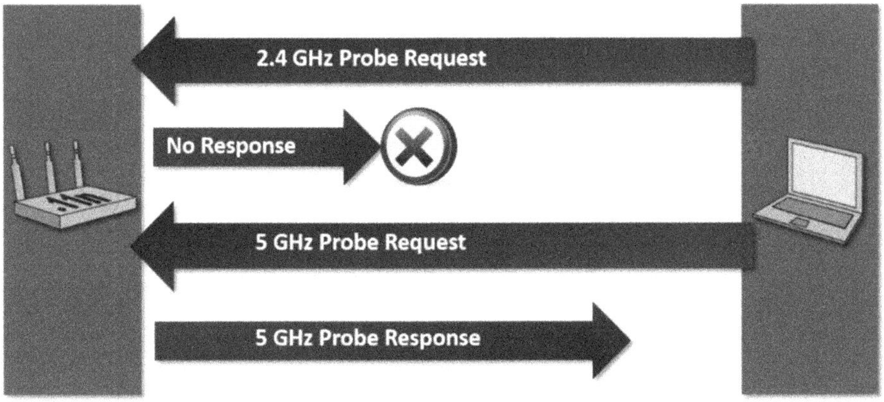

Figure 5.16: Band Steering

5.9: Using WLAN Design Software

Now that we've addressed some of the factors to consider before entering the WLAN design tool of your choice, we can begin to look at design software and how it operates. Throughout this section, we will see four independent tools (non-AP vendor-specific) that can be used to design a WLAN:

- TamoSoft TamoGraph Site Survey
- Ekahau Site Survey Pro
- AirMagnet Survey Pro
- iBwave Wi-Fi

In addition to these independent tools, many vendors provide integrated design capabilities in their controllers, management systems, or cloud. Additionally, some provide software for download. However, most such solutions lack the power of the more advanced independent WLAN design tools.

Before we look at design tools, however, we should evaluate the use of a spectrum analyzer in the design information gathering process.

Spectrum Analysis

Some design tools include spectrum analysis capabilities. Spectrum analysis is not really part of design in most modern design scenarios, which are performed after a proper site survey, but it does prove useful in those design-as-you-survey scenarios. Even if an AP-on-a-stick type of survey is not performed, a spectrum sweep of the target requirement areas can be beneficial as discussed in previous chapters. In this section, we will explore the use of a spectrum analyzer and its reporting capabilities in this process. AirMagnet Spectrum XT will be used, but other spectrum analyzers could be used as well.

Capturing Spectrum Data

Spectrum captures are essential because the RF space is used by many different devices. Thankfully, modern spectrum analysis tools provide both the capturing of the spectrum data and reporting of it. In this section, you will learn to use a spectrum analysis tool to capture and report on the data gathered.

With the software installed and the spectrum adapter inserted you can launch the spectrum analyzer on a laptop computer and walk through the facility. To use Wi-Fi integration, make sure the proper WLAN adapter is selected, as shown in Figure 5.17. Using Wi-Fi integration will allow the spectrum analyzer to gather Beacon frames and other frames to identify WLAN devices specifically. With some spectrum analyzers, you will need to configure them to record the session, such as AirMagnet Spectrum XT. In this application, simply click File > Record to do so.

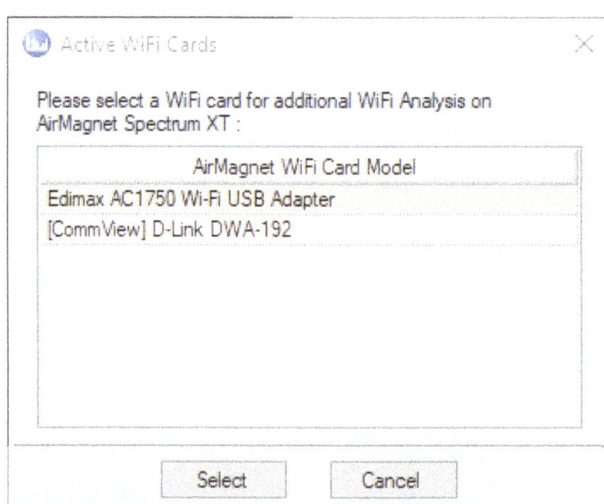

Figure 5.17: Selecting the WLAN Adapter

While walking through the facility, should you notice an interferer or interesting RF signature, you can add notes to the recording. In Spectrum XT, choose the Add Note button on the toolbar. Figure 5.18 shows this interface.

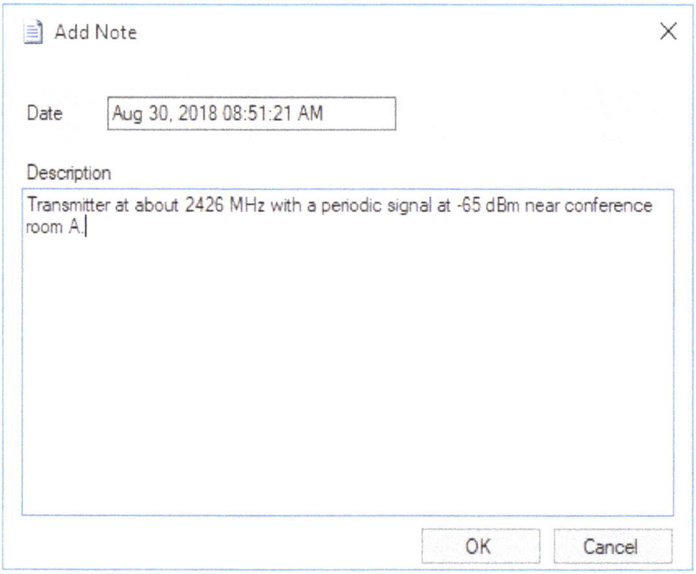

Figure 5.18: Adding Notes to the Spectrum Capture

If the spectrum analyzer does not sweep all the Wi-Fi bands (2.4 GHz and all 5 GHz bands) concurrently, be sure to switch between bands at each paused location to record the information. When interferers are detected, you may capture a screenshot of the Real-Time FFT and other views to document it as shown in Figure 5.19.

The software may offer a data export function for interferers. Spectrum XT allows you to export the interferer data. Select the Spectrum Graphs > Non-Wi-Fi Interferers view in one of the view panes. You will see the center frequency of the interferer as well as the power levels of that interferer and a Wi-Fi channel where it was seen. Once exported, given that the list could be very long, you can open it in Excel and format it for viewing and analysis. This CSV document will be valuable as you design the WLAN. In the Excel document, as your tracking interferers, consider adding a note about the location and, if possessed, add a photo of the interferer if it was physically located. Figure 5.20 shows the Non-Wi-Fi Interferers view in Spectrum XT and Figure 5.21 shows the exported data formatted in Excel.

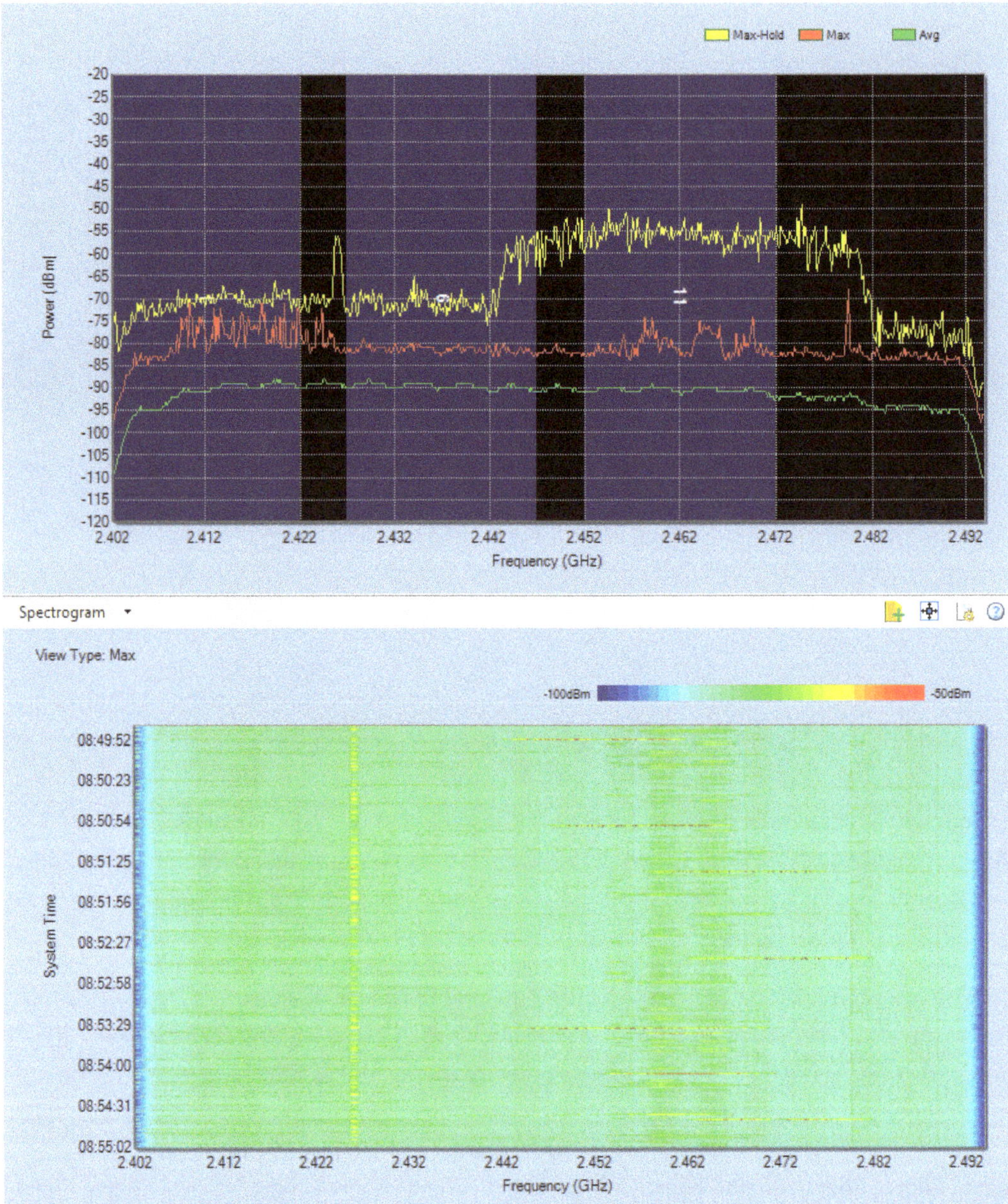

Figure 5.19: Capturing of an interferer around 2426 MHz

Name	Peak Powe...	Avg Power dBm	Last Seen Channel	Affected Channels	Center Frequency GHz	Duty Cycle
Type: Possible Interferer (1)						
Possible Interferer (Id 1)	-69	-71	2	1..14	2.416	100.00

Figure 5.20: Non-Wi-Fi Interferers

	A	B	C	D	E	F	G	H	I	J	K	L	M	N
1	Non-WiFi Interference													
2	View By	Active Interference Summary												
3														
4		Name	Peak Power dBm	Avg Power dBm	Last Seen Channel	Affected Channels	Center Frequency GHz	Duty Cycle	First Seen Time	Last Seen Time	Event Count	Active	Hopping	Custom
5	Type:Possible Interf	1												
6		Possible Interferer (Id 1)	-70	-72	5	1..14	2.431	100	8/30/2018 8:46:43 AM	8/30/2018 8:58:12 AM	58	True	False	False

Figure 5.21: Exported Interferers Data

During your walkthrough with the spectrum analyzer, you can also note existing Wi-Fi networks and signal strengths in use. Instead of just looking at spectrum data, you can use views like the Wi-Fi Devices view shown in Figure 5.22. This data can also be exported and used in Excel or another analysis or reporting tool as shown in Figure 5.23. Once again, the software does not know where you are in the facility or at the location, so you must document notes about the location where the information was gathered as you export, or screen capture this data. The data is important, but the location source of the data is most important when you begin designing around existing Wi-Fi or non-Wi-Fi interference.

Another useful view is Channel Occupancy as shown in Figure 5.24. This view shows MAC addresses (or names, if you've entered or gathered them) and indicates the signal strength. Note, in Figure 5.24, that we have moved channels 1, 6, and 11 to the left in the view for easier observation of those channels in 2.4 GHz.

We will observe one final view, and that is Channel Utilization. This is a very useful view in post-validation but can also be valuable in Wi-Fi-congested areas during pre-design surveys. Figure 5.25 shows the view. Note that it reports the channel utilization while revealing the data rates used during utilization. The report on data rates used is very useful in post-validation. In Figure 5.25, you can see that utilization is actually very low on these channels.

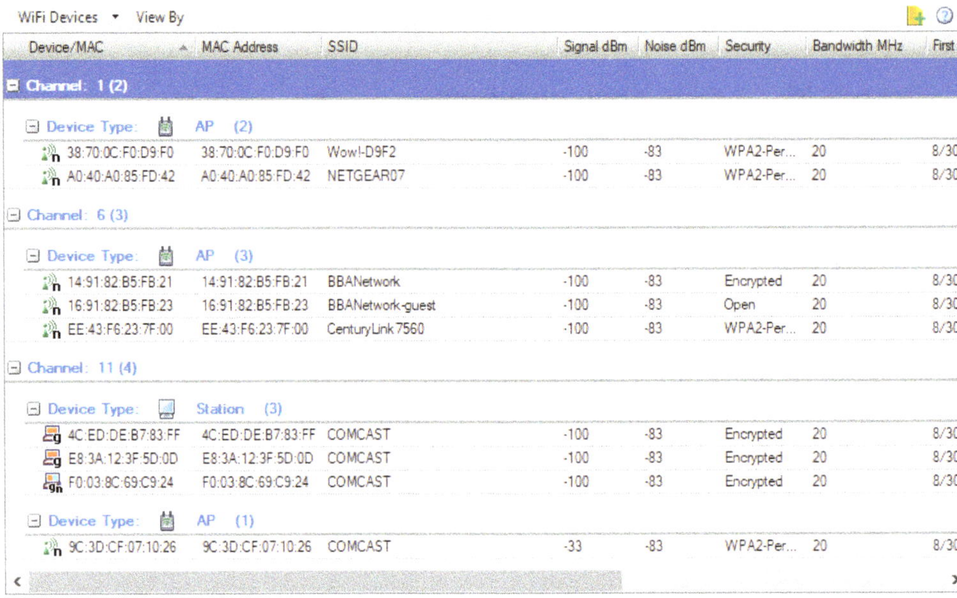

Device/MAC	MAC Address	SSID	Signal dBm	Noise dBm	Security	Bandwidth MHz	First F
Channel: 1 (2)							
Device Type: AP (2)							
38:70:0C:F0:D9:F0	38:70:0C:F0:D9:F0	Wow!-D9F2	-100	-83	WPA2-Per...	20	8/30/
A0:40:A0:85:FD:42	A0:40:A0:85:FD:42	NETGEAR07	-100	-83	WPA2-Per...	20	8/30/
Channel: 6 (3)							
Device Type: AP (3)							
14:91:82:B5:FB:21	14:91:82:B5:FB:21	BBANetwork	-100	-83	Encrypted	20	8/30/
16:91:82:B5:FB:23	16:91:82:B5:FB:23	BBANetwork-guest	-100	-83	Open	20	8/30/
EE:43:F6:23:7F:00	EE:43:F6:23:7F:00	CenturyLink7560	-100	-83	WPA2-Per...	20	8/30/
Channel: 11 (4)							
Device Type: Station (3)							
4C:ED:DE:B7:83:FF	4C:ED:DE:B7:83:FF	COMCAST	-100	-83	Encrypted	20	8/30/
E8:3A:12:3F:5D:0D	E8:3A:12:3F:5D:0D	COMCAST	-100	-83	Encrypted	20	8/30/
F0:03:8C:69:C9:24	F0:03:8C:69:C9:24	COMCAST	-100	-83	Encrypted	20	8/30/
Device Type: AP (1)							
9C:3D:CF:07:10:26	9C:3D:CF:07:10:26	COMCAST	-33	-83	WPA2-Per...	20	8/30/

Figure 5.22: Wi-Fi Devices View

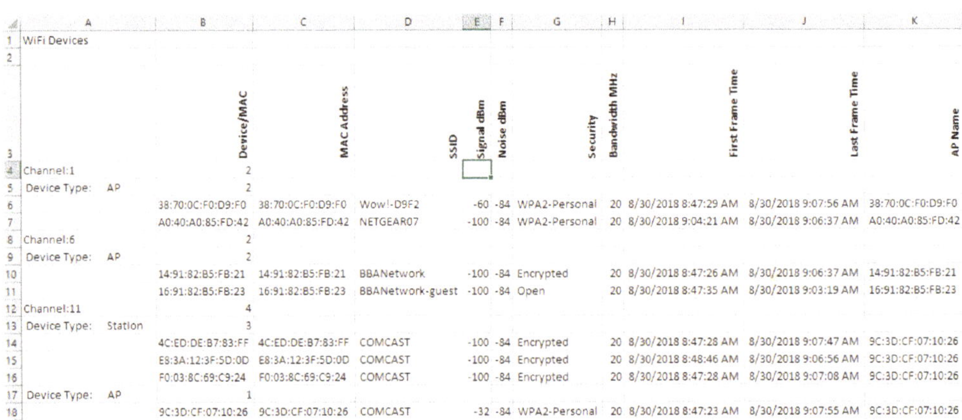

	A	B	C	D	E	F	G	H	I	J	K
1	WiFi Devices										
2											
3		Device/MAC	MAC Address	SSID	Signal dBm	Noise dBm	Security	Bandwidth MHz	First Frame Time	Last Frame Time	AP Name
4	Channel:1	2									
5	Device Type: AP	2									
6		38:70:0C:F0:D9:F0	38:70:0C:F0:D9:F0	Wow!-D9F2	-60	-84	WPA2-Personal	20	8/30/2018 8:47:29 AM	8/30/2018 9:07:56 AM	38:70:0C:F0:D9:F0
7		A0:40:A0:85:FD:42	A0:40:A0:85:FD:42	NETGEAR07	-100	-84	WPA2-Personal	20	8/30/2018 9:04:21 AM	8/30/2018 9:06:37 AM	A0:40:A0:85:FD:42
8	Channel:6	2									
9	Device Type: AP	2									
10		14:91:82:B5:FB:21	14:91:82:B5:FB:21	BBANetwork	-100	-84	Encrypted	20	8/30/2018 8:47:26 AM	8/30/2018 9:06:37 AM	14:91:82:B5:FB:21
11		16:91:82:B5:FB:23	16:91:82:B5:FB:23	BBANetwork-guest	-100	-84	Open	20	8/30/2018 8:47:35 AM	8/30/2018 9:03:19 AM	16:91:82:B5:FB:23
12	Channel:11	4									
13	Device Type: Station	3									
14		4C:ED:DE:B7:83:FF	4C:ED:DE:B7:83:FF	COMCAST	-100	-84	Encrypted	20	8/30/2018 8:47:28 AM	8/30/2018 9:07:47 AM	9C:3D:CF:07:10:26
15		E8:3A:12:3F:5D:0D	E8:3A:12:3F:5D:0D	COMCAST	-100	-84	Encrypted	20	8/30/2018 8:48:46 AM	8/30/2018 9:06:56 AM	9C:3D:CF:07:10:26
16		F0:03:8C:69:C9:24	F0:03:8C:69:C9:24	COMCAST	-100	-84	Encrypted	20	8/30/2018 8:47:28 AM	8/30/2018 9:07:08 AM	9C:3D:CF:07:10:26
17	Device Type: AP	1									
18		9C:3D:CF:07:10:26	9C:3D:CF:07:10:26	COMCAST	-32	-84	WPA2-Personal	20	8/30/2018 8:47:23 AM	8/30/2018 9:07:55 AM	9C:3D:CF:07:10:26

Figure 5.23: Wi-Fi Devices Exported to Excel

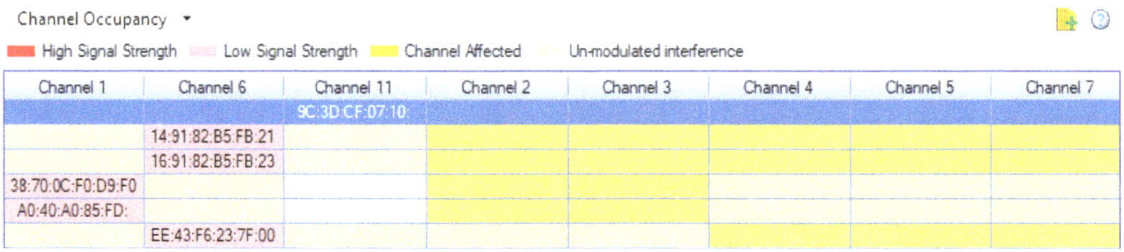

Figure 5.24: Channel Occupancy

If you plan to use the Spectrum XT reporting engine, instead of exporting and screen capturing along the way, you can click the Add to Report button for any view. This button, for example, is shown in Figure 5.25 in the upper right corner left of the Help button. When you click the button, you will be presented with a screen like the one in Figure 5.26. Be sure to note the location, as stated previously, to provide documentation value during the design process.

You can repeat the Add to Report function dozens or hundreds of times as required. After walking through the area that is the target of the WLAN design, you can generate a report that assists you in the design process. Spectrum analysis is your friend in the pre-design survey.

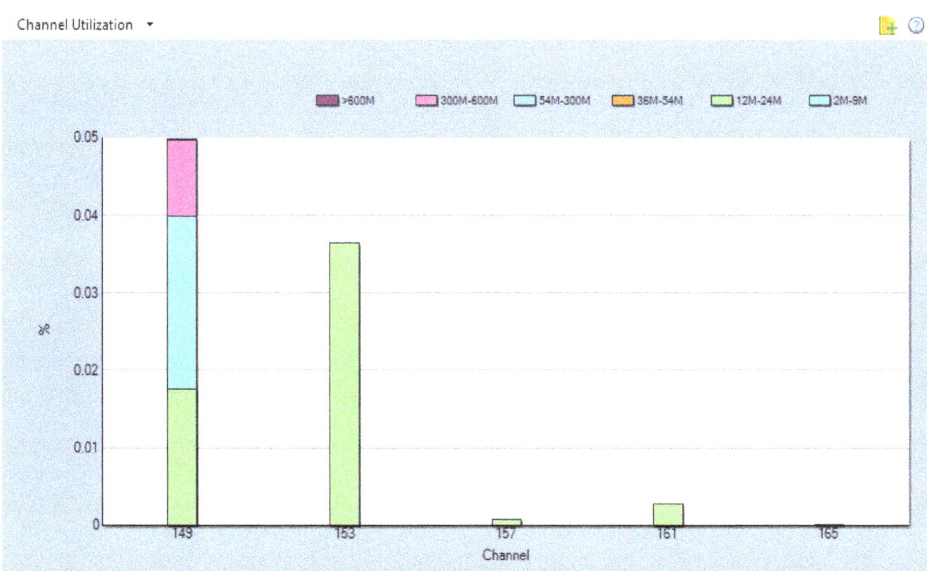

Figure 5.25: Channel Utilization

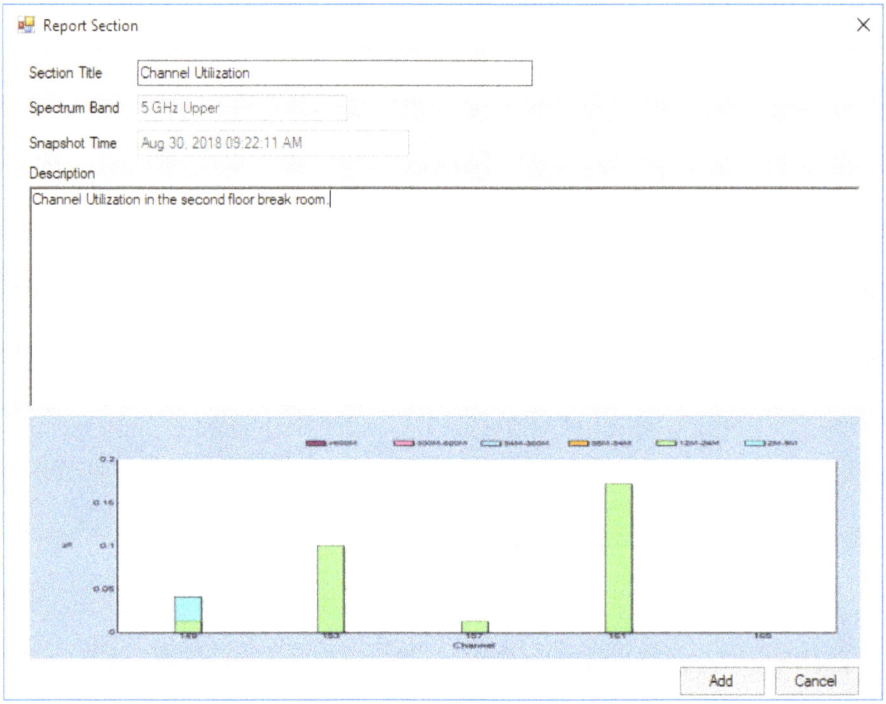

Figure 5.26: Adding a Section to the Report

Reporting on the Capture

Most spectrum analysis tools allow you to export data or in some way gather information to generate a report. The good news is that Spectrum XT allows you to generate reports and export them to Crystal Reports, PDF, and other formats. If you used the Add to Report feature while performing the spectrum walkthrough and added notes about locations, you could generate a report like the one in Figure 5.27. Notice the noted included in the report document.

We've addressed spectrum capture here to ensure you understand the basic steps you will take during a spectrum sweep walkthrough of a target design location. The basic process is:

1. Move to a location.
2. Capture spectrum data in each band.
3. Export data, gather screen captures or use the reporting features. Be sure to note locations.

4. If interferers are discovered, locate them whenever possible and document your findings.
5. Repeat.

You may perform this process five times in some design projects and five hundred times in others.

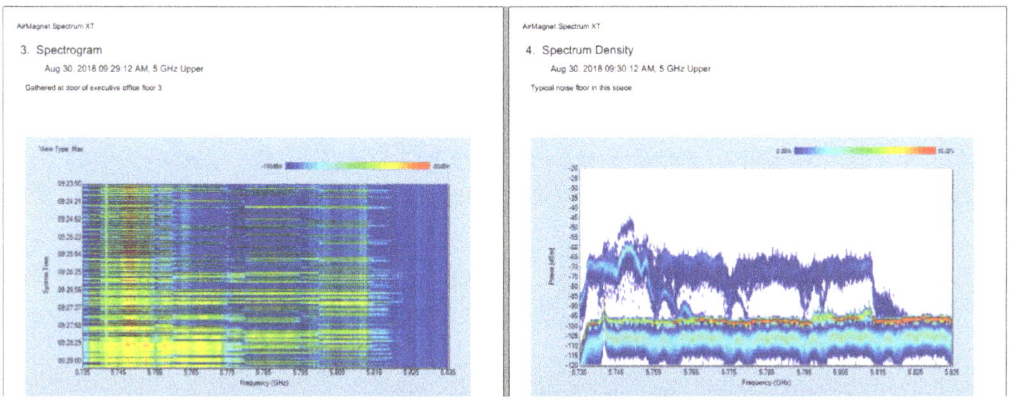

Figure 5.27: Report Generated from the Spectrum Capture

Common Tools

With everything you've gathered up to this point, you are ready to design the WLAN. You have gathered documents, including floor plans, that will be used in this process. Hopefully, you've also performed or had someone perform a spectrum analysis walkthrough. This data will be used to make design decisions using your WLAN design software of choice. You have also considered the important factors in the design.

In this section, we will cover common features of WLAN design software and then briefly explore the features of four popular, independent WLAN design tools.

Common Features

WLAN design software has a minimum set of features to allow you to design a WLAN and see (through simulation) how the WLAN will "look" in the target environment. You see how it "looks" with different views of the design WLAN, such as signal strength, expected data rates, number of APs, and more. Most WLAN design applications will have the following basic features:

- Floor plan import: Allows you to import either a flat floorplan, which will require drawing building materials or a CAD drawing that may already have the materials defined.
- Calibration: Used to set the dimensions of the floor plan. This is very important. It is used in the propagation algorithms to simulate coverage.
- Building material definition: Used to define walls, floors, and other materials based on attenuation factors. Also, very important for the propagation algorithms.
- Passive and active surveys: Allows you to gather data either passively (just gathering signal strength) or actively (sending/receiving data to/from APs).
- AP placement: Placing APs on floor plans to accomplish the desired design.
- AP configuration: Configuring AP parameters, such as output power, antennas in use, channel, channel width, and other parameters.
- Coverage zone (requirement area) definition: Allows you to draw areas that can have unique requirements, what we call requirement areas. Some tools call these coverage zones or capacity zones. Figure 5.28 shows the specifications for a requirement area defined in Ekahau Site Survey Pro.
- Simulation: This is the primary feature that brings the most value in these tools. It simulates propagation of RF signals based on physics and the materials you've defined in the space as well as AP placement.

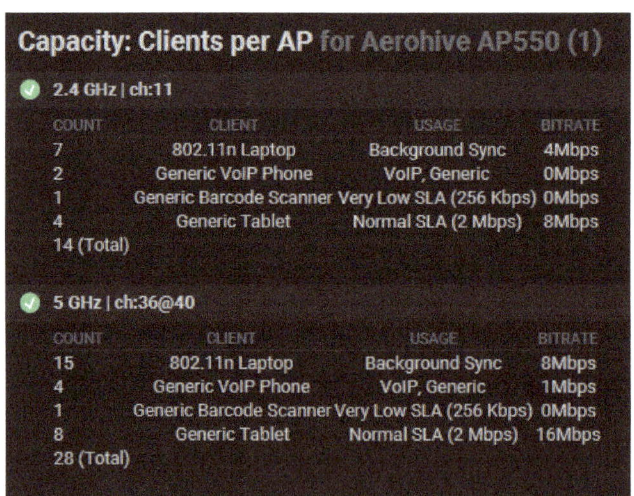
Capacity: Clients per AP for Aerohive AP550 (1)

2.4 GHz | ch:11

COUNT	CLIENT	USAGE	BITRATE
7	802.11n Laptop	Background Sync	4Mbps
2	Generic VoIP Phone	VoIP, Generic	0Mbps
1	Generic Barcode Scanner	Very Low SLA (256 Kbps)	0Mbps
4	Generic Tablet	Normal SLA (2 Mbps)	8Mbps
14 (Total)			

5 GHz | ch:36@40

COUNT	CLIENT	USAGE	BITRATE
15	802.11n Laptop	Background Sync	8Mbps
4	Generic VoIP Phone	VoIP, Generic	1Mbps
1	Generic Barcode Scanner	Very Low SLA (256 Kbps)	0Mbps
8	Generic Tablet	Normal SLA (2 Mbps)	16Mbps
28 (Total)			

Figure 5.28: Requirement Area Specifications in Ekahau Site Survey Pro

TamoGraph Site Survey

TamoSoft makes TamoGraph Site Survey. Like their protocol analysis software (CommView for WiFi), TamoGraph has one of the simpler interfaces in the site survey and design toolset. However, for most WLAN projects, it provides what is needed to implement basic designs. According to TamoSoft, TamoGraph Site Survey (see Figure 5.29) supports the following features:

- Simple and fast data collection
- Passive, active, and predictive surveys
- Spectrum analysis (uses Metageek wiSpy DBx)
- Comprehensive WLAN analysis with easy-to-understand visualization of signal level, interference, access point coverage areas, data rates, network issues, etc.
- Automatic access point location
- Detailed information about every access point: channel, maximum data rate, vendor, encryption type, etc.
- Full support of 802.11ac, as well as 802.11a, 802.11b, 802.11g, and 802.11n networks
- GPS-assisted outdoor surveys
- Dozens of antenna patterns allow you to model APs by major WLAN vendors: Cisco, Aruba, Ruckus, Aerohive, Ubiquiti, Meraki, and many others
- Detailed reporting in PDF and HTML formats
- Attractive, competitive pricing

> A demonstration of performing a basic design with TamoGraph is available for viewing as a value-add to this book. You will find it at cwnp.link/TamoGraph.

Ekahau Site Survey Pro

Ekahau Site Survey Pro is among the leading WLAN design tools in popularity. It has advanced features allowing for complex designs. Additionally, in 2017, Ekahau released the Ekahau Sidekick, which is an all-in-one Wi-Fi adapter and spectrum analysis adapter for use in surveys with Ekahau Site Survey Pro. It offers very high-resolution spectrum analysis but lacks the advanced features of a tool like Spectrum XT with its many views because it is designed to work with site survey software and not as a general-purpose spectrum analyzer.

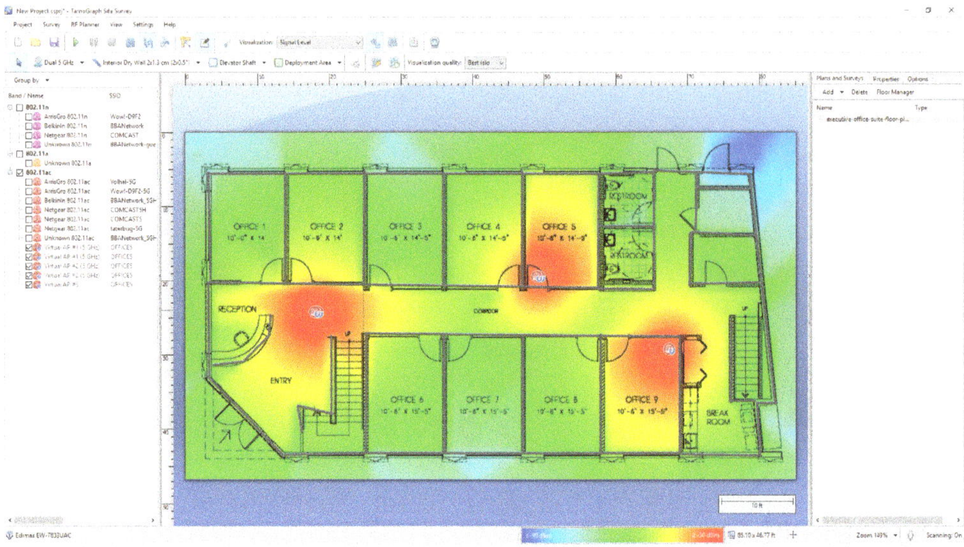

Figure 5.29: TamoGraph Site Survey

According to Ekahau, Ekahau Site Survey Pro (see Figure 5.30) supports the following features:

- Industry's fastest 3D prediction engine
- Multi-floor and multi-building planning
- Wall materials, wall height
- Convert CAD drawings into walls
- Antenna direction, uptilt, downtilt
- Outdoor planning for urban areas
- Automatic AP placement & channel assignment based on coverage and capacity requirements
- Manual one-click AP placement option
- Crystal-clear heatmaps
- Hundreds of Wi-Fi access points and antennas
- Expansion planning combines measurements & simulation
- Fully supports 802.11ac, 802.11n, and 802.11abg
- Simultaneous 2.4 and 5GHz surveys
- Simultaneous passive & active & spectrum surveys

- Throughput (iPerf) support
- Multi-adapter support
- GPS enabled outdoor surveys
- Touch screen optimized surveying
- Locates all APs
- Fills in the gaps to places you didn't visit
- Automatic channel optimization
- Supports 802.11a/b/g/n/ac
- Easy-to-read, crystal-clear heatmaps
- Signal strength coverage
- Signal-to-noise ratio
- Data rate
- Throughput
- Packet loss
- Jitter
- Round-trip times
- Maximum Channel Bandwidth
- Number of access points (overlap)
- Co-channel / adjacent channel interference
- Network capacity problems
- Capacity issues, such as excessive VoIP calls per radio, AP overload, ...
- Spectrum Channel Power
- Spectrum Utilization
- 802.11ac/n channel width
- Built-in requirements for VoIP, RTLS, video...
- Customizable requirements
- Customizable requirement profiles
- Locate all access points
- Analyze for all APs, per SSID, channel or AP
- Visualize data for all or selected surveys
- Visualize measured/simulated/both
- Simulate how the network will perform using different Wi-Fi devices
- Fully supports 802.11a/b/g/n/ac
- Survey Inspector: detailed, combined Wi-Fi and spectrum analysis
- And many more

A demonstration of performing a basic design with Ekahau Site Survey Pro is available for viewing as a value-add to this book. You will find it at cwnp.link/Ekahau. This video is in place of a lengthy step-by-step procedure in the book.

Figure 5.30: Ekahau Site Survey Pro

AirMagnet Survey Pro

AirMagnet Survey Pro is one of the oldest site survey and design tools in the industry. It supports spectrum integration like Ekahau Site Survey Pro and TamoGraph Site Survey, if you have a Spectrum XT adapter. According to NETSCOUT, AirMagnet Survey Pro (see Figure 5.31) supports the following features:

- Design and deploy the most accurate indoor & outdoor wireless LAN network (802.11n/a/b/g/ac) correctly the first time and prevent costly rework & IT complaints
- Collect real-world data by performing unique true end-user experience (wireless LAN throughput, data rates, retries, losses) measurements
- Minimize expensive wireless 802.11n/a/b/g/ac LAN performance impact due to RF interference sources by performing simultaneous spectrum analysis walkthroughs

- Single-click wireless LAN network readiness verification for VoWiFi and Location Services
- Confidently certify the wireless network for any design/application requirements using customer-ready pass/fail assessment reports
- Customize reports enabling efficient hand over of results based on WLAN design/application requirements

A demonstration of performing a basic design with AirMagnet Survey Pro is available for viewing as a value-add to this book. You will find it at cwnp.link/AirMagnetSP. This video is in place of a lengthy step-by-step procedure in the book.

Figure 5.31: AirMagnet Survey Pro

iBwave Wi-Fi

The final tool referenced in this chapter is iBwave Wi-Fi. These tools have not been presented in any order of priority or preference. The authors of this book use multiple tools and each author has her or his own preference. CWNP makes no recommendations of specific tools and believes that all of these tools can be used for design, and the designer must select the tool having her or his required features and fitting within the purchasing budget.

iBwave Wi-Fi comes from an extensive background in RF modeling from previous RF design tools created by the company. The modeling in iBwave Wi-Fi is some of the best available in the industry. According to iBwave, iBwave Wi-Fi (see Figure 5.32) supports the following features:

- Active Surveys with an integrated server. Assess a network's performance without using iPerf or another third-party tool. Simply launch the 'Active Survey' server from your desktop and then survey with either the mobile app or PC version.
- A Complete Bill of Materials. With the ability to design your entire network with APs, cabling and network equipment, you can quickly generate an accurate bill of materials for your customer.
- Auto-AP Placement and Auto-Channel Assignment. Use the optional AP placement in either the mobile app or PC version to quickly start your Wi-Fi network design. Freely move, delete or add APs around the floor plan to get the design just right. Quickly assign each AP in your design to a channel with the auto-channel assignment functionality - then update channels if/where needed.
- Smart Antenna Contouring. View live signal strength predictions as you design, so you do not have to run full prediction results multiple times.
- Automatic Cable Routing Alignment. Model cable trays and risers in iBwave Wi-Fi Mobile or iBwave Wi-Fi and then save hours of time with the automatic cable alignment functionality that automatically snaps them into place.
- Cloud-connectivity to easily share files. Easily share files with your team and with your customers by using the free 10GB of storage in the iBwave cloud to save and access files from anywhere, anytime.
- Multi-adapter Passive Survey. Use of several Wi-Fi adapters to collect survey data. Each of them can be configured to scan a certain band, specific channels or groups of channels. Decrease time to collect data when tasks are distributed among the adapters.

A demonstration of performing a basic design with iBwave Wi-Fi is available for viewing as a value-add to this book. You will find it at cwnp.link/iBwaveWi-Fi. This video is in place of a lengthy step-by-step procedure in the book[85].

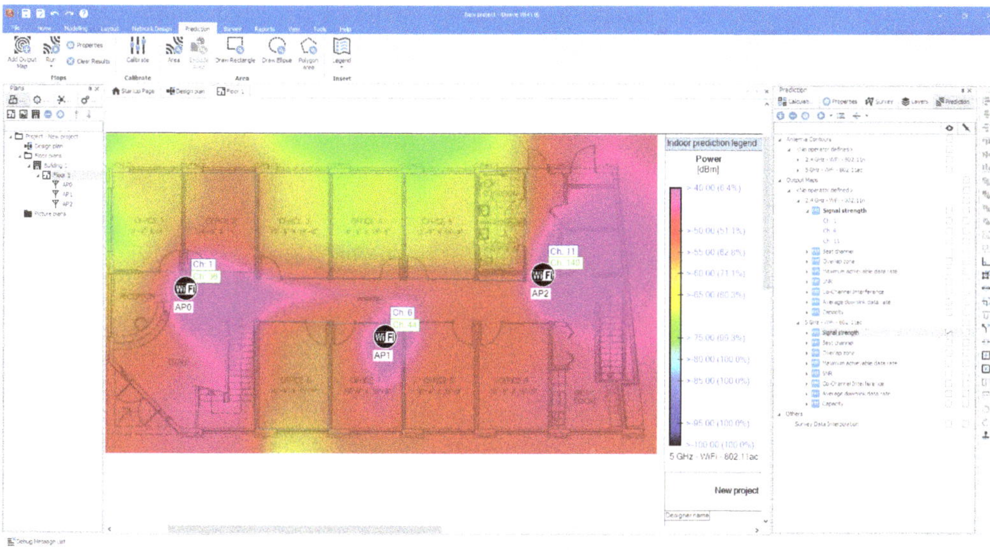

Figure 5.32: iBwave Wi-Fi

Be sure to watch the companion videos referenced throughout the preceding section. One hundred pages of book materials can be covered in 10-15 minutes in videos like these, and they are freely available on CWNPTV YouTube channel.

Best Practices

A few best practices can go a long way toward success when using WLAN design tools. The following guidelines should be followed at a minimum:

- Import and calibrate floor plans: Don't forget to calibrate the floor plan. Use something with known dimensions in the floor plan for the calibration. For

[85] Newer versions of the tools are available today; however, the important basic concepts of using the tools are still the same and these videos provide useful demonstrations today.

example, if a room is known to be 10 feet by 14 feet, the 10-foot width can act as your calibration. Some use doors and assume they are a common width, but doors vary. If you are unsure and not onsite, have someone that is onsite measure the width of a room that is referenced on the floor plan and use that. When you are accurate in your floor plan calibration, you are more likely to have an accurate design.

- Set project parameters: the project parameters define basic information about the project and vary by tool. iBwave Wi-Fi seems to have the most project parameters, and whatever tool you use, you must understand the meaning of the parameters and the impact they will have on your design. Common project parameters include:
    - Project Name
    - Project Description
    - Environment parameters (office space, outdoors, high-density, etc.)
    - Measurement units (feet vs. meters)
    - Channels to scan for site surveys
    - Floor plans
    - Noise level for predictive surveys (design)
    - Clients and their capabilities
    - Measurable requirements (minimum signal level, minimum SNR, the minimum number of APs required, minimum PHY rate). These are shown in Figure 5.33 from TamoGraph Site Survey.
- Select and place APs and antennas manually or using automated placement tools and define configuration parameters: Auto placement tools can be helpful in gaining a quick perspective on what might be needed in a WLAN design; however, the designer must validate the placement (the software doesn't know about constraints that you know about) and ensure proper AP configuration specifications.
- Adjust AP settings to accommodate design requirements: This includes adjusting channels, output power, channel widths, spatial streams, SSIDs, and other parameters that vary by design application.
- Define appropriate requirements areas using software features: As you've learned, all areas within a design target are not the same. Be sure to specify the WLAN requirements for each requirement area.

- Document cabling requirements: This is an area where some tools are more powerful than others. For example, some WLAN design tools will allow you to show the cable runs on the floor plan, so your installers know exactly how it works. Others will simply require that you document it in the physical installation guide documents.

5.10: Documentation

The final step of the design is documentation. If you have to create all of the documentation manually, it will prove to be very time-consuming. This is where good WLAN design software saves the day. With it, you can often generate the four common documents required: Bill of Materials, Site Survey Report, Design Report, and Physical Installation Guide.

Bill of Materials (BoM)

The BoM is the list of hardware and software that must be purchased or used to perform the WLAN installation based on the design. If your chosen software has an exhaustive and updated database of AP models, antennas, and cables, it may be able to generate a details BoM for you. Even if it does not, it is likely to have the capability at least of listing the APs required for the installation as shown in Figure 5.34.

Generally, the BoM you create must be more detailed and include AP models and possibly pricing information.

Site Survey Report

The site survey report will report on the results of an actual site survey. This is different from a design report. The design report shows how you plan to implement the network. The site survey report shows what you found when surveying within the target areas. Site survey reports typically include heatmaps, detected device lists, and other custom information you may want to include.

Figure 5.35 shows the report generation feature in Ekahau Site Survey Pro and Figure 5.36 shows a portion of the generated report.

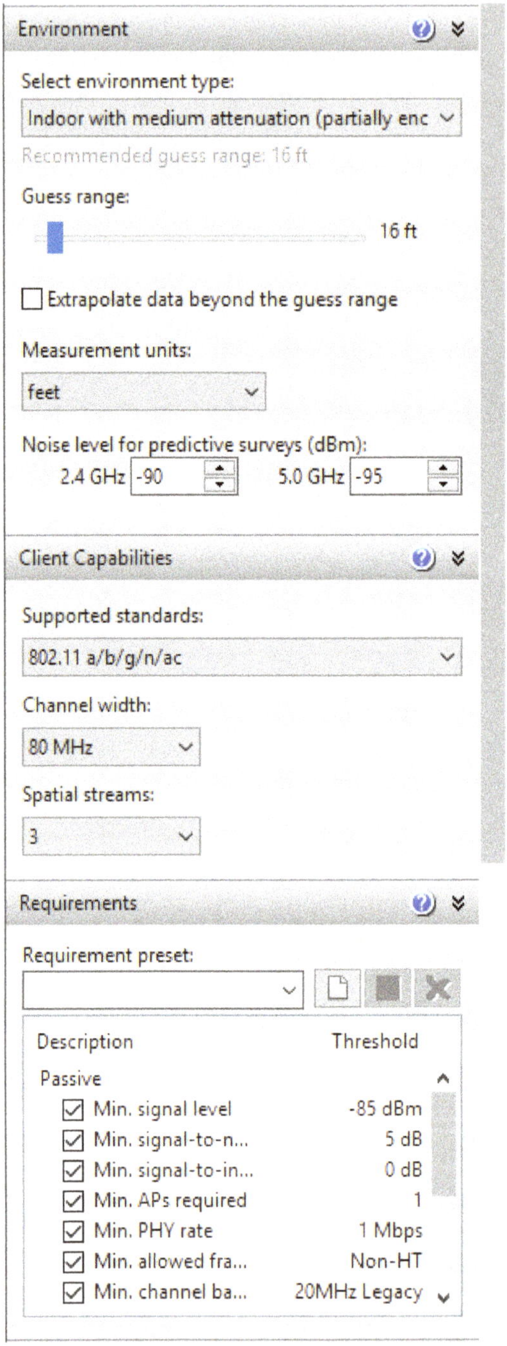

Figure 5.33: Project Parameters in TamoGraph Site Survey

List of APs

Name	SSID	MAC	Vendor	Channel	Max Rate	Encryption	Max Signal
Virtual AP #2 (2.4 GHz)	SSID 2	FF:FF:FF:00:00:01		11	144.4	None	N/A
Virtual AP #3 (5 GHz)	SSID 3	FF:FF:FF:00:00:02		44 (44-48@40)	600.0	None	N/A
Virtual AP #2 (5 GHz)	SSID 2	FF:FF:FF:00:00:01		149 (149-153@40)	600.0	None	N/A
Virtual AP #1 (5 GHz)	SSID 1	FF:FF:FF:00:00:00		36 (36-40@40)	600.0	None	N/A
Virtual AP #1 (2.4 GHz)	SSID 1	FF:FF:FF:00:00:00		1	144.4	None	N/A
Virtual AP #3 (2.4 GHz)	SSID 3	FF:FF:FF:00:00:02		6	144.4	None	N/A
Virtual AP #6	SSID 6	FF:FF:FF:00:00:05		157 (157-161@40)	600.0	None	N/A
Virtual AP #7	SSID 7	FF:FF:FF:00:00:06		44 (44-48@40)	600.0	None	N/A
Virtual AP #5	SSID 5	FF:FF:FF:00:00:04		149 (149-153@40)	600.0	None	N/A
Virtual AP #4 (5 GHz)	SSID 4	FF:FF:FF:00:00:03		157 (157-161@40)	600.0	None	N/A
Virtual AP #4 (2.4 GHz)	SSID 4	FF:FF:FF:00:00:03		1	144.4	None	N/A

Figure 5.34: List of Required APs

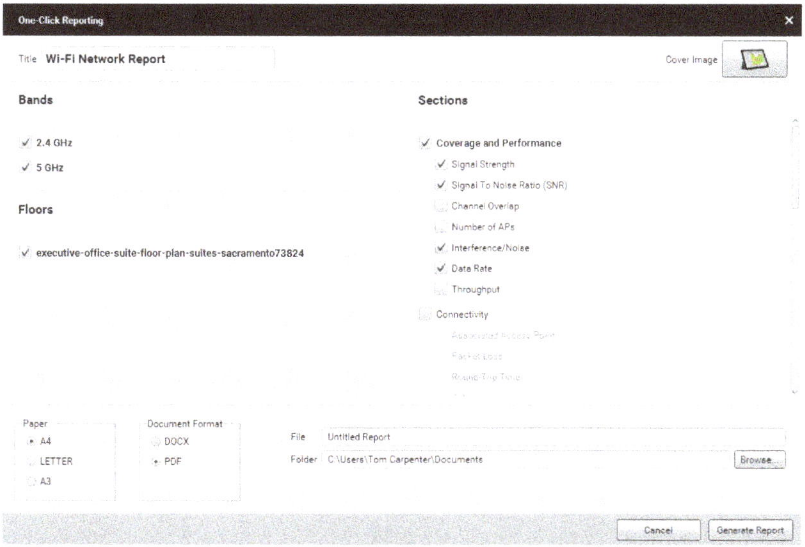

Figure 5.35: One-Click Report Generator (Ekahau)

Some tools also support custom report templates. This allows you to add logos, images, and other information you may want to include in your reports and is particularly useful when generating reports for a customer. You can include legal information in the document that is standard for your contracts or reports.

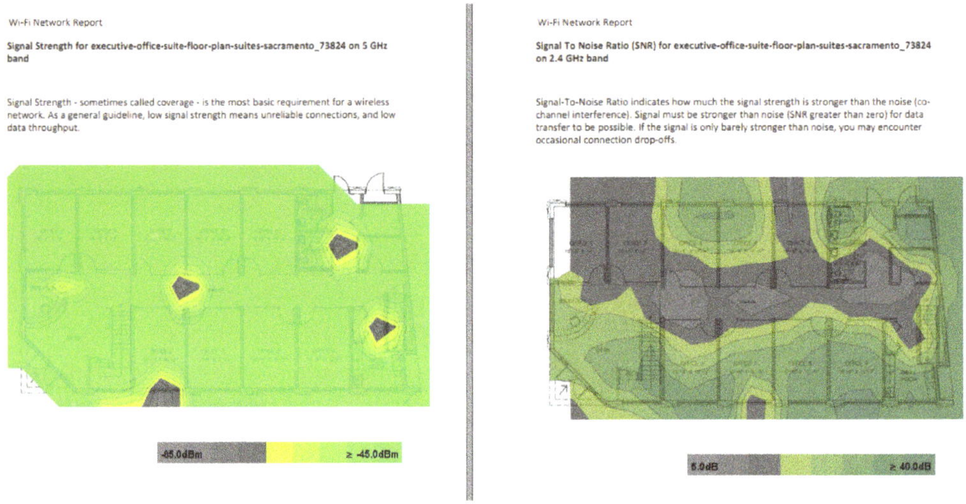

Figure 5.36: Report Generated by Ekahau Site Survey Pro

Design Report

The design report can be generated from the same software that generates the site survey report. Again, the difference is that you are generated the report based on your planned design rather than actual survey data. In the post-validation survey, you will perform a site survey after installation, and generate the survey report again. This latter report should be very similar to your design report, assuming the pre-design survey was effective.

Physical Installation Guide

Finally, the physical installation guide will be a plan for install technicians to use (who may be CWT or CWNA certified) to perform the installation. The installation guide should include the following at a minimum:

- Controller, WNMS, cloud-based configuration recommendations: Indicate locations for installation in the network structure, licensing required for APs, special features requiring licensing, WLAN profile settings, auto-channel selection settings (if used), etc.
- AP mounting locations and methods: Include detailed instructions on the mounting location down to the inch/centimeter. For example, AP1 should be mounted on the ceiling T-rail using a T-rail mounting kit 420 centimeters from the west wall and 860 centimeters from the south wall. We know this can be painful, but if you are not performing the installation, this greatly improves the likelihood of success when your design is implemented.
- AP configuration settings (if not configured by a controller): Include channel, output power, SSIDs, security settings, and any other required parameters. In some cases, you may be asked to provide complete step-by-step instructions. If so, leave nothing out. It's easy to miss steps in your documentation when you're an expert.
- Required cables and PoE parameters: List cables (including category type) and lengths. Provide details on the PoE power that should be provided to each AP so the switch administrator can properly provision the port or so that the install technician can use the appropriate PoE injector.
- Antenna orientation, if using external antennas: The design success will depend on the antennas being properly oriented. If external antennas are used, how should the be mounted or positioned to comply with your design?
- Any tools and equipment needed by the installers (cable tools, pliers, screwdrivers, ladders, tape measure, gloves, etc.): Informing the install technicians of the tools needed is always best. Without this provision, much time can be lost "going back for tools."

5.11 Chapter Summary

In this chapter, you learned to design a typical WLAN. You explored items you must consider and evaluated WLAN pre-design tools (spectrum analyzers) and design tools (site survey software, which includes predictive design capabilities). You also explored the required documentation that should be deliverables from the design phase. In the next chapter, you will learn about WLAN design for special use cases.

5.12: Facts to Remember

- Controller-based architectures offer both centralized and distributed data forwarding.
- Some cloud-based architectures now offer local servers or appliances that provide centralized data forwarding to enable centralized filtering for WLAN traffic.
- Software Defined Radios (SDR) or Software Configurable Radios (SCR) are usually implemented with one fixed 5 GHz radio and one configurable radio that can operate in either 2.4 GHz or 5 GHz.
- CCI is caused by both APs and clients and must be considered in the design.
- The CCI boundary is not the same as the association boundary. Disabling low data rates does not reduce the size of the CCI boundary.
- 40 MHz channels should never be used in 2.4 GHz and, in most deployments, only channels 1, 6, and 11 will be used.
- If deploying dual-5 GHz APs, configure as much separation between the radio channels as possible to avoid interference.
- Lightweight APs can locate controllers using DNS, DHCP, broadcasts, or pre-staged configuration.
- A spectrum walkthrough of a target design area is essential to understanding the RF activity that exists in the space.
- It is essential to calibrate the floor plan before beginning a site survey or design in the software.
- Project parameters should be set accurately in the design software to ensure propagation modeling is accurate.
- Remember to draw the walls (if they cannot be imported) before generated heatmaps.
- Requirement areas should be defined in the design using the features of your selected design tool.

5.13: Review Questions

1. What is a primary benefit of centralized data forwarding?
    a. Reduced latency
    b. Centralized filtering
    c. Increased delay
    d. Enhanced wired performance

2. What is included in the BoM?
    a. Heatmaps
    b. Spectrum captures
    c. Hardware lists
    d. Mounting locations

3. What is the first thing you should do after importing a floor plan into design software?
    a. Calibrate it
    b. Place APs
    c. Draw walls
    d. Render heatmaps

4. What tool may be able to create a BoM for you automatically?
    a. Spectrum analyzer
    b. Protocol analyzer
    c. Design/Site survey software
    d. PoE switch

5. What report includes captured data from a real environment?
    a. Design report
    b. Site survey report
    c. BoM
    d. Physical installation guide

6. When performing a spectrum analysis walkthrough of a target design area, you notice a source of potential interference in an area. What should you do?
    a. Remove yourself from the project; it will be too hard
    b. Document the location in notes
    c. Remove the interfering device
    d. Cancel the entire project

7. Why is it important to define building materials in your WLAN design software?
    a. So the customer will know the cost of erecting the building
    b. So the software can properly model RF propagation
    c. So the software can recommend the best vendor APs
    d. So the software can inform you of walls that must be removed

8. What is not typically defined for requirement areas within WLAN design software even though you may consider documenting it in your design report or physical installation guide?
    a. Types of clients
    b. Number of clients
    c. Desired signal strength
    d. PoE requirements

9. What is the term used for the colorful overlay used in design software to show various metrics related to the WLAN?
    a. Heatmaps
    b. Floor plans
    c. Peer-maps
    d. Reports

10. What is not typically included in project parameters within WLAN design software?
    a. AP locations
    b. Project name
    c. Measurement units
    d. Clients

5.14: Review Answers

1. The correct answer is B. A benefit of centralized data forwarding is WLAN traffic filtering. Because the traffic passes through the controller, the controller can allow or disallow it based on filtering rules.
2. The correct answer is C. The Bill of Materials (BoM) will include the list of hardware and software required to implement the WLAN.
3. The correct answer is A. Before performing any design tasks on an imported floor plan, you should calibrate it, which means you are defining its dimensions.
4. The correct answer is C. WLAN design software may offer a BoM function in the reporting engine. This feature is most useful when the software has a large and frequently updated vendor hardware database.
5. The correct answer is B. The site survey report includes the captured data. The design report may also include it, but it is not the intention of the design report.
6. The correct answer is B. When you locate sources of interference, you should document the location where it was detected in your notes. Additionally, you may need to locate and identify the source.
7. The correct answer is B. If the building materials are not defined, the software is unaware of the attenuation and reflection factors of that material.
8. The correct answer is D. While you may desire to define PoE requirements for your requirement areas, most WLAN design software does not directly link PoE budgets to the requirement, coverage, deployment, or capacity area/zones.
9. The correct answer is A. These overlays are called heatmaps and often include signal strength, SNR, expected data rates, throughput rates, number of APs, CCI, and more.
10. The correct answer is A. AP locations are defined on the floor plans and not in the project parameters.

Chapter 6: Designing for Specific Scenarios

Objectives Covered:

1.4.6 Understand common vertical markets

2.2.1 Select and use the appropriate design tools

In the preceding chapters, information was provided for designing standard enterprise WLAN deployments. This chapter goes into detail for more varied applications, with a focus on understanding the different types of wireless backhaul options, defining the components of a link budget, and diving into the requirements and additional networking technologies that are usually required in specific types of applications and vertical markets.

The design is as much an art as a science, making design one of the most challenging activities in any engineering enterprise, including and especially in wireless. Design requires both following established engineering principles as well as creativity in adapting those principles to specific problems.

There is an inherent Catch 22 with the process of design: you need experience to do design successfully, but you need to perform design to gain experience. Engineers will often talk about best practices, but these are simply just codified rules based on experience, usually lessons learned after bad experiences when things went drastically wrong. While generally applicable in a wide variety of situations, specific best practices may not always be relevant in specific situations.

So how does one close the experience gap? There is a general process for design, and axioms to follow. There are several academic theories on the best way to perform generic engineering design, and these generally can be summarized as follows:

- Understand your requirements: The biggest error that most designers make is to jump into generating a solution before fully defining and understanding the problem. Functional requirements are what the system must do, and it is critical to capture these properly prior to generating a design solution. In Wi-Fi, the most common functional requirements are as follows:
    - Usage: What kind of devices and applications will be connecting to the network? Are there differences in types of users (e.g., guests, staff, appliances, etc.)? What kind of user access control and security is required for each type of user or device?
    - Coverage: What areas of the facility require Wi-Fi signal? Are there areas of the facility that should not get a Wi-Fi signal?
    - Capacity: What are the quantities of devices that shall use the network simultaneously, and what are the differences in terms of usage levels and/or client density across different coverage areas

    - Control: What are the requirements for monitoring and managing the network over time? Is centralized control or cloud-monitoring required?
    - Integration: What other network appliances are required to be integrated to provide full functionality (e.g., switches, routers, modems, WIDS/WIPS, beacons, etc.)?
- Understand your constraints: Constraints are what the system must work around. Constraints do not directly relate to system functionality, but generally, have a huge impact on the design solution. In Wi-Fi applications, the most common constraints include budget, aesthetics, cable path restrictions, and mounting / mounting location restrictions (e.g., where electrical power is available).
- Understand your design parameters: The design parameters are the degrees of freedom that can be manipulated to satisfy the functional requirements and constraints. In Wi-Fi applications, the most common degrees of freedom are
    - The selection of a particular AP make and model, and it associated antennas
    - The location of the APs
    - The channel settings and width per AP per band
    - The transmit power per AP per band

As a CWDP, it is critical to understand that different AP vendors are suited to particular applications, and specific AP models are usually intended for specific applications. One size does not fit all!

- Maintain functional independence: It is unfortunately quite common for requirements and constraints to change during a project, necessitating changes to the design. This is usually termed scope creep. The better that the design parameters are kept independent, the easier it is to make changes that don't cause unanticipated negative impacts elsewhere. Ideally, each functional requirement should be both satisfied and influenced independently by one design parameter, so as to easily accommodate changes in requirements. Such independence is virtually impossible to achieve in practice, though minimizing the coupling by having each design parameter primarily influence only a single functional requirement generally makes design changes simpler, less disruptive, and less iterative. Common changes in Wi-Fi applications include new types and/or increased quantities of client devices to be supported, additional areas of coverage, etc.

    - Also, remember that the expectations of the system are likely to change over its lifetime after deployment. Typical Wi-Fi refresh cycles are 2-3 years in large enterprise deployments, though are 5-7 years or more in most SMB applications. During that time, new client devices will emerge, and generally, more client devices will be connecting simultaneously. Thus, the demand on the network will increase while the network itself is static. It is therefore critical to build *slack* (i.e., margin) into the network design up front, as it will be consumed over time.

- Minimize complexity: Colloquially called the KISS rule (Keep It Simple, Stupid), it is important to avoid making the design unnecessarily complex. There is naturally a careful balance between designing for margin vs. over-designing (i.e., building in functionality and capacity that will never actually get used), as over-designing will drive up costs and reduce functional independence. Unfortunately, it is all-too-common for networks to get over-designed in practice. In Wi-Fi, the most common over-design problem occurs in the selection of the AP itself. AP vendors and models are often selected based on the comfort level, preferences, and commercial interests of the system integrator, as opposed to selecting the AP vendor and model best suited to the application. This tendency leads to APs being deployed with features that are not needed for the application. Those features are naturally still being paid for in the cost of the APs. Furthermore, enabling these unnecessary features can cause their own functionality issues that take the focus off of satisfying the actual system requirements. Pick the best AP (i.e., the simplest AP) suitable for the application.

With this context in mind, the remainder of this chapter will discuss common wireless applications and the best ways to understand the requirements and generate appropriate design solutions for them.

6.1: Wireless Backhaul

All Wi-Fi applications are a form of *wireless backhaul*, which is a method of connecting devices wirelessly to a wired network, and in most cases ultimately to the Internet.

In a conventional Wi-Fi network, as covered in previous chapters, each access point (AP) requires a wired Ethernet connection to provide backhaul to the wired network infrastructure and ultimately the Internet. The Ethernet connection typically supplies both data connectivity and electrical power to the AP via Power over Ethernet (PoE), with CAT5e / CAT6 / CAT6a cabling connecting the AP to a network of wired switches. This type of connection is known, somewhat hyperbolically, as a *wired AP*. This is depicted in Figure 6.1.

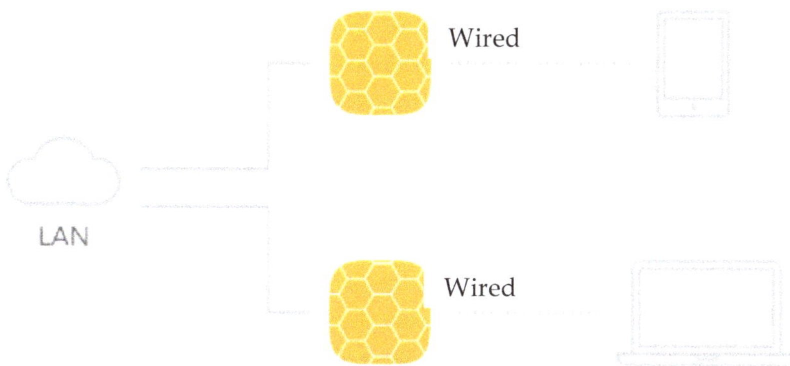

Figure 6.1: Schematic diagram of a conventional wired AP network. (Source: LigoWave Networks Inc.)

In many environments, however, it is either impossible or prohibitively expensive to run an Ethernet cable from a switch to at least some of the APs. This is especially common in outdoor applications, though can also occur indoors where it is impractical or cost prohibitive to run a cable. In such cases, Wi-Fi itself can be used to provide wireless backhaul from the AP (or other network appliance, such as a remote IP camera) to the wired network. Each wireless backhaul link is referred to as a *hop,* and it is possible to have a chain of multiple hops between the remote wireless AP to the root wireless AP that has a wired connection to the network.

By this definition, a wireless client device always has one hop to the access point via its Wi-Fi connection. Keep this in mind when multiple wireless hops are involved in the path of connecting that client device to the Internet.

There are multiple options for providing wireless backhaul to remote APs. Naturally, each option has both benefits and limitations. Most critically, each wireless hop introduces latency, which adds in a linear fashion with the number of hops. Repeaters and mesh APs, because they are sharing the same radio for both client access bd wireless backhaul, also intrinsically lower the throughput and user capacity, often as a square of the number of hops.

It is critical to understand your functional requirements and constraints, as well as the benefits and limitations of each wireless backhaul option when designing a Wi-Fi network and selecting a specific wireless backhaul approach.

Option 1: Client Bridge

An access point operating in Client Bridge mode provides Wi-Fi connectivity for a wired client device. A Client Bridge is intended to connect an individual wired client device to a Wi-Fi network. This is depicted in Figure 6.2.

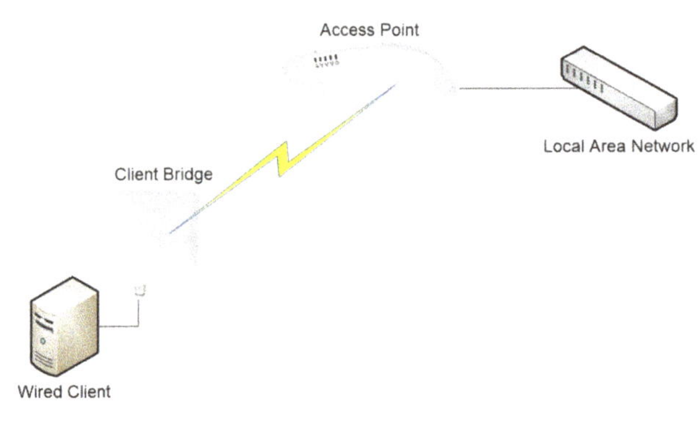

Figure 06.2: Example of a network utilizing a client bridge. (Source: LigoWave Networks Inc.)

The typical application is to connect a single client device with a wired network interface, such as an appliance, PC, or IP camera, to a Wi-Fi network. For several years, many game console vendors sold client bridges as "gaming adapters" to connect game consoles with Ethernet ports to a home Wi-Fi network.

From the Wi-Fi network's perspective, the connected client device is the client bridge, not any wired Ethernet device(s) behind it. The client bridge serves as a transparent pass-through, to make the wired client device appear to be a wireless client device. When multiple wired client devices are connected through a single Client Bridge (i.e., via a network switch), they share the same MAC address on the network, namely the WLAN MAC address of the Client Bridge. While multiple wired client devices can still be configured with different Layer 3 static IP addresses, having multiple client devices all sharing the same Layer 2 MAC address on the network can lead to unusual problems in some circumstances. For example, each wired device sharing the same WLAN MAC address may or may not be able to obtain an independent Layer 3 DHCP address, depending on the DHCP server.

When using an AP in Client Bridge mode, it is best practice to connect only one wired client device. There are better wireless backhaul options when multiple wired devices need to be connected.

For typical applications, Client Bridge mode is only utilized on single-band APs. For dual-band access points, one radio (typically 5 GHz) will be configured to operate in client bridge mode, while the other radio (typically 2.4 GHz) will be used for providing Wi-Fi connectivity on an independent SSID to client devices. Client Bridge mode is generally only available on APs operating in standalone mode, meaning that each AP must be configured individually and cannot be managed or monitored from a centralized controller.

Option 2: Repeaters

An access point operating in repeater mode provides both Wi-Fi connectivity to client devices as well as providing a wireless backhaul connection to one or more wired APs. This is depicted in Figure 6.3. Repeaters are intended for very small networks (e.g., home environments), where individual repeater APs are used to fill in particular coverage gaps. Individual client MAC addresses are preserved, though the VLAN (if any) is defined by the main access point's SSID that is being repeated.

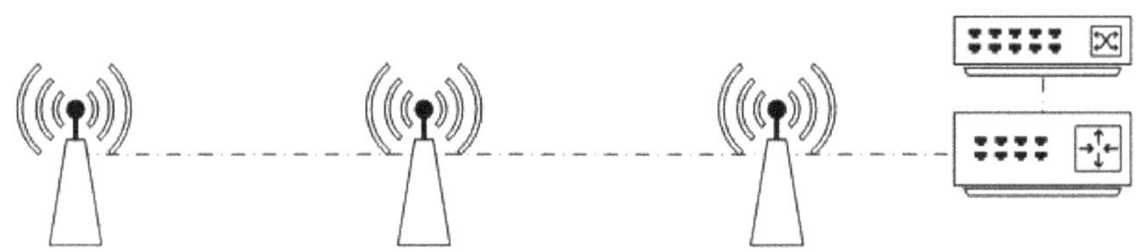

Figure 6.3: Example of a network utilizing wireless repeaters

For dual-band access points, one radio (typically 5 GHz) will be configured to operate in repeater mode, while the other radio (typically 2.4 GHz) will be exclusively for providing Wi-Fi connectivity to client devices. Note that both Wi-Fi bands depend upon the repeater radio for backhaul. Since the repeater radio must spend half its time providing Wi-Fi connectivity to client devices and half its time providing wireless backhaul, the data capacity of a repeater radio for both backhaul and for Wi-Fi client connectivity is reduced by 50%. When there are multiple hops, the data capacity is reduced by 50% at each hop. Thus, for two hops, the total data capacity is only 1/4, for three hops it is 1/8, for four hops it is 1/16, and so forth.

Repeater mode is generally only available on APs in standalone mode, meaning that each AP must be configured individually and cannot be managed or monitored from a centralized controller.

Option 3: Point-to-Multi-Point Wireless Bridge Links

A dedicated pair of APs, usually with integrated directional antennas, are configured to operate in Wireless Bridge mode to create a point-to-point (PTP) or point-to-multipoint (PTMP) link to provide wireless backhaul. The wireless bridge link on the remote end is connected to the remote AP via its wired Ethernet interface. From the perspective of the rest of the network, this wireless connection looks like a wired connection; the wired

Ethernet frame is encapsulated and encrypted in a Wi-Fi packet on one end, transmitted across the wireless link, and then de-encapsulated and decrypted on the other end. Thus, all wired Layer 2 information (i.e., client MAC addresses, VLANs, etc.) are preserved across the Wireless Bridge link. This is depicted in Figure 6.4.

Point-to-multipoint (PTMP) wireless bridge links operate similarly, but instead of one remote station, there are multiple remote endpoint stations (a.k.a. subscribers) connected to a single base station. This is shown in Figure 6.5 Note that unlike parallel PTP links which are normally placed on independent channels, in PTMP the APs are all on the same channel by necessity. Thus, remote endpoints collectively share the total available airtime bandwidth of the link.

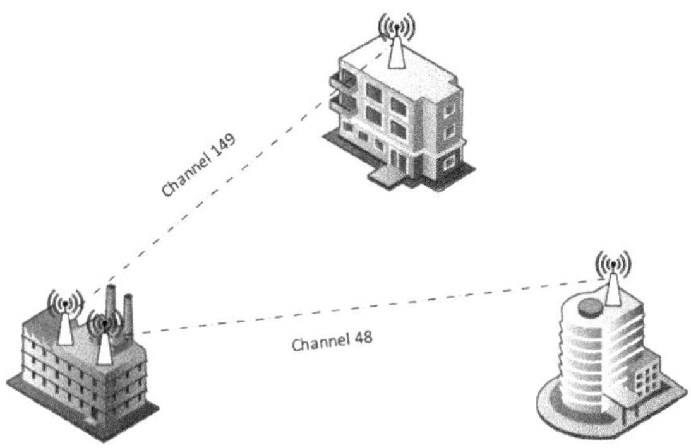

Figure 6.4: Examples of parallel point-to-point (PTP) wireless bridge links

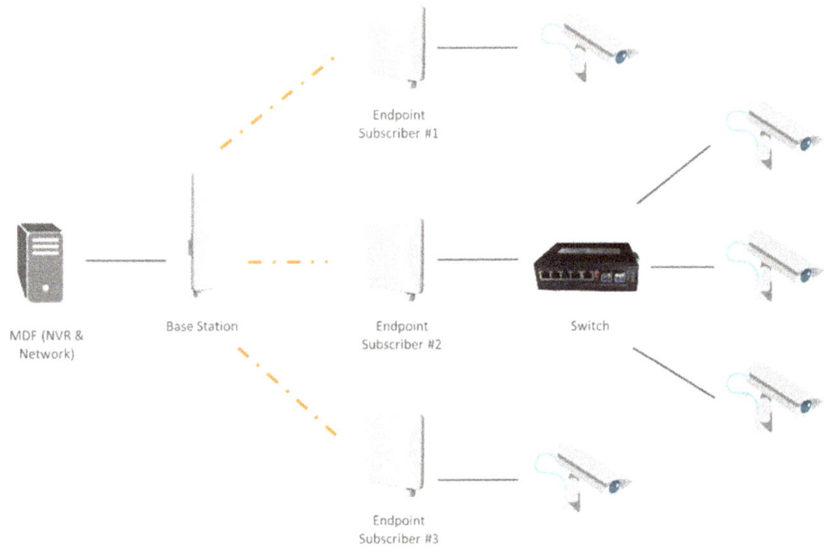

Figure 6.5: Example of a point-to-multipoint wireless bridge link, for surveillance or Wi-Fi applications. (Source: LigoWave Networks Inc.)

The most common wireless bridge link technology is known as a wireless distribution system (WDS). WDS uses a variation of the 802.11 protocol to communicate between two wireless radios. Wireless client devices cannot connect directly to a wireless bridge link. Depending on the vendor, the APs can be configured in two different modes:

- WDS Bridge: In WDS Bridge mode, the base and endpoints are configured identically, but each AP must explicitly specify the MAC address of the "remote" APs allowed to communicate with it. An encryption key is usually specified, and the SSID is hidden. In this manner, the links are statically established, so that each WDS Bridge AP only accepts connections from pre-defined APs. This mode can be useful for smaller PTP/PTMP networks, but it doesn't scale to larger networks, as generally, the base can only have up to 4 (prior to 802.11ac) or 8 (802.11ac wave 1 and later) remote endpoints. Adding an additional endpoint or changing out an AP also requires reprogramming MAC addresses on the other side of the link.
- WDS Access Point / WDS Station: This mode looks more like conventional Wi-Fi, where the base station is configured as the WDS Access Point and the endpoints are configured as WDS Stations. For PTP, which AP is which doesn't matter,

though by convention the upstream AP is generally defined as the base station. The SSID is defined by and broadcast from the base station.

Wireless Bridges usually require dedicated single-band hardware at each location, though some APs allow for one radio (typically the 5 GHz) to be in WDS bridge mode and the other radio (typically the 2.4 GHz) to be in AP mode to provide Wi-Fi service client devices.

It is best practice to use wireless bridge links with dedicated access points for networks requiring both wireless backhaul and high bandwidth and/or high user capacity Wi-Fi. While each hop adds latency, there is no throughput or user capacity degradation, since the point-to-(multi)point wireless bridge links are solely dedicated to wireless backhaul, with Wi-Fi access for client devices being handled by separate access points. This serves to decouple the functional requirement to provide Wi-Fi from the functional requirement to provide wireless backhaul.

A network of multiple wireless bridges operating in the same area requires its own design effort to ensure appropriate bandwidth capacity and channel utilization.

There are AP vendors that specialize in PTP and PTMP applications. Such vendors generally offer the following:

- Dedicated AP Models: Specialty PTP/PTMP vendors will have entire product lines of AP models for such applications. These APs are single-band and usually have integrated directional antennas. It is not uncommon to see a series of APs in a product family with the same core hardware and firmware, where the differences are primarily in the antenna, usually having a low gain patch and high gain patch models for endpoints and sector, omni, and/or external antenna port models for PTMP base stations.
- Hardware Acceleration: High-end AP models for PTP/PTMP may incorporate QoS hardware acceleration, so that QoS processing is offloaded to dedicated hardware and not software, thus increasing the total processing power available to handling data traffic and increasing the effective capacity of the link.
- Proprietary Wireless Protocols: Wi-Fi was not originally designed for wireless bridging, and while WDS enables that capability, it still uses CSMA/CA at its core, thus introducing a lot of unnecessary overhead for wireless links. Accordingly, some vendors have developed proprietary wireless protocols to

utilize the airtime more efficiently. These proprietary protocols are generally based on time-division multiplex access (TDMA) techniques, thus performing collision avoidance by having the base station dictate which radio can transmit in assigned time slots. Such protocols are generally optimized to both load balance based on traffic demand per radio and generally will also include QoS algorithms to optimize PTP and PTMP links for voice and video applications.

- Design and Installation Tools: PTP/PTMP vendors generally provide either online or offline design tools for planning links and aligning links during installation.
- Centralized Cloud Management: Conventionally, wireless bridging was generally only available on standalone APs, meaning that each AP must be configured individually and cannot be managed or monitored from a centralized controller. At best, monitoring and management functionality of such links could be managed via SNMP to a WNMS, though this option required a lot of network expertise by the network operator to configure and maintain. Specialty PTP/PTMP vendors are now introducing cloud-management platforms for monitoring and managing PTP/PTMP networks, like the cloud management platforms offered by many vendors for managing Wi-Fi APs. For vendors that provide both PTP/PTMP and Wi-Fi networking AP products, this is a very compelling commercial offering, as multiple sites that integrate both PTP/PTMP and Wi-Fi can have all their APs monitored and managed from a single cloud interface.

Point-to-Multipoint Wireless Bridge Network Example

Figure 6.6 shows an example of an outdoor surveillance network at a garden-style apartment complex utilizing point-to-multipoint links to provide wireless backhaul to cameras mounted on either light poles or on the sides of buildings. The network video recorder (NVR) is in the main distribution frame (MDF) with the property's Internet feed and router. For apartment complexes like this, the MDF is usually in the clubhouse, and there is no Ethernet or fiber cabling that interconnects the clubhouse to the other residential buildings on the property.

The pins on the map indicate the buildings or light poles that contain surveillance cameras. Cameras are usually positioned where they need the field of view for surveillance and where they have access to electrical power. For residential properties like this, cameras are usually positioned to provide surveillance of all parking areas and building entryways. High-resolution cameras are also quite common at gates and vehicular entrances, so the license plates of all entering cars can be recorded. A building with multiple cameras may have co-located cameras pointed in different directions, and/or cameras on different sides of the same building. Typically, the cameras attached to a building are connected and powered via Ethernet cabling to a network switch located in an *intermediate distribution frame* (IDF), which may be a utility room, laundry room, or even a NEMA enclosure on the side of the building.

The colored lines indicate the point-to-(multi)point wireless bridge links. Since all the cameras in a building terminate at an IDF, the wireless bridge links need to provide backhaul from the various IDFs to the MDF. For applications where only one camera is located in a particular spot (e.g., a light pole), a switch may not be used, and instead, the camera will be connected directly to the AP. Typically, one tries to follow the roadways or to establish links between buildings where there is empty space and hopefully little foliage. Ideally, single hops would be established from the MDF to all the IDFs, but it is extremely common for the MDF not to have visual line of sight, let alone RF line of sight, to all of IDFs, even when the MDF is at or near the center of the property, as in this example. When the MDF is located in one corner of the property, which is all too common, such projects can be even more challenging.

Figure 6.6: Example of a wireless network utilizing point-to-(multi)point links for backhaul to outdoor wireless APs. (Source: LigoWave, Inc.).

Accordingly, it, therefore, becomes necessary to *relay* links to reach all IDFs. One generally picks relay points where there are going to be cameras, and thus IDFs, that require wireless backhaul anyway. The relay point will generally have multiple wireless bridge APs, with one pointed upstream towards the MDF, and one or more APs pointed downstream towards additional IDFs. These types of relay links are often referred to as *branches*. Note that each hop adds latency, and retransmissions due to interference will contribute to jitter.

In this example, there is one building for which that is not possible, so a relay point needed to be established on a light pole on the left side of the example. For this PTMP link, the base station is on the pole, and one of the endpoints is downstream whereas the other is upstream. Fortunately, the wireless bridge links self-optimize based on the volume of traffic, so the wireless bridge links themselves don't care which direction is upstream vs. downstream.

When doing relay links, the total amount of bandwidth on each link needs to be calculated. The upstream links carry both the data traffic from the IDF it is directly connected to plus the data traffic from all downstream branches. A very common mistake in such designs is to aggregate too much traffic on the initial link from the MDF, causing that entire branch with all sub-branches to have performance issues. Load balancing is critical in such applications so that each branch coming from the MDF has approximately the same level of traffic.

Each *PTP/PTMP group* is usually depicted in a different color, so as to visualize them better. Each AP in a PTP/PTMP group needs to be on the same channel in order to communicate, but neighboring PTP/PTMP groups need to be on independent channels to prevent self-interference between PTP/PTMP groups. Since the APs are directional, PTP/PTMP groups that are not co-located with each other will generally not see each other (or see each other only at very low energy levels), so channel reuse is readily possible and usually necessary.

For channel re-use purposes, it is often convenient to not use the maximum channel sizes available from the APs, but rather select smaller channel sizes based on the total amount of traffic flow. In cases with multiple sub-branches, the links downstream (towards the endpoints) may use smaller channel widths, whereas the links upstream (towards the MSF) may use larger channel widths.

To maximize the available airtime and to minimize the number of hops, one generally wants to have as many PTP / PTMP groups as possible emerging from the MDF, while minimizing the total number of hops to minimize the effects of latency, for which video surveillance and QoS Wi-Fi applications are particularly sensitive.

It is a best practice is to keep the total number of wireless bridge hops along each branch to three or less. If there were also Wi-Fi access points in this application, then the link between the client device and the Wi-Fi AP itself represents an additional hop in the chain.

Option 4: Mesh Networks

In a *mesh* network, the Wi-Fi AP uses its own radio to provide a wireless backhaul to other Wi-Fi APs on the network, eventually reaching a Wi-Fi AP with a wired Ethernet connection to the wired backhaul infrastructure and the network. In this sense, a mesh network is a network of repeaters, though mesh is designed to operate automatically and more intelligently on a large scale.

A mesh network creates a set of "dynamic wireless bridge" links, using routing algorithms to automatically calculate the most optimal wireless path through the network back to a wired root node. This makes mesh networks relatively robust to the failure of an individual AP; in a process referred to as *self-healing*, the routing algorithms will automatically calculate the "next best" path through the network if an AP in the path goes offline. Since the routing functions are done automatically within the mesh software, mesh networks are straightforward to set up and are thus scalable to cover large geographic areas. All wired Layer 2 information (i.e., client MAC addresses, VLANs, etc.) are preserved across the mesh link. A conceptual example of a mesh network is shown in Figure 6.7

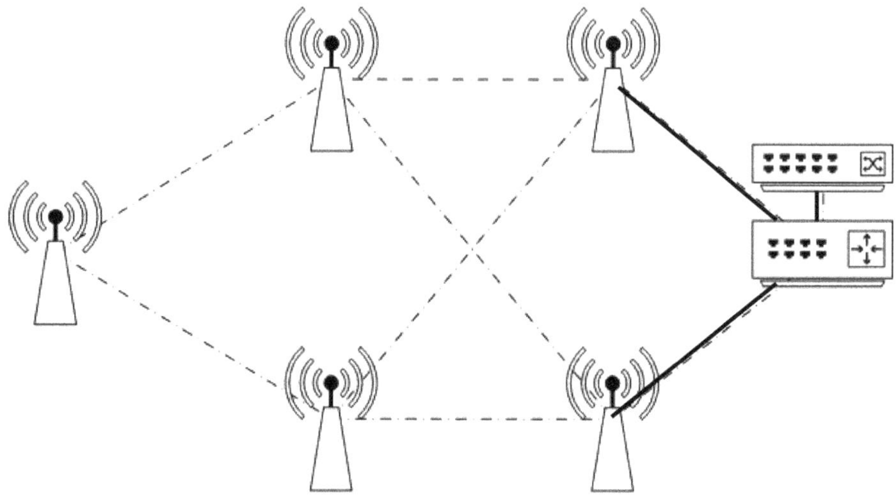

Figure 6.7: An example of a mesh network, demonstrating the possible redundant paths. The two right-most APs are root APs, as they have wired Ethernet connections back to the wired network.

The mesh network control architecture can either be *centralized* or *distributed*. With a centralized mesh architecture, an AP controller is required to calculate and coordinate the mesh parameters for each AP. This architecture has the advantage of centralizing all the data for decision making and optimization. However, the scalability of the mesh network is ultimately limited by the capacity of the AP controller. In a distributed mesh architecture, each AP operationally acts like a router, continuously sharing information about its connection status with its neighbors, and each AP uses this information to compute its own optimal mesh path. The distributed mesh architecture operates at Layer 2 but functionally behaves very similarly to how Internet routers compute optimal paths through the Internet at Layer 3.

In a distributed mesh architecture, an AP controller is not required, though is generally extremely useful in providing centralized real-time monitoring of the mesh network, as well as establishing the core initial mesh network configuration parameters, such as mesh ID, encryption, etc.

Unfortunately, mesh networks have significant limitations, most notably in the loss of both throughput and user capacity, which scales geometrically as the number of wireless hops increase, as well as the increases in latency, which scales linearly as the number of wireless hops increase.

Accordingly, mesh networks are not suitable for high bandwidth or latency-sensitive applications. Because of these performance limitations, it is generally recommended that mesh networks be *avoided* unless no other viable wireless backhaul options are available. Mesh networks should only be used in environments where providing Ethernet data wiring to access points or cameras is impossible or cost-prohibitive, and when bandwidth performance is not critical.

Mesh networks were originally trendy in the mid-2000s[86], as a way of both providing metropolitan Wi-Fi coverage as well as coverage for large outdoor properties where wiring was prohibitively expensive, such as RV parks, garden-style apartment complexes, marinas, etc. While many mesh networks were successfully deployed, most of these efforts ultimately failed, especially metropolitan Wi-Fi. Early mesh networks relied upon single-radio APs on 2.4 GHz using 802.11g. When dual-band APs were introduced, only 802.11a was available on the 5 GHz band, which still led to very low throughputs as the number of hops increased.

With the wide adoption of dual-band access points with 802.11ac, there has been renewed interest in mesh for both Wi-Fi access and surveillance applications. Accordingly, several startup companies have introduced mesh Wi-Fi products utilizing 802.11ac for the consumer market to extend coverage in private homes. Since such networks are limited in scope to only a few APs and one or two hops, and since the bandwidth of the mesh links generally exceeds the bandwidth coming into the home, the mesh is surprisingly suitable for this type of application. Nonetheless, while the data rates of 802.11ac are approximately 25 times larger than the 802.11a data rates of a decade ago, the number of client devices and their bandwidth demands have also grown exponentially during that time. The fundamental limitations of mesh networks are therefore still the same, and thus mesh may ultimately once again prove to be a passing fad.

---

[86] Mesh networks are very common in wireless IoT deployments. Many of these deployments use protocols other than Wi-Fi, such as 802.15.4. Time will tell how popular Wi-Fi becomes for IoT now that the 6 GHz band has opened. With the massive number of 20 MHz channels available, I expect it to be used quite heavily for IoT in the 6 GHz band.

Nonetheless, mesh networks are sometimes the only viable option. If mesh needs to be deployed, it is best to design it properly to maximize its performance and mitigate its inherent limitations.

6.2: Mesh Network Terminology and Best Practices

The access points in a mesh network are categorized as either *root APs* or *node APs*. Note that AP vendors tend to use their own variations on terminology, but functionally they will fit into these categories.

- Root AP: This is an access point with a wired connection to the wired switch infrastructure. The remote node APs establish wireless backhaul connections to the root AP. Note that the wired connection utilized by a root AP can either be (1) a direct Ethernet or fiber-optic connection to the wired switch infrastructure or (2) a wired connection to a separate wireless point-to-(multi)point bridge link on an independent channel.
- Node AP: This is an access point without a wired Ethernet connection. Wireless backhaul to the network is established via a wireless connection to a root AP or to other node APs. Note that the node AP still requires electrical power, so an Ethernet connection to a PoE injector is common, even though the "network" end of the PoE injector may not be connected at all, or may serve to interconnect a wired client device, such as an IP camera.

The path from a particular node APs back to a particular root AP can require connections via multiple intermediate remote node APs. Each wireless link in this chain is referred to as a *hop*. The mesh routing algorithm selects the most optimal route through the network. The optimization function used by the mesh APs is generally proprietary to each AP vendor, but typically the mesh algorithm attempts to balance several (often conflicting) parameters, such as the following:

- **Minimize the number of hops**, so as to minimize the total wireless latency and throughput penalty of the network
- **Maximize the signal strength of each hop**, to maximize the achievable Wi-Fi data rates between the mesh radios on each hop. For maximum data rates in 802.11ac, the received signal strength indicator (RSSI) would ideally be in the -40 dBm to -50 dBm range, though this is usually unachievable in practice since

omnidirectional antennas are typically used to create the widest field of view to neighboring APs. RSSIs should be greater than -65 dBm for decent data rate performance between hops.

- **Balance the load on each AP**, to account for the number of associated client devices and the total throughput consumption on each AP. The throughput load stacks as the number of hops increase, so intermediate remote nodes that are heavily utilized with client traffic will not give as many resources to remote downstream nodes.

Because of the competing tradeoffs in this optimization process, mesh networks can often result in counter-intuitive and/or sub-optimal topologies.

As a best practice, the network design should cluster the mesh APs into groups consisting of up to four remote nodes that are only one hop away from a root node. Thus, at least 20% of the mesh APs, distributed roughly evenly throughout the property, should be root APs. Each remote node AP is therefore nominally only one hop away from a root AP. In the event of a failure of a root AP, the nearby node APs will then only be 2-3 hops away from another root AP. This approach generally requires creating additional root APs, which can be done either by running Ethernet or fiber-optic cable to the particular remote locations or by establishing dedicated point-to-(multi)point wireless bridge links to create "wireless wires" from the root AP back to the wired network.

Thus, mesh network design involves the following sequence of steps:

1. Place the necessary quantity of APs to optimize the area of Wi-Fi coverage, based on the mounting location constraints
2. Group the APs into clusters and select a reasonably centrally-located root AP for each cluster
3. Design a wireless bridge network to connect the desired root APs back to the MDF.

It is a best practice to set the root AP on a static independent channel, and each remote node AP should be set to "auto channel." This is done to maximize the airtime capacity of the overall network so that multiple neighboring root nodes do not create self-interference. The remote node APs are set to auto-channel so that they can fail over to a different root APs in the event of the failure of their primary root AP. When utilizing point-to-(multi)point wireless bridge links to establish root nodes, the wireless bridge links must also be on static independent channels, and thus must be accounted for in the overall channelization plan.

Most vendors that support mesh will allow root APs and node APs can operate in one of two modes. Again, terminology for this feature is usually vendor-specific:

- Mesh AP Mode: In this mode, the wireless radio acts like a repeater, providing both Wi-Fi connectivity to client devices as well as providing a backhaul connection to one or more node APs. For single-band mesh access points, this is the only operational mode available. For dual-band access points, one of the bands (typically 5 GHz) will be configured to operate in this mode. The other band (typically 2.4 GHz) will be exclusively for providing Wi-Fi connectivity to client devices. Note that both Wi-Fi bands depend upon the mesh radio for backhaul. Since the mesh radio must spend half its time providing connectivity to client devices and half its time providing backhaul, the data capacity of the mesh radio for both backhaul and for Wi-Fi client connectivity is reduced by 50%. When there are multiple hops, the data capacity is ***reduced by 50% per hop.*** Thus, for two hops, the total data capacity is only 1/4, for three hops it is 1/8, for four hops it is 1/16, and so forth.
- Mesh Point Mode: In this mode, available only for dual-band APs, the wireless mesh radio (typically 5 GHz) only provides wireless backhaul, and the other radio (typically 2.4 GHz) only provides Wi-Fi connectivity to client devices. Operationally, the mesh radio operates like a dynamic wireless bridge link, so while each hop still introduces latency which adds linearly, there is no 50% throughput penalty per hop, since the mesh radio is not also servicing client devices on the same radio and can be devoted exclusively to wireless backhaul. Since Wi-Fi access to client devices is restricted to only one radio (typically 2.4 GHz), the overall client capacity of the AP is that of a single-band AP. Furthermore, even dual-band 802.11ac client devices will only be able to connect at 802.11n data rates on the 2.4 GHz radio.

It is best practice to have mesh networks configured to operate in Mesh Point mode. The loss of bandwidth capacity from lacking wireless 5 GHz wireless connectivity tends to pale in comparison to the loss of bandwidth capacity from losing 50% of bandwidth per hop. This also allows for the transmit power of the mesh radios to be set at their maximum value, to provide the maximum signal strength between nodes, without being imbalanced with the low transmit power capability of most 5 GHz client devices.

In both operational modes, the overall data capacity of a mesh AP is reduced as compared to the same AP operating in a conventional configuration with a wired Ethernet connection to a wired switch infrastructure. ***Accordingly, a mesh Wi-Fi network will never have the same level of throughput and client capacity of a conventional Wi-Fi network.***

Mesh Network Example

Figure 6.8 shows an example mesh network. This is an RV park with 437 spaces spread across a roughly 2000′ x 1000′ area. The main distribution frame (MDF) is in the southwest corner of the property, and trees in parts of the property preclude direct line-of-sight to many locations.

Figure 6.8: Example of a mesh network, utilizing point-to-multipoint links to create additional root nodes. (Source: Imperial Network Solutions, LLC .)

The red links and bubbles indicate wireless bridge links from the MDF to each of the root APs. In some cases, multiple relay wireless bridge links need to be established to reach each mesh cluster. The point-to-(multi)point links are designated by the base station ("Master") or endpoints ("Slave") with a letter and number index. For example, the wireless bridge link going between the MDF and G8-R is designated Link D, with Master D connected to Slave D1. The other colors and bubbles represent the root APs and node APs in Mesh Point mode, and the nominal mesh links between the node APs and the root APs in each cluster. In the figure, each cluster is designated with a group number and an index to indicate that it is a root AP or remote AP. For example, in the right, the root node is designated G8-R, and the nominal remote APs are designated G8-1 to G8-4.

6.3: Building a Link Budget

As you learned in your CWNA studies, a link budget is simply the aggregation of all signal gains and losses as a signal is communicated from a transmitter to a receiver. Link budgets are commonly only calculated explicitly for long-distance point-to-(multi)point applications, though the principle is applicable to all forms of wireless backhaul communication.

Link budgets are typically calculated in decibels (dB), as the use of logarithmic units allows for simple calculation by simply adding the gains and losses as the signal makes its way between the transmitter and receiver.

The components of a link budget are as follows:

- Transmitter Effective Isotropic Radiated Power (EIRP)
    - Transmit power
    - Transmitter cabling losses between radio and antenna
    - Transmitter antenna gain
- Free space path loss
- Attenuation
- Receiver antenna gain
- Receiver cable losses

Note that there is no "receiver power"; listening is purely passive in Wi-Fi communications.

When all of these components of the link budget are added and subtracted, the result is the *received signal strength indicator* or *RSSI*. The RSSI, therefore, indicates how well the receiver hears the transmitter.

For troubleshooting wireless networks, and for many activities such as roaming and mesh optimization, it is important that wireless stations (i.e., access points, client devices, wireless bridge links, repeaters, or mesh nodes) measure and report on the RSSI and use that information for decision-making. Many APs will provide the RSSI of connected stations directly in dBm, though some will provide the RSSI as a percentage, based on the use of an 8-bit number in most AP chipsets to store the RSSI. RSSI was defined arbitrarily in the IEEE 802.11 spec and thus left to chipset vendors and wireless manufacturers to interpret. To further complicate matters, stations are generally not calibrated to a fixed standard. Thus, the measured RSSI can vary by approximately 3-6 dB between two stations of the same make and model under identical conditions, let alone between different models from different vendors. Accordingly, the measured RSSI is a poor metric to use in practice, but alas it is the primary metric available. For the purposes of discussion in this chapter, RSSI will be referenced as a known quantity, but keep in mind that this number is not as accurate in practice as we like to pretend. This variability in measured RSSI is one of many reasons why good WLAN designers always ensure that extra margins are always built into their designs.

In the IEEE 802.11 protocol, we define a series of standard modulation and coding scheme (MCS) rates to determine how much data can get pushed in a single transmission. A station will indicate its *receive sensitivity* for each MCS rate in its specification, which is the lowest RSSI at which the station can successfully receive and interpret a transmission at that MCS rate. In general, a higher MCS rate leads to a worse receive sensitivity.

For example, Table 6.1 shows the receive sensitivity scale for a particular 802.11 b/g AP.

The difference between the actual RSSI and the receive sensitivity is known as the *fade margin*.

Fade margin is simple enough conceptually, but hard to work with in practice without memorizing the MCS receive sensitivity for every AP and every client device on the network. It is thus more common to use *signal-to-noise ratio (SNR)* instead of *fade margin* in Wi-Fi network design. The *signal-to-noise ratio (SNR)* is the ratio between the RSSI and the noise floor. When we define both the RSSI and the noise floor in logarithmic measurements of dBm, the SNR is the difference in these two values. The formula for calculating SNR is simple:

```
SNR = RSSI (dBm) - noise floor (dBm)
```

For example, if the RSSI is measured at -70 dBm and the if the noise floor is rated at -95 dBm, the SNR is (-70 dBm) – (-95 dBm) = +25 dB.

dBm Power Level	Data Rate
-94 dBm	1 Mbps
-93 dBm	2 Mbps
-92 dBm	5.5 Mbps
-86 dBm	6 Mbps
-86 dBm	9 Mbps
-90 dBm	11 Mbps
-86 dBm	12 Mbps
-86 dBm	18 Mbps
-84 dBm	24 Mbps
-80 dBm	36 Mbps
-75 dBm	48 Mbps
-71 dBm	54 Mbps

Table 6.1: Example receive sensitivity scale for an 802.11g station

Know how to calculate SNR. Be warned that the signal strength may be provided to you in mW, and you will therefore first need to convert power levels to dBm using the rules of 10s and 3s. You should recall this from your CWNA studies as well.

The *thermal noise floor* is the RSSI at which a radio signal cannot be distinguished from the surrounding environment. Every object gives off energy at very low levels across the entire RF spectrum. Consider this to be the "background hum of the universe." The amount of energy present in the environment across a given portion of the spectrum is a function of absolute temperature and the channel width. For temperatures at which human beings operate (i.e., roughly -40°C to +60°C), the temperature component varies by < 1 dB and is thus generally treated as a constant. Note, however, that with the use of 40 MHz or 80 MHz channels in IEEE 802.11n / 802.11ac, this background noise doubles or quadruples, increasing by 3 dB or 6 dB, respectively.

The universe itself, however, isn't the only potential source of RF noise. Neighboring Wi-Fi or other RF sources can increase the ambient noise level at the receiver. Even for signals that are not directed at the receiver, reflections, scattering, and other attenuation effects will serve to increase the noise floor.

A useful analogy for understanding and/or explaining SNR and MCS rates is the experience of trying to have a conversation in a noisy restaurant with poor acoustics (e.g. marble walls or a tin roof) that lead to a lot of reflections (i.e., echo). If the restaurant is empty and you and your dinner companion are the only people having a conversation, you can still hear each other clearly when talking at a normal or even at a low volume. As the restaurant fills up with more patrons, each having their own conversations, the level of noise increases, so your volume (i.e., transmit power) needs to increase and/or the speed of your words (i.e., your MCS rate) needs to decrease, in order to maintain the conversation. Eventually, when the restaurant is packed, you are probably screaming at the top of your lungs (i.e., at max transmit power) and speaking very slowly (i.e., at minimum MCS rate), and you still may not be heard by your dinner companion. At this point, get the check and find a different restaurant!

In Wi-Fi design, it is typical to want the transmissions to occur at the highest possible MCS rates, so as to achieve the best possible data throughputs. The higher the SNR, the stronger the signal, and thus the higher the MCS rate that can be used. Table 6.2 below shows the achievable MCS rate based on channel width, SNR, and RSSI. This is a very useful tool for all CWDPs.

To understand the achievable SNR in a Wi-Fi or wireless backhaul scenario, it is important to understand the impact of each component of the link budget:

As your noise floor increases, your signal (RSSI) needs to increase by the same amount to maintain a given SNR. The SNR determines the actual data rate achieved and not the signal strength alone.

802.11ac - VHT MCS, SNR and RSSI

VHT MCS	Modulation	Coding	20MHz Data Rate 800ns	20MHz Data Rate 400ns	20MHz Min. SNR	20MHz RSSI	40MHz Data Rate 800ns	40MHz Data Rate 400ns	40MHz Min. SNR	40MHz RSSI	80MHz Data Rate 800ns	80MHz Data Rate 400ns	80MHz Min. SNR	80MHz RSSI	160MHz Data Rate 800ns	160MHz Data Rate 400ns	160MHz Min. SNR	160MHz RSSI
1 Spatial Stream																		
0	BPSK	1/2	6.5	7.2	2	-82	13.5	15	5	-79	29.3	32.5	8	-76	58.5	65	11	-73
1	QPSK	1/2	13	14.4	5	-79	27	30	8	-76	58.5	65	11	-73	117	130	14	-70
2	QPSK	3/4	19.5	21.7	9	-77	40.5	45	12	-74	87.8	97.5	15	-71	175.5	195	18	-68
3	16-QAM	1/2	26	28.9	11	-74	54	60	14	-71	117	130	17	-68	234	260	20	-65
4	16-QAM	3/4	39	43.3	15	-70	81	90	18	-67	175.5	195	21	-64	351	390	24	-61
5	64-QAM	2/3	52	57.8	18	-66	108	120	21	-63	234	260	24	-60	468	520	27	-57
6	64-QAM	3/4	58.5	65	20	-65	121.5	135	23	-62	263.3	292.5	26	-59	526.5	585	29	-56
7	64-QAM	5/6	65	72.2	25	-64	135	150	28	-61	292.5	325	31	-58	585	650	34	-55
8	256-QAM	3/4	78	86.7	29	-59	162	180	32	-56	351	390	35	-53	702	780	38	-50
9	256-QAM	5/6			31	-57	180	200	34	-54	390	433.3	37	-51	780	866.7	40	-48
2 Spatial Streams																		
0	BPSK	1/2	13	14.4	2	-82	27	30	5	-79	58.5	65	8	-76	117	130	11	-73
1	QPSK	1/2	26	28.9	5	-79	54	60	8	-76	117	130	11	-73	234	260	14	-70
2	QPSK	3/4	39	43.3	9	-77	81	90	12	-74	175.5	195	15	-71	351	390	18	-68
3	16-QAM	1/2	52	57.8	11	-74	108	120	14	-71	234	260	17	-68	468	520	20	-65
4	16-QAM	3/4	78	86.7	15	-70	162	180	18	-67	351	390	21	-64	702	780	24	-61
5	64-QAM	2/3	104	115.6	18	-66	216	240	21	-63	468	520	24	-60	936	1040	27	-57
6	64-QAM	3/4	117	130.3	20	-65	243	270	23	-62	526.5	585	26	-59	1053	1170	29	-56
7	64-QAM	5/6	130	144.4	25	-64	270	300	28	-61	585	650	31	-58	1170	1300	34	-55
8	256-QAM	3/4	156	173.3	29	-59	324	360	32	-56	702	780	35	-53	1404	1560	38	-50
9	256-QAM	5/6			31	-57	360	400	34	-54	780	866.7	37	-51	1560	1733.3	40	-48
3 Spatial Streams																		
0	BPSK	1/2	19.5	21.7	2	-82	40.5	45	5	-79	87.8	97.5	8	-76	175.5	195	11	-73
1	QPSK	1/2	39	43.3	5	-79	81	90	8	-76	175.5	195	11	-73	351	390	14	-70
2	QPSK	3/4	58.5	65	9	-77	121.5	135	12	-74	263.3	292.5	15	-71	526.5	585	18	-68
3	16-QAM	1/2	78	86.7	11	-74	162	180	14	-71	351	390	17	-68	702	780	20	-65
4	16-QAM	3/4	117	130	15	-70	243	270	18	-67	526.5	585	21	-64	1053	1170	24	-61
5	64-QAM	2/3	156	173.3	18	-66	324	360	21	-63	702	780	24	-60	1404	1560	27	-57
6	64-QAM	3/4	175.5	195	20	-65	364.5	405	23	-62			26	-59	1579.5	1755	29	-56
7	64-QAM	5/6	195	216.7	25	-64	405	450	28	-61	877.5	975	31	-58	1755	1950	34	-55
8	256-QAM	3/4	234	260	29	-59	486	540	32	-56	1053	1170	35	-53	2106	2340	38	-50
9	256-QAM	5/6	260	288.9	31	-57	540	600	34	-54	1170	1300	37	-51			40	-48

Table 6.2: Reference table showing MCS rates, SNR, and RSSI (courtesy Andrew von Nagy)

Effective Isotropic Radiated Power (EIRP) and Antennas

The effective Isotropic radiated power (EIRP) is the power coming out of the transmitter antenna. It is the sum of the radio transmit power, the gain of the transmit antenna, and any losses from the RF cabling between the radio and the antenna.

Each regulatory domain will generally specify the maximum EIRP from a transmitter in an unlicensed band. For example, in the United States of America, the maximum EIRP on the 2.4 GHz band is 36 dBm.

CWNP will never test on the specific requirements for different regulatory domains. However, a CWDP is expected to be fully aware of these regulatory constraints for his or her region and abide by them. When designing wireless networks, make sure you both know and follow the maximum EIRP constraints imposed by the country in which the network is operating.

For wireless stations with either internal antennas or external antennas directly mounted to the antenna port, the cable lengths inside the device are so short that their losses are generally negligible. The losses from antenna cabling only need to be accounted for when the antenna is not co-located with the radio and is therefore connected with an external 50-ohm coaxial cable. The RF losses are a function of the resistance of the cable and its length. In applications with external antennas, therefore, it is generally advisable to use the best quality (i.e., lowest resistance) cable available and to keep the cable lengths as short as possible to minimize RF losses from antenna cabling.

As RF coaxial cable quality increases (i.e., lower resistance), it tends to be harder to work with, as it is generally more expensive, thicker, and stiffer. Thus, if there is margin available in the wireless backhaul link, it can be acceptable to specify a lower quality antenna cable to keep your material and installation costs down.

An antenna shapes the radio signal. It acts like a lens, passively amplifying both the radio's ability to talk and to listen in particular directions. Antenna gains are defined relative to an isotropic radiator, which is a theoretical point source that radiates energy equally in all directions (i.e., spherically). An antenna's gain, therefore, is a measure of its deviation from a perfect sphere. The gain of an antenna is usually measured in "dBi," which literally means "decibels relative to an isotropic radiator." An isotropic radiator cannot be physically constructed but serves as a useful mathematical reference. An isotropic radiator has a gain of 0 dBi.

In older Wi-Fi applications, some antenna gains were defined in "dBd," which is gain relative to a standard dipole antenna. A standard dipole antenna has a gain of 2.15 dBi, so if you ever encounter the use of dBd on an antenna spec, just remember to subtract out this factor when calculating the antenna gain and link budget.

Every AP and antenna vendor provides polar plots showing the shape of the antenna signal coverage. These are usually depicted in two 2D antenna charts known as azimuth (horizontal—the view from the top down on an omni antenna) and elevation (vertical—the view from the side on an omni antenna). The beamwidth is a measurement of how broad or narrow the focus of the RF energy is as it propagates from the antenna along the main lobe. The main lobe is the primary RF energy coming from the antenna. The side-lobes are the other lobes of radiation in the non-intended directions. Beamwidth is measured both vertically and horizontally, so do not let the term width confuse you into thinking it is a one-dimensional measurement. Specifically, the beamwidth is a measurement taken from the center of the RF signal to the points on the vertical and horizontal axis where the signal decreases by 3 dB, or half power, as shown in Figure 6.9. The vertical and horizontal beamwidth measurements are generally stated in degrees.

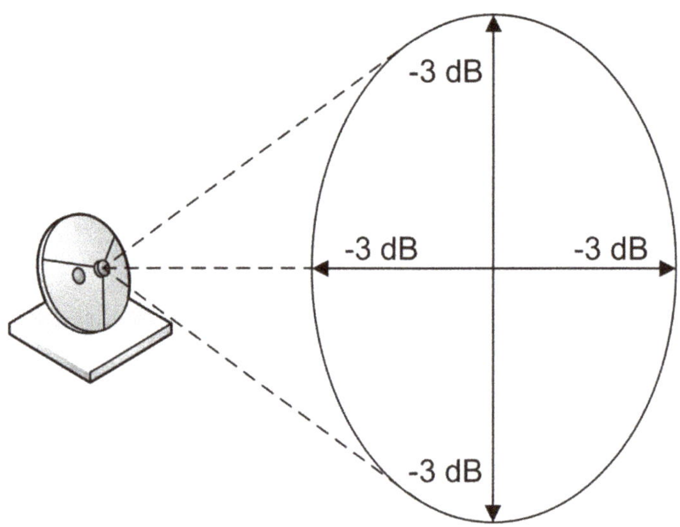

Figure 6.9: Illustration of beamwidth

Remember that the beamwidth is calculated where the signal reaches half power or -3 dB or half power. Remember, however, that coverage does not stop at this point. For more useful visual representations, reference the Azimuth and Elevation charts provided by the AP vendor or antenna manufacturer.

Figure 6.10 shows an example of an antenna polar plot for an AP with an integrated 90° sector antenna, with Azimuth (H) chart, and Figure 6.10 shows an example of an Elevation (V) chart. The 3D pattern for this antenna is rendered in Figure 6.11.

> The marketing department of at least one major AP manufacturer has recently acquired the bad habit of defining their beamwidths over -6 dB vs. -3 dB. This is intended to make the beamwidths appear larger on the spec sheets than they are in fact. While you can find this reclassification in the fine print, it is not readily obvious just by looking at the spec sheet. Caveat Emptor!

The difference between an Azimuth and Elevation chart is simple: the Azimuth chart shows a top-down view of the propagation path (to the left, in front, to the right, and behind the antenna), and the Elevation chart shows a side view of the propagation path (above, in front, below, and behind the antenna). Think of these charts in terms of an antenna that is positioned vertically upright. If you are standing directly above it and looking down on it, you see the perspective of an Azimuth chart. If you are beside it and looking at it from a horizontally level position, you see the perspective of an Elevation chart.

As with most Azimuth charts, the direction of propagation is represented in the upward direction; however, the actual direction will depend on how you position the antenna. The chart is reporting the different signal strength you can expect at different degrees from the antenna. For example, at 90° and 270 ° (to the immediate left and right of the antenna's intended propagation direction) you will see a loss of approximately 5-6 dB. Directly behind the antenna, at 180 °, you will see a loss of approximately 9 – 12 dB. This is a sector antenna and is intended to propagate its energy in one direction but in a fairly wide pattern.

For dual-band APs, the azimuth and elevation charts will be different for each band. Most antenna designers try to make the coverage area reasonably close on both bands to keep the coverage areas the same. However, antenna design often involves compromises, especially when keeping antennas to limited sizes and/or using the same antenna elements for dual-band coverage. It is important to understand both the gain and the coverage area differences when designing for a dual-band AP.

Table 6.3 provides a table of common beamwidths for various antenna types.

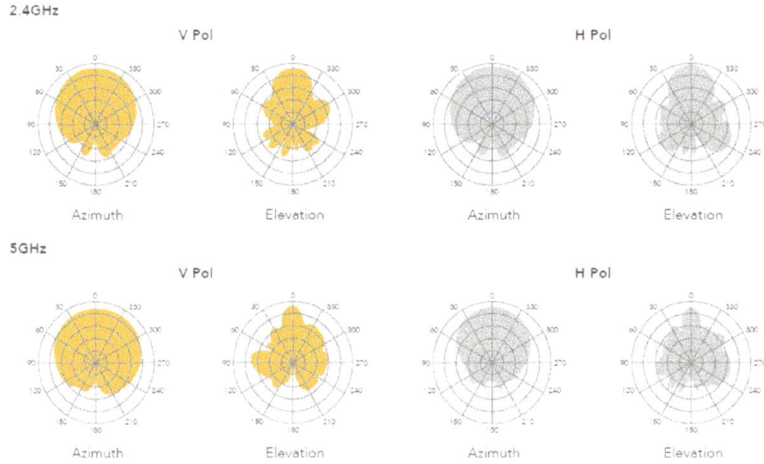

Figure 6.10: Antenna polar plot for LigoWave NFT Blizzard 2ac-90. The plot shows both azimuth and elevation readings for vertical and horizontal polarization. (Source: LigoWave Networks Inc.)

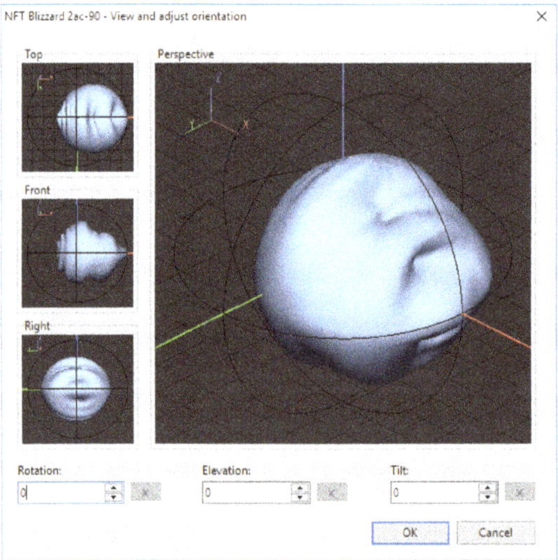

Figure 6.11: 3D Rendering of NFT Blizzard 2ac-90 antenna (5 GH GZz). Image rendered by Tamograph v. 5.0 (build 222)

Antenna Type	Horizontal Beamwidth	Vertical Beamwidth
Omni-directional	360 degrees	**7 to 80 degrees**
Patch/panel	30 to 180 degrees	**6 to 90 degrees**
Yagi	30 to 78 degrees	**14 to 64 degrees**
Sector	60 to 180 degrees	**7 to 17 degrees**
Parabolic dish	**4 to 25 degrees**	**4 to 21 degrees**

Table 6.3: Common beamwidths for various types of antennas.

Antennas are a key parameter to be controlled in Wi-Fi designs. APs with internal antennas are often selected for applications for the following advantages:

- Aesthetics: people generally don't like seeing antennas
- Cost: If the antenna is included, one doesn't need to purchase separate antennas and antenna cabling
- Ease of Installation: One component vs. several components to install.
- Maintaining MIMO / MU-MIMO Performance: For APs that utilize a series of dipole antennas for MIMO and/or MU-MIMO, the antennas need to be at the appropriate phase offsets to work optimally. If the antennas are misaligned when installed (or become so later), MIMO / MU-MIMO performance will degrade.

When an AP model with an internal antenna is selected, the AP manufacturer has dictated the antenna to be used. Nonetheless, some vendors offer multiple models in a product family where the only material difference between products is the embedded antenna. This is very common for both outdoor Wi-Fi as well as PTP/PTMP applications.

Propagation of the Signal Between the Transmitter and Receiver

As the signal leaves the transmitter and heads towards the receiver, it is subject to two major effects, known as free-space path loss and attenuation.

Free-space path loss (FSPL) is the geometric effect of a signal weakening as it gets further away from the transmitter. FSPL is defined relative to an isotropic radiator since the effect of the antennas at both the transmitter and receiver to focus the signal on each end is already accounted for separately in the link budget. As the signal propagates farther away from a transmitter using an isotropic radiator, the EIRP remains constant but is spread out over a greater area, i.e., the circumference of the isotropic sphere. The receiver (with its own isotropic radiator) is effectively exposed to only one small portion of the circumference of that sphere. Thus, the further away from the receiver, the more spread out the energy of the transmission, and the smaller sample of that energy that the receiver encounters.

The formula for calculating FSPL is as follows:

$$\mathrm{FSPL(dB)} = 20\ \mathrm{LOG10}(d) + 20\ \mathrm{LOG10}\ (f) + 20\ \mathrm{LOG10}\ (4\pi/c)$$

Even though FSPL is a geometric effect, for the purposes of calculating a link budget, FSPL is defined relative to an isotropic radiator at a specific frequency. Accordingly, there is a wavelength component in the FSPL calculation, as it is a convenient place to account for the physical differences in wavelengths when it comes to signal propagation. As a result, FSPL is generally more pronounced as frequency increases. This partially explains why higher frequency signals do not propagate as far as lower frequency signals, given the same transmit power and antenna gain.

FSPL is the largest single source of loss in a link budget, even for a client device relatively close to an AP. You should recognize the formula for FSPL, though you will not be required to define or calculate it.

Attenuation is a general term for the loss of signal as it encounters and interacts with an object in the environment. For conventional indoor Wi-Fi applications, the two main forms of attenuation encountered are *absorption* and *reflection*.

Absorption is the amount of energy that is physically absorbed by an object, such as a wall, as the RF energy propagates through it. The absorption is measured in dB, as it is a multiplicative loss.

The Laws of Thermodynamics indicate that the RF energy absorbed by a wall or other object is ultimately converted to and released as heat into the environment. At the energy levels at which Wi-Fi operates, however, one would need extremely sensitive equipment in a highly controlled environment to measure this effect.

Different materials have different rates of absorption. Generally, most common building materials absorb more energy at higher frequencies, which also effectively serves to limits the range of higher frequency signals as compared to lower frequency signals.

For example, standard drywall has a loss of 3 dB at 2.4 GHz and 6 dB at 5 GHz. To put this in perspective, this means in an indoor environment where a transmitting dual-band AP is in the next room from client devices, 50% of the 2.4 GHz transmission energy is lost, and 75% of the 5 GHz transmission energy is lost, going through a single piece of drywall!

When designing for dual-band APs, which is fairly standard in most Wi-Fi deployments, it is common to do an offset in the transmission power settings between the 2.4 GHz and 5 GHz radio, so that the coverage area is roughly equal on both bands. For APs where the antenna gains are equal on both bands (most vendors seem to have indoor AP models with either 3 dBi at 2.4 GHz / 3 dBi at 5 GHz or 5 dBi at 2.4 GHz / 5 dBi at 5 GHz internal omni antennas), an offset in transmit power of 6 dB between the bands is a typical recommendation. This offset should be adjusted in cases where the antenna gains in a dual-band AP are not matched (e.g., an AP with internal 5 dBi at 2.4 GHz / 7 dBi at 5 GHz).

For indoor deployments with a lot of structure in relatively small areas (i.e., several walls, including closets, bends in hallways, several little rooms such as a medical clinic, etc.), just using transmit power offsets to equalize the dual-band coverage area can prove insufficient. In such cases, the recommended approach is to lay out the APs for enough 5 GHz coverage, and then selectively disable the 2.4 GHz radios on specific APs to prevent too much coverage on the 2.4 GHz band. Sometimes, transmit power on the 2.4 GHz band needs to be increased by a few dB to provide sufficient coverage. Such is the art of Wi-Fi design!

One potential scenario soon is "multi-band" APs, e.g., 900 MHz / 2.4 GHz / 5 GHz tailored for IoT applications or 2.4 GHz / 5 GHz / 60 GHz for high-density applications. If such AP products should come to pass, the use of transmit power offsets and disabling certain radios on specific APs will become even more important, and trickier, in optimizing Wi-Fi designs.

The other significant attenuation effect of indoor Wi-Fi deployments is *reflection*, where a portion of the signal is reflected by the object instead of passing through it. If the receiver is on the other side of the object, any energy reflected away is now heading in the wrong direction and will not get to the receiver. Reflectivity is typically indicated as a percentage. As with absorption, different building materials will have different reflectivity factors, though unlike absorption, reflectivity may or may not differ substantially as a function of wavelength. For rough surfaces, the reflection will get spread out in several directions (think of a highly faceted mirror), which is an effect referred to as scatter.

Predictive design modeling tools such as Ekahau, AirMagnet, iBwave, and TamoGraph have "standard" values for both absorption and reflectivity for common building materials. An example from TamoGraph is shown in Table 6.4.

The absorption and reflection of actual building materials in a particular facility may not be "standard," so when possible, it is always best to measure the attenuation during a site survey and adjust the predictive design model accordingly.

Building Material	2.4 GHz		5 GHz	
	Absorption	Reflection	Absorption	Reflection
Brick 3.5"	6 dB	6%	10 dB	13%
Brick 10"	10 dB	6%	25 dB	13%
Cubicle Divider	1 dB	12%	2 dB	0%
Concrete 8"	10 dB	40%	13 dB	30%
Concrete 18"	18 dB	40%	30 dB	30%
Concrete 27"	30 dB	40%	45 dB	30%
Drywall	3 dB	6%	6 dB	7%
Glass (interior)	3 dB	7%	6 dB	32%
Glass (exterior)	7 dB	7%	6 dB	32%
Glass (exterior coated)	13 dB	7%	20 dB	32%
Steel Fire Door 1.75"	13 dB	90%	25 dB	90%
Steel Fire Door 2.5"	19 dB	90%	32 dB	90%
Wood Door (hollow)	4 dB	12%	7 dB	0%
Wood Door (solid)	6 dB	2%	10 dB	3%

Table 6.4: Typical material absorption and reflection values for common building materials. (Source: TamoGraph Version 4.2, Build 185).

Reality is naturally more complicated. Signals may reflect multiple times, especially in environments with highly reflective materials, such as metal walls. While reflections serve to increase multipath which is good for increasing MIMO, a lot of reflections can ultimately serve to increase the noise floor. There are real-world applications in designing for highly reflective environments, such as point-to-multipoint links in elevator shafts to enable surveillance or Wi-Fi inside each elevator cab.

For outdoor deployments, especially for point-to-(multi)point applications, there tend to be few if any objects directly in the path, and thus there isn't a lot of absorption or reflections to worry about.

The lack of objects in an outdoor environment also limits MIMO. Two-stream MIMO for PTP/PTMP links is achieved via polarization, i.e., one antenna is vertically polarized, and the other antenna is horizontally polarized. Thus, modern APs designed for point-to-(multi)point applications are always dual-stream, i.e., 2x2:2. This practical limitation of physics, however, has not stopped some AP vendors from manufacturing and peddling 3x3:3 MIMO and 4x4:4 MU-MIMO Wi-Fi access points for use in outdoor environments. Another example of the power of marketing.

There are two other attenuation effects that must be accounted for when designing outdoor point-to-(multi)point links: diffraction and earth bulge.

Diffraction occurs when an RF signal travels around an object, and the object serves to bend the signal in a different direction. As diffraction becomes more pronounced, more of the RF energy simply gets diverted away from the receiver, causing lower RSSIs. In radio communications, diffraction occurs with objects in the RF line of sight.

If you stand on top of a tall building, you can see for a very great distance. You may even be able to see for several kilometers / miles on a clear day. If you can physically see something, it is said to be in your visual line of sight (visual LoS) or more colloquially your line of sight (LoS). This visual LoS is actually the transmission path of light waves from the object you are viewing (transmitter) to your eyes (receiver). Visual LoS is an apparently straight line from your perspective, but light waves are subject to similar behavior as RF waves like refraction and reflection. Therefore, the actual visual LoS may not be straight. Consider viewing an object in a mirror. The object is not directly in front of you, yet it appears to be so, showing that visual LoS is not necessarily a straight line between two objects.

Since RF is part of the same electromagnetic phenomenon as visible light, RF behaves similarly to visual LoS. However, because the wavelengths for Wi-Fi (2-6 in / 50-300 mm) are substantially longer than the wavelengths for visible light (300-700 nm), more space is needed for the RF waves to be seen at each end of the connection. This extra space is calculated as the Fresnel Zone (pronounced frah-nell).

Fresnel Zones, named after the French physicist Augustin-Jean Fresnel, are a theoretically infinite number of ellipsoidal areas around the LoS in an RF link. The impact on the wireless link from the first Fresnel zone dominates; the impact of objects located in the second and successive Fresnel Zones collectively total a small percentage of that of the first Fresnel zone. Accordingly, only the first Fresnel Zone is usually considered when planning a point-to-(multi)point link. By popular convention, this text will use the term "Fresnel Zone" to refer to the first Fresnel Zone and will ignore the effects from the successive Fresnel Zones.

The Fresnel Zone has been referenced as an ellipsoid-shaped area, an American football-shaped area, and even a Zeppelin-shaped area. Figure 6.12 shows how a Fresnel Zone is typically depicted.

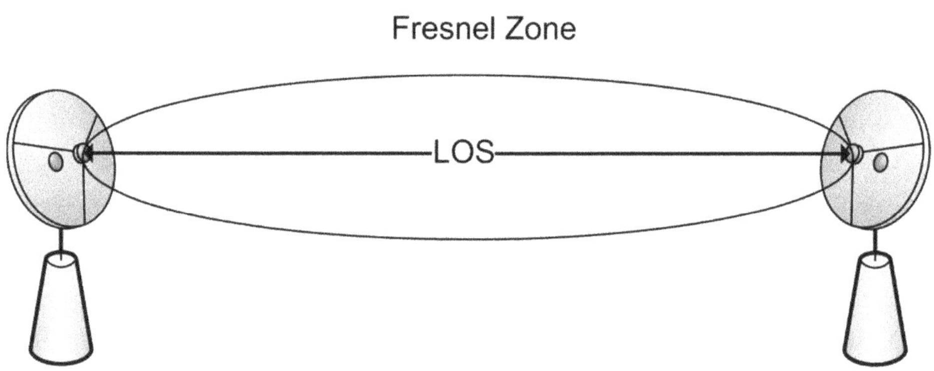

Figure 6.12: Illustration of the Fresnel Zone.

The Fresnel Zone is an area surrounding the visual LoS and must remain at least 60% free of obstructions to have a minimal material impact on the link. Wireless link designers are encouraged to try and keep the area 100% free of obstructions. This is usually done by raising the height of the links, but that solution is sometimes impractical if not impossible for specific applications, especially when shooting under a tree canopy line in a wooded campground or RV park.

Also, remember that some level of margin is always a good idea in wireless design. For instance, when erecting towers and shooting above the existing tree line, trees have this nasty tendency to grow into your Fresnel Zone over time. The only ways of fixing that issue are to either (a) raise the links higher every few years, or (b) get a hold of a chainsaw and do some tree trimming. Having personally encountered this problem on more than one deployment, neither option is particularly attractive, if even acceptable at all, to the customer.

Know that minor obstructions into the Fresnel Zone won't kill your link, but you should always try to minimize them. Also, always leave yourself comfortable margins when designing point-to-(multi)point links.

The formula to calculate the radius of the Fresnel Zone radius *R* at any distance along the path is as follows. Here, the Greek lambda is the wavelength used for transmission, *d1* is the distance from the transmitter to the point of interest, and *d2* is the distance from the point of interest to the receiver, such that *d1* + *d2* = *D*, which is the total distance of the link.

$$R = \sqrt{\frac{\lambda d_1 d_2}{d_1 + d_2}}$$

To calculate the maximum radius of the Fresnel Zone, which occurs halfway between the transmitter and receiver where d1=d2=D/2, this formula reduces to the following:

$$R_{max} = \frac{1}{2}\sqrt{\lambda D}$$

Especially for short distance links which will be aligned by visual LoS, it is usually much easier only to calculate the maximum Fresnel Zone radius, and simply tell your installers to keep that cylindrical area clear, which is usually much easier to visualize. It's also a good idea to round up your Fresnel Zone calculations, to ensure some additional margin.

The chart in Figure 6.13 shows you the maximum Fresnel Zone radius for 5 GHz links going out to a distance of 1000 ft / 300 m.

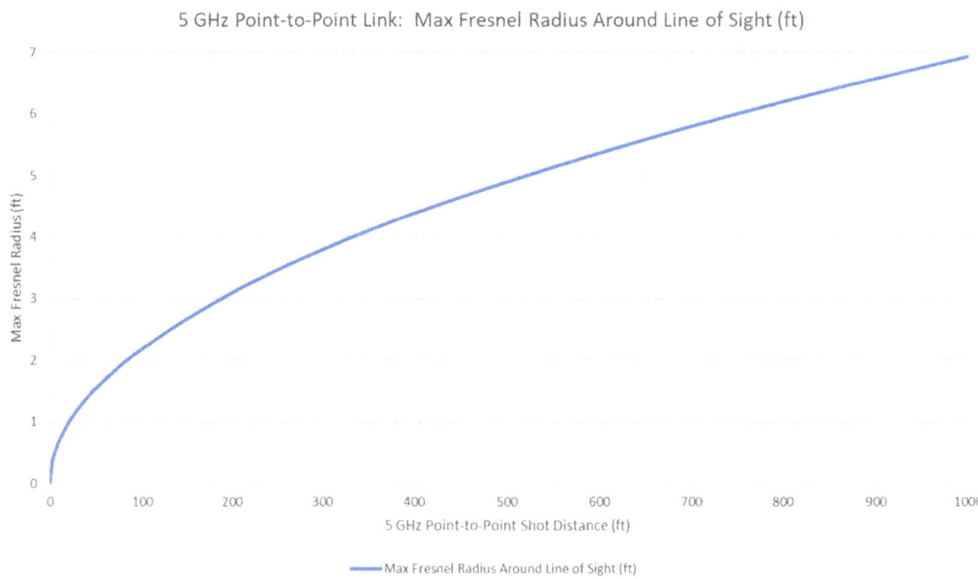

Figure 6.13: Fresnel Zone calculation for 5 GHz links under 1000 feet

In many point-to-(multi)point applications, the Fresnel Zone blockage might not be coming from below. When doing PTP/PTMP links indoors or in a wooded area with a trimmed under the canopy at a fixed height, the blockage may come from above (e.g. light fixtures, rafters, trees, etc.). Such links tend to be reasonably short, so the Fresnel Zone isn't necessarily that big, but you don't necessarily want to deploy the PTP / PTMP links near the ceiling. In dense urban environments or even garden-style apartment complexes, it's very common to need to shoot between buildings, so Fresnel Zone obstructions can come from the sides.

> Remember the lesson of Star Trek II: Always think three dimensionally in your wireless designs. Always think four-dimensionally when trees are involved!

Another factor that must be considered for long-distance wireless links is the curvature if the Earth itself. Because the Earth is round, the surface of the planet can itself project into the Fresnel Zone. Generally, the only way to avoid this is to raise the height of your APs.

This is illustrated in Figure 6.14. When creating wireless links over distances greater than 7 miles, *earth bulge* will need to be accounted for.

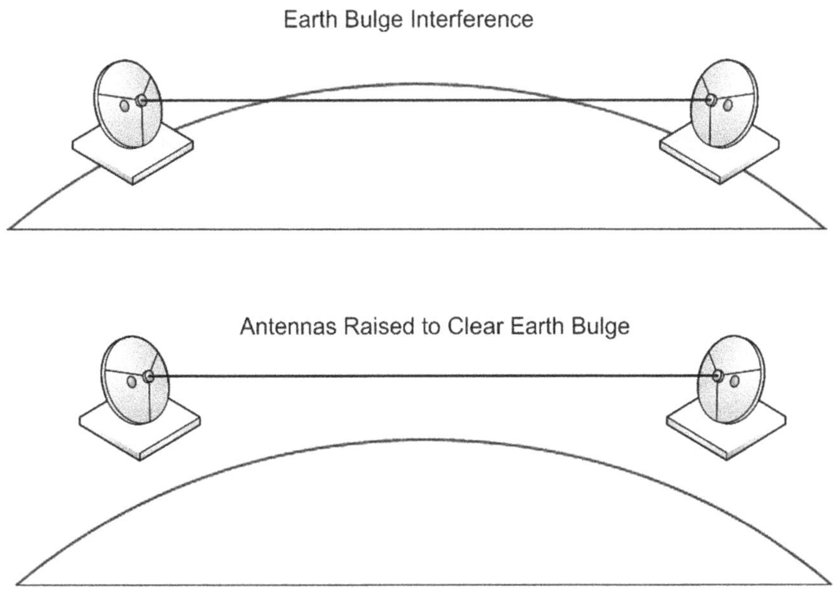

Figure 6.14: Illustration of earth bulge

After aligning the antennas for a long-distance link, they will appear from the ground as if they are pointed downwards, which, in fact, they are. Such links are shooting over the horizon. If the antennas were mounted perpendicularly to the vertical tower or building, the antenna would literally be shooting into space at a tangential angle.

The formula for calculating the extra height your antennas will need to compensate for earth bulge is as follows:

$$H = D^2/8$$

Where *H* is the height of earth bulge in feet and *D* is the distance between antennas in miles. For example, an 11-mile link would require the following additional height to:

$$11^2/8 = 15.12 \text{ feet}$$

Rounding up, one would need to raise the antenna height by 15.5 to 16 feet to accommodate for earth bulge.

Note that these formulas are additive, so a long-distance link with objects in the middle would have to be raised both to keep the Fresnel Zone clear and to account for the curvature of the Earth.

You will not need to memorize the formulas provided here for the CWDP examination; however, they will prove a useful reference for you when you need to create PTP / PTMP links in the future. Some vendors that specialize in point-to-(multi)point applications provide free online calculators for working out Fresnel Zones and minimum antenna heights, in addition to bandwidth requirements.

Tying Everything Together at The Receiver with the Link Budget

The antenna at the receiver, along with any incurred cable losses, works the same way as it does at the transmitter. The significant difference is that there is no receiver power.

There are radio systems that do active amplification at the receiver. This technique is not used in Wi-Fi, however, as it would provide no practical benefit. Remember that the quality of a wireless link is always ultimately defined by achievable data throughput, which is driven by the achievable MCS rate, which is ultimately achieved by maximizing the signal-to-noise ratio (SNR). A power amplifier at the receiver would serve to amplify both the signal and the noise. Thus, a power amplifier at the receiver would increase the RSSI but also increase the noise floor, thus leaving SNR constant. (The power amplifier is actually likely to introduce some level of noise by itself, so realistically SNR would likely be a tad worse.)

Long-distance PTP links are conventionally done with licensed microwave frequencies, such as 10 GHz, so as to avoid interference from other sources. However, links that are mounted very high and/or in rural areas likely don't have external interference sources, and it comes down to your requirements and your link budget.

The receive sensitivity of most access points is typically significantly better than the receive sensitivity of client devices. Essentially, APs have better hearing than client devices. This becomes extremely important when setting transmit power levels on APs. Some vendors publicly advocate matching the transmit power on the AP to the transmit power of the worst client device, so that the link budgets between the AP and client device will be balanced in both directions. Thus the same MCS rates can be achieved in both directions. However, doing this usually results in very low power level settings on the AP, and thus the need to deploy more APs for a given coverage area. It is fine for the APs to speak "a little louder" (i.e., 3-6 dB) than the client devices, while still maintaining a balanced connection (i.e., the same or similar MCS rates) in both directions.

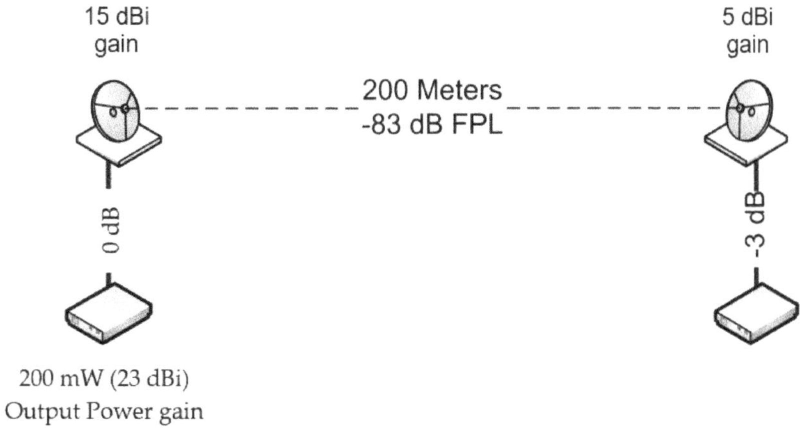

Effective Isotropic Radiated Power (EIRP)		38 dB
- Transmitter Power	23 dBm	
- Transmitter Cable Losses	-0 dB	
- Transmitter Antenna Gain	15 dBi	
Free Space Path Loss (FSPL)		-83 dB
Attenuation		-0 dB
Receiver Antenna Gain		5 dBi
Receiver Cable Losses		-3 dB
RSSI at Receiver		-43 dBm

Figure 6.15: Sample link budget calculation

Furthermore, depending on the application, it may also be perfectly fine for the MCS rates to widely imbalanced; specifically, it is often acceptable to be slower in the upstream direction (client device to AP). In a typical indoor Wi-Fi deployment primarily servicing low-power mobile client devices, most traffic is downstream (e.g., streaming real-time video from services like YouTube or Netflix), so it is important that SNR and MCS rates be as high as possible. The traffic upstream is requests or email. Even if the user is uploading videos to social media, that is not a real-time application, and thus it is acceptable that such transmissions may take longer.

The counter-argument to this is the growth of video chat applications like FaceTime and Skype. However, such chats are usually not at overwhelmingly high video resolutions, as compared to streaming an HD or 4K movie. As always, understand your actual usage requirements before jumping into a design solution, else you risk deploying a very clever and expensive solution to a problem that doesn't exist in practice.

For applications like wireless video surveillance, the bulk of the traffic is upstream from the camera to the NVR server in the MDF. Most wireless IP cameras use cheap 2.4 GHz only chipsets, and serious surveillance integrators have been burned by those devices too often to use them. Wireless surveillance is therefore typically done using one or more wired IP cameras connected to a wireless bridge, thus the antenna and transmit power level settings of that wireless bridge AP are controllable aspects of the wireless design.

Antenna Alignment

Antenna alignment is critical for ensuring the strongest RSSI on each side of a point-to-point link. The procedures you follow will depend on the link distance and the fade margin you have in your link budget.

The goal is to maximize the SNR to the point where the link is reliably and consistently achieving the highest MCS rate. For 5 GHz 802.11ac links at 80 MHz, an SNR of at least 37 dB is required. Assuming no interference of the channel and thus a noise floor of -94 dBm, you would need an RSSI of at least -94 dBm + 37 dB = -57 dBm. There will generally be some low level of background noise, and the extra margin is always good, so usually it is best to target an RSSI in the -40 dBm to -50 dBm range. For short distance links, this is usually achievable easily. For long-distance links, this can be challenging but achievable. For very long-distance links, FSPL will dominate, making this RSSI target impossible; such applications, however, are usually focused on reliable data connections and not quite as obsessed with maximum data throughputs, making lower MCS rates acceptable.

Most PTP/PTMP vendors will have built-in alignment tools. This may include LEDs on the AP, as well as screens in the GUI to show the RSSI from the other side. These should always be used for *fine alignment* of the APs in making those minor adjustments to the horizontal and vertical tilt of each antenna to maximize RSSI before locking down their mounts. The methodology used for *coarse alignment* (i.e., finding the other side of the link so that they see each other, to begin with) will depend on the link distance and the beamwidths of the antennas. Many PTP / PTMP link applications are quite short (< 1000 ft / 300 m), and thus alignment need not be very challenging. Short links are common in surveillance applications, mesh applications, and when connecting neighboring buildings on a property. In such applications, the link budget is fairly forgiving, and it is common to use APs with internal directional patch antennas that have broad beamwidths (e.g. > 30°). This is done on purpose so that it is sufficient to just align them by "eye" (i.e., visual LoS). For slightly longer shots, a rifle-like sight is useful, in conjunction with using levels. Some vendors have levels and/or angular protractor marks on their mounting hardware to facilitate this.

For really short links (e.g., less than 100 feet or 30 meters), it is possible to get very strong RSSIs in excess of -30 dBm, depending on the antenna gains used. An RSSI above approximately -35 dBm can actually degrade performance because it is "too loud" for the receiver. In such circumstances, it's appropriate to purposely misalign the antennas to keep the RSSI below -35 dBm.

In point-to-multipoint applications, the base station (i.e., the "point") must cover a wide area, so it is common to use a broad patch, sector, or even omnidirectional antennas at the base station. That decision will be dictated by the area of coverage required and the capacity. In high capacity applications, it's common to use multiple bases to provide 360° coverage, such as 4 APs each with 90° sector antennas on independent channels. Alignment is therefore typically one-sided, where a high gain / small beamwidth antenna is used at each of the multi-points to point back to the base. Point-to-multipoint applications typically have a combination of short distance and longer distance links, so care needs to be taken in both transmit power settings and antenna selection at each endpoint to maximize the RSSI. In cases where particular endpoints are outside the field of view of the base station, but the distances are short, one can use a high gain antenna at the endpoint and aim into a side-lobe of the base antenna and maintain a good link.

For long-distance links, Visual LOS is typically not an option. These types of alignments generally require coordination on both ends of the link to adjust the horizontal and vertical tilts of each high gain / small beamwidth antenna to achieve coarse alignment. For shots of up to a few miles, laser sights or even GPS systems can be used. When shooting over the horizon, some vendors also have tools to help indicate the points on the horizon that the antenna should be aiming for.

Interference

A link budget does not account for interference from neighboring Wi-Fi or other sources. However, interference is usually the largest bane of any Wi-Fi network. At best, interference will raise the noise floor and lower SNR. At worst, interference causes retransmissions resulting in lower effective data throughputs and spotty performance.

Interference always occurs at the receiver, when it receives multiple transmissions on the same or overlapping channels simultaneously. The receiver is unable to distinguish the messages from each other, and this receives a garbled combination of the transmissions. The receiver, therefore, doesn't understand the message and doesn't send an ACK, forcing the transmitter to retransmit the same message again until it gets through.

You can think of interference like being at a cocktail party when multiple people are talking. Despite the surrounding conversations, you can generally hold a conversation with someone by focusing in on the pitch of his or her particular voice (i.e., channel) as opposed to the other voices in the room (i.e., on independent channels). However, if someone near you is also speaking in a separate conversation it has the same or similar pitch to their voice (i.e., overlapping or identical channel), you hear both conversations at once and will be unable to distinguish them, forcing the person who is speaking to you to repeat himself or herself.

The symptoms of interference are typically complaints of slow speeds, inconsistent speeds, and even frequent disconnects. Interference gets worse as traffic levels on the APs increase. Two neighboring APs set to the same channel will not interfere with each other if there are no clients connected to them. Even under light usage conditions (i.e., a small number of clients on each AP that are only each passing small amounts of data traffic), interference effects will be small; with relatively little traffic, the statistical likelihood of two transmissions simultaneously occurring is very low. However, as usage increases, transmissions back and forth are happening more frequently, which raises the likelihood that two transmissions associated with the two different APs will either simultaneously occur or that the physical carrier sense for stations on one AP will trigger because of activity from the neighboring AP, effectively slowing down that station. Under high utilization conditions, it is extremely likely that multiple transmissions will occur simultaneously, and physical carrier sense will get triggered quite often, grinding down actual traffic on both neighboring APs to a slow crawl as many of the stations contend simultaneously for access to the channel.

Interference can be classified into two categories. Self-interference is interference from your own APs on your own network, whereas external interference is interference from neighboring APs on different networks or non-Wi-Fi sources. The effects on the network are the same. The difference lies in control. As the Wi-Fi designer, you have total control over the settings on your own APs. Furthermore, the closest APs to you are always your own, which means interference from your own APs will generally be stronger than interference from external sources (with some notable exceptions, discussed below).

As a CWDP, one of your primary design goals is to minimize self-interference. You have complete control over the channel and transmit power settings of your own APs, so you have no excuse to need to tolerate self-interference. Note, with only three channels on 2.4 GHz band and relatively low signal attenuation, it is usually impossible to eliminate self-interference on the 2.4 GHz band, but the right combination of channel and transmit power settings can minimize self-interference to acceptable levels. For 5 GHz, especially when using smaller channel widths, self-interference can usually be avoided.

There is an ongoing debate within the wireless engineering community as to whether or not *auto-channel* should be used in deployments vs. assigning static channels to each AP in an alternating pattern. Auto-channel is intended to enable the AP to adapt to uncontrolled changes in the RF environment (i.e., sources of external interference). Auto-channel is not a part of the IEEE 802.11 standard, and thus every AP vendor has a proprietary approach, some of which are quite elaborate. While it can certainly be observed that some AP vendors perform auto-channel "better" than others, none of them actually do it right. Automatic channelization turns out to be a highly complex nonlinear optimization problem that changes with every network. In short, auto-channel introduces complexity into your network, and from design theory, added complexity usually serves to degrade performance.

Where auto-channel is useful, or perhaps more correctly unavoidable, is in environments infested with external Wi-Fi networks. Such networks are either immediately adjacent to (or, even worse, co-located with) your own networks. For indoor Wi-Fi deployments, immediately adjacent Wi-Fi networks are commonplace in multi-tenant residential and commercial buildings. For example, if your business occupies the 10th floor of a 20 story office building, it is very common to see the Wi-Fi signals from neighboring businesses on surrounding floors, even from businesses that are several floors away, especially on the 2.4 GHz band. Apartment buildings are often even worse. A typical apartment is usually around 800-1200 square feet. The cable or telco company providing Internet service typically deploys a consumer-grade Wi-Fi access point / router designed to cover a 2500 square foot house, naturally blasting at full transmit power, and all-too-frequently set to use 40 MHz channels on the 2.4 GHz band and 80 MHz channels on the 5 GHz band, because they market their service on theoretical speed, not actual usability. It is therefore quite common to see many neighboring transmissions from numerous adjacent apartments, and for 2.4 GHz and at least the UNII-1 and UNII-3 bands of 5 GHz to be completely saturated.

Outdoor networks suffer the same or even worse external interference effects, especially in urban or suburban environments. For outdoor Wi-Fi applications like a public park in a downtown area, all of the residential and commercial buildings are blissfully blasting out their own Wi-Fi for all the neighborhood to hear. For the surveillance example above, the use of directional antennas helps to filter out some of the surrounding interference, though such environments can still be very challenging.

There is, unsurprisingly, a lot of interest in using alternative frequencies. For example, IEEE 802.11ad opens up the use of the 60 GHz band (57 GHz – 66 GHz, depending on region) for Wi-Fi applications. With 2 GHz channels and very high attenuation through virtually all common building materials, it is considered to be attractive for very high bandwidth, very short distance applications. At 60 GHz, even normal drywall will highly attenuate the signal, making it only practical for single room Wi-Fi. Some PTP/PTMP vendors have also started introducing 60 GHz PTP/PTMP links to the market for gigabit data throughputs over a few hundred feet. Unfortunately, weather effects such as rain, snow, and fog will also attenuate the 60 GHz signal, which means signal degradation or loss as a function of weather. Nonetheless, 60 GHz may wind up being a more attractive technology from an interference-avoidance standpoint vs. just a raw speed perspective.

There is also a lot of interest in the 6 GHz band (5.9 GHz – 7.1 GHz), especially for point-to-multipoint applications, to avoid the saturated 5 GHz channels. This band is lightly used and is unlicensed in some regulatory domains, including Russia, Africa, and parts of the Middle East. Another attractive aspect of this band is that it can be accessed with existing 802.11ac 5 GHz chipsets, using some dedicated electronics to shift the operating frequency upwards and antennas optimized for use at 6 GHz.

The best practices for 60 GHz wireless designs will clearly be different than what we are used to from 2.4 GHz and 5 GHz. At the time of this writing, 60 GHz seems to have a place in the PTP/PTMP arena. For Wi-Fi applications, the success or failure of 802.11ad will ultimately depend on whether a critical mass of client devices support the technology, which didn't happen with 802.11ac wave 2 and appears dubious for 802.11ax – at least in the near future. Always understand your "usage" functional requirement in selecting your AP technology.

At the time of this writing, the FCC is evaluating whether to open this band for Wi-Fi applications or not, and many worldwide regulatory domains follow the lead of the FCC. That said, it is likely that the FCC would require some type of DFS protections for these bands if they are opened for Wi-Fi bands, which will likely necessitate the need for new chipsets. Time will tell.

The good news is that, despite the trend of commoditization, Wi-Fi is still a growing market with interesting and exciting technology developments.

6.4: Non-Wireless Solutions that Impact WLANs

There are several companion networking technologies that are used in conjunction with Wi-Fi and/or wireless backhaul deployments. While detailed expertise in each of these areas is not required for the CWDP, as a wireless engineer you will likely to also be responsible for implementing several of these technologies throughout your career.

The intent of this section is, therefore, to give a broad introduction to the most common technologies and applications you are likely to encounter.

VLANs and Client Isolation

VLANs, or **V**irtual **L**ocal **A**rea **N**etworks, are one of the most powerful technologies in wired and wireless networking. While in use for many years in large enterprise deployments, it is only now beginning to see use in Wi-Fi networks for small-to-medium-business (SMB) applications. VLANs are usually not documented or explained well, making VLANs both misunderstood and underutilized. This section explains VLANs from a practical perspective and the applications where they should be used.

At its simplest, VLANs enable you to transform one physical Local Area Network into multiple, isolated, logical Local Area Networks. Thus, you literally have multiple wireless and wired LANs with different purposes and intents that are co-located physically, without the expense of multiple sets of hardware and multiple sets of cabling. VLANs are illustrated in Figure 6.16.

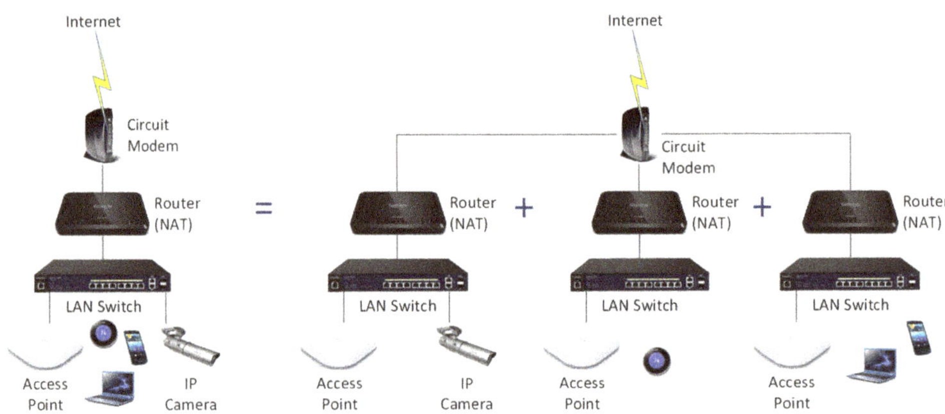

Figure 6.16: Conceptual hardware view of VLANs. (Source: Imperial Network Solutions, LLC.)

This is extremely useful for even small network applications, especially with the growth of wireless surveillance, IoT, and the proliferation of general network appliances measuring and controlling our environment.

Even in the simplest case of a private home, where only one AP may be needed to provide full Wi-Fi coverage, there is generally at least the need for two distinct and isolated networks. First and foremost, a network is needed for the residents of the home, allowing access to all PCs, network printers, multimedia devices such as Apple TV or SONOS, IP cameras, NEST thermostats, and the like. However, a separate network is needed by *guests*, as the homeowner undoubtedly doesn't want their teenager's friends to hack into any of the computers or network appliances in the home. In more complex

SMB applications, such as coffee shops, restaurants, doctor's offices, and multi-dwelling units such as apartment buildings, dormitories, or hotels, there are also usually at least two distinct and isolated networks needed. There is the network needed by the business staff for operations, such as point-of-sale, IP surveillance, access control, HVAC control, multimedia, etc. There is also the need for the businesses' customers to be able to simply and easily access the Internet, but not have any access to any network devices used for operations.

Generically, most Wi-Fi networks need to be segmented at least into an operations network and a visitor network. The requirements for who is allowed access to these networks and how they get used are quite different.

- Operations: Typically, access to this network needs to be restricted to authorized personnel, as it contains data that is confidential to the business and critical to day-to-day operations, especially financial and security data. For a small network, typically WPA2 Personal (i.e., an AES encrypted pre-shared key or passphrase) is sufficient security: the code must be known in order for the wireless client to connect to this network. In larger networks, using 802.1X with RADIUS is appropriate. Once connected to the network, the client device likely needs access to other wireless devices and appliances on the network, so client devices need the ability to intercommunicate with each other. In some environments, breaking up the operations network into multiple separate VLANs by function is appropriate, especially to separate out applications like security (i.e., cameras and access control), facilities (HVAC, lighting control), and voice over IP / voice over Wi-Fi.
- Visitors: Also, commonly called a "hotspot," access to this network needs to be unrestricted, as the facility has no control over what devices their customers are bringing in and using. Thus, typically no encryption is used to facilitate access, though a captive portal may or may not be used to capture email / social media information or set out terms and conditions. Once connected, visitors should only have access to the Internet, and have no access to any network devices on the operations network, nor access to any other client devices on the visitor network. You don't want a hotel guest hacking into the device in a different room.

Most access points support multiple SSIDs. Multiple SSIDs are most often used with VLANs to segment user traffic by type of client. Once segmented, the flow of the traffic can be controlled on the network.

Many SMBs tend to provide a "visitor" network that requires a WPA2 Personal passphrase. Doing this creates more overhead for the staff, who must provide the customers the passphrase, and it only creates a false sense of security. The logic of using a pre-shared key is that a hacker sniffing unencrypted radio frequency transmissions can intercept data traffic. While this is technically true, using standard WPA2 Personal doesn't solve this problem. A hacker who possesses the passphrase and who captures the unencrypted association exchange between a client device and an access point when it first connects to the network can use the collective information to decrypt the client device traffic. Furthermore, most security issues on visitor / hotspot networks do not require sniffing the radio frequency, but rather come from the "wired side" of the network; all Wi-Fi encryption only occurs between the client device and the access point, as the access point decrypts all data traffic before passing it on to the wired network infrastructure. If client isolation on the network is not set up properly (an all-too-common problem), a wireless hacker can simply connect to another wireless client device through the wired network. Accordingly, the only thing that WPA2 Personal adds to your guest network is increased overhead on the staff, who have to give out the passphrase to all of the visitors. Often, this passphrase is publicly posted for convenience, which is totally antithetical to the goal of providing security. If you want to remain secure when using a hotspot / visitor network, make sure you are using application-level security, such as https for web surfing and SSL for your email service. Personal or corporate VPNs are also eminently appropriate and effective.

Many consumer wireless router appliances come with the ability to set up a "guest network" with a separate subnet and DHCP range. The access point, therefore, creates a Layer 3 (IP) barrier between the guest network and the operations network to isolate them from each other. Using this feature, however, is only appropriate for single AP networks. On networks consisting of multiple APs, there is no mechanism to allow roaming between APs, since each AP is creating a separate and functionally isolated guest network. Any client attempting to roam on the "guest network" will have to re-establish a Layer 3 connection, thus interrupting any streaming applications. Thus, on all multi-AP networks, VLANs should always be used to provide visitor / guest networking.

Client devices don't know, and generally shouldn't know, anything about the VLAN configuration of a network. Accordingly, all VLAN configuration is done on the network router, switch(es), and access point(s). When a client device sends data, each packet is "tagged" as it enters the network, so it can be routed to the correct destination. When the data reaches its intended destination, the tag is removed, which is commonly referred to as either "untagging" or "stripping the tag."

A good analogy for understanding VLANs is what happens to your luggage when you check in at the airport. The luggage (i.e., Layer 2 data frame) is tagged when you check in at the airport for a flight. This makes sure the luggage is routed on to the correct plane(s) to travel with you and available for retrieval at your destination.

In networking, the VLAN tag is a 4-byte element inserted into the MAC header of the Ethernet packet. This element contains a 12-bit number indicating the VLAN ID (or VID), meaning that, in theory, a network can have $2^{12} = 4096$ tags. The "all zeros" and "all ones" tags (i.e., VLAN 0 and VLAN 4095) are not used per the 802.1q specification. Furthermore, VLAN 1 is reserved for "untagged traffic," meaning that any data traffic in a network that does not have a VLAN tag is considered to be on VLAN 1. This is why all switch and access point VLANs default to VLAN 1.

By default, each port on a network switch will drop VLAN traffic, so any VLAN traffic that is allowed through a switch port must be ***explicitly defined*** in the switch configuration. Trunk ports are used to interconnect switches (and access points), where each VLAN in use on the network is explicitly defined as a *tagged VLAN*, meaning that the switch will pass traffic on that VLAN without touching the VLAN tag.

The tagging / untagging mechanism in switches and access points are a bit different between a wired client and a wireless client, but functionally they are identical:

- Wireless: A wireless client associates with a particular SSID, and in the configuration of the access point, the SSID is associated with a particular VLAN. All traffic coming from a wireless client is tagged with the VLAN ID associated with the SSID. Similarly, the access point strips the tag associated with the SSID for all data traffic being transmitted to a wireless client. Thus, from the perspective of the switch, all traffic coming from or going to an access point is tagged.

- Wired: A wired client is connected to a particular port on a particular switch. This port has two settings associated with it, which unfortunately are commonly put on different screens. The first is the *PVID*. This setting indicates the VLAN ID that should be tagged onto all traffic coming into the port (i.e., from the wired client). Since each port must have all allowed VLANs explicitly defined on it, an *untagged VLAN* can be defined, such that any traffic on that particular VLAN gets its tagged stripped before the traffic leaves the port. By definition, a wired port connected to a client can have only one PVID and should have only one untagged VLAN, and these two should match in order for the connected wired client to communicate in both directions on that VLAN.

The router configuration similarly becomes slightly more complex. Each VLAN on the network is considered to be a sub-interface of the LAN interface (since multiple VLANs exist on the same physical wire / network interface card). Thus, instead of defining an IP address, subnet, and DHCP range for a single LAN, each VLAN is treated as a separate LAN and requires an independent subnet, IP address, and DHCP range. By convention, some people like matching the second or third octet of the subnet to the VLAN ID. For example, VLAN 8 could be given the subnet 10.8.0.0/16 or 192.168.8.0/24, VLAN 16 could be given the subnet 10.16.0.0/16 or 192.168.16.0/24, etc. The VLAN ID and the subnet settings are independent (VLANs are defined at Layer 2 whereas subnets are defined at Layer 3), so no correlation between the VLAN ID and the subnet is required, but it is often done for convenience and ease of maintenance.

Typically, VLANs are used to keep the various LAN subnets isolated, so the router is generally just doing WAN to VLAN routing. Cross-VLAN routing can be done in specific instances, and usually requires explicit rules to be set up on the firewall to allow particular exceptions. One common example would be a hotel with a printer in the lobby that they want both staff and guests to be able to use. This printer would be placed on a dedicated "shared device VLAN," and router rules can be defined to route traffic from either the operations VLAN or the visitor VLAN to the shared devices VLAN.

Most consumer routers, including routers supplied by cable and telco companies to consumers, do not support VLANs. This limitation is not a hardware limitation. All consumer routers have radios, processors, memory, and Ethernet ports. The hardware is there. The limitation is in the software. With fewer complicated features included in the software, support costs can be reduced for the system vendor.

There is also a technology known as dynamic VLANs, supported by some AP vendors, which enable multiple users to connect to a single SSID but be assigned to different VLANs. This is deployed in conjunction with WPA2 Enterprise security utilizing an authentication server that stores a database of allowed users and devices. When a user gets authenticated via RADIUS, the authentication server database can also provide the appropriate VLAN tag for that user to the access point. The AP then knows that all data traffic from that particular client device MAC address destined for the wired network is to be placed on that VLAN.

Just as a facility's operations and visitors are segmented on two (or more) VLANs to separate the network traffic, it is best practice to use a separate VLAN for the web and CLI interfaces for the network equipment, such as the network router, switch(es), and access point(s). This way, no users on any internal user VLAN can access (and therefore hack) the network equipment. The management VLAN, therefore, is the VLAN for the network equipment.

By default, all network equipment comes with the management VLAN set to 1, and all managed and smart switches are configured such that every port is PVID 1 / untagged VLAN 1. Strictly speaking, if your operations users and visitor users are on their own VLANs, then the LAN is isolated from the VLAN users and acts as a separate VLAN (i.e., VLAN 1). However, any user connecting to a misconfigured access point or a non-configured switch port can access your network equipment. Therefore, it is best practice to explicitly define a VLAN for management and configure all of the network equipment to require traffic to be on the management VLAN in order to access its configuration.

Since switches and access points in their default configurations are on management VLAN 1, it can be tricky to implement a management VLAN on a network and have all of the switches and APs accessible on the management VLAN. If the process is not performed in the correct sequence, the risk is run of completely losing access to some or all of the network equipment.

VLANs are a powerful tool and are usually going to be an integral part of more complex Wi-Fi network designs. Can you get yourself into trouble using VLANs? Absolutely. VLANs are quite easy to misconfigure. That said, aside from just "fat fingering" a wrong setting, VLAN mistakes are most commonly made on the network switches. When using VLANs, be sure to understand how traffic flows through your network from your client device through the AP and switches to the router.

The usual problems encountered with VLANs are as follows, along with some guidelines for avoiding and/or troubleshooting them:

- The device is on the wrong VLAN: This is usually due to traffic being put on to the wrong VLAN as it enters the network. Make sure all SSID settings and PVID / untagged VLAN switch settings are correct. Fortunately, this is usually easy to catch, especially if a client device is configured for DHCP. One look at the IP address on the client device will indicate if it is not getting a DHCP address or is getting a DHCP address on the wrong subnet. For static clients, an arping or nmap on the wrong VLAN will reveal the presence of the client.
- Data traffic doesn't flow: This is usually due to either traffic being put onto the wrong VLAN as it enters the network, or switch ports not being properly and explicitly configured to pass traffic on that VLAN. Remember that all network equipment ports on a switch, defined here as ports connected to either the router, backhaul to other switches, or access points, should be trunk ports and be configured for tagged VLANs for all VLANs used in the network, including management VLANs. All ports connected to client devices and/or network appliances should only be configured for the PVID / untagged VLAN that the client should be connecting to.
- Lose access to network configuration or equipment: This is virtually always a VLAN mismatch between the PC being used to configure the network devices and the management VLAN set up on the device. Management VLANs should generally be configured last; once you configure a network device to use a management VLAN, you will typically lose access to that device until changing the port of your PC to be on the same VLAN as the management VLAN. If you have a switch configured to be on management VLAN 4000, but none of the switch ports are configured for tagged or untagged access on VLAN 4000, you are cut off from the switch and have no way to access the configuration, short of a serial interface or a hard reset. It is advisable that each network switch have at least one PoE port designated as a management port, and that it be configured to be PVID / untagged VLAN on the management VLAN.

High-Density Applications

In the early days of Wi-Fi, the primary design focus was on providing coverage in all desired areas. As Wi-Fi has become more prevalent, however, the design focus has shifted towards handling a higher capacity of devices on the network. High capacity Wi-Fi designs come into play whenever you have many people / client devices in a relatively small space. Examples of areas where Wi-Fi design gets driven by capacity vs. coverage include stadiums, arenas, classrooms, lecture halls, conference centers, large ballrooms, auditoriums, concert halls, theaters, school cafeterias, etc.

To illustrate the issue, think of a university lecture hall that measures approximately 75 ft x 75 ft (25 m x 25 m) that can readily hold 500 or more students. From a coverage standpoint, one AP in the middle of the ceiling would provide excellent signal throughout the lecture hall. However, despite having an excellent signal everywhere, this network would fail. Assuming an average of 1.5 Wi-Fi devices per student (e.g., laptops, tablets, and smartphones), approximately 750 client devices are simultaneously trying to get online simultaneously during the lecture. Depending on whether the course has online content (or if the professor is inescapably boring), each device may be passing a large amount of traffic. Even the most expensive and sophisticated AP available would struggle to hand the onslaught of 750 clients at all, let alone trying to pass sizable amounts of data. Obviously, more APs are needed in the space to handle the high capacity of client devices.

Naturally, however, just adding extra APs is likely to cause its own set of problems, including co-channel interference and unnecessary client roaming. There are several strategies for adding APs intelligently to minimize such effects. A typical high capacity application will use many of these in tandem, though not necessarily all are appropriate for a given application. Most of these strategies are intended to segment the space into smaller coverage zones, where each coverage zone has a manageable number of client devices.

- Stick to static channels with small channel widths: Channel planning and control are crucial in every high capacity application. The goal is managing large numbers of client devices, not in providing each client device with the maximum possible speed. Use a static channel plan and use 20 MHz channels on both 2.4 GHz and 5 GHz.
- **Limit bandwidth per user:** Most AP platforms support limiting the upstream and downstream bandwidth per client device. While this may have little practical effect on managing the airtime, this is primarily done to manage the total Internet bandwidth of the site, to prevent one or a small number of users

from consuming most of the bandwidth with aggressive applications like BitTorrent.

- Turn down transmit power settings: Good coverage usually comes for free in a high capacity design, so virtually all of the clients will be in close range to at least one AP and be able to achieve high MCS rates. To minimize excessive client roaming, it is desirable to minimize the overlap areas between neighboring APs. Turning down the transmit power to low or minimum levels will shrink the size of the coverage area, thus shrinking the number of client devices covered by that AP.
- Use directional antennas: This approach is common in large venues like stadium and arena deployments, where specialty antennas that are both low gain and highly directional are used to create a precise and limited coverage area.
- Implement Band Steering: Most AP manufactures have the feature to encourage dual-band client devices to connect to the 5 GHz band. Band steering is not part of the IEEE 802.11 spec, and thus vendors have proprietary methods of performing this. The usual approach involves delaying responses to probe requests on the 2.4 GHz band, or even ignoring probe requests, authentication requests, or association requests on the 2.4 GHz band if the AP also sees the same client device MAC address on the 5 GHz band.
- Implement Load Balancing: Some AP vendors offer the ability to load balance across APs. (This is done automatically in SCA deployments.) Load balancing is not part of the IEEE 802.11 spec, and thus vendors have proprietary methods of performing this. The usual approach is that an AP that gets to a certain client threshold will delay responses to (or just ignore) probe requests and authentication requests, to encourage new clients to seek out different APs on the same network. Some vendors can also do load balancing in conjunction with band steering for high capacity environments, where some percentage of dual-band clients are encouraged to connect on the 2.4 GHz band to utilize that radio more fully.
- Deploy 5 GHz only APs: Most dual-band APs allow you to disable the 2.4 GHz radio. Since there are more channels available on 5 GHz vs. 2.4 GHz, it's not unusual to design 2.4 GHz for coverage to handle any 2.4 GHz only clients, and 5 GHz for capacity, meaning that a significant portion of the APs may have their 2.4 GHz radios disabled.

- Use tri-band APs: Some vendors have introduced APs with one 2.4 GHz radio and dual 5 GHz radios, where one 5 GHz radio is on the lower UNII-1 and UNII-2 band, and the other is on the higher UNII-2e and UNII-3 bands. Such APs can talk to two 5 GHz clients both simultaneously and bi-directionally, and don't require any support by the client aside from 5 GHz capability.

MU-MIMO in IEEE 802.11ac wave 2 was intended to expand capacity by enabling the AP to communicate unicast messages downstream with several client devices simultaneously. MU-MIMO requires support from the client device to provide appropriate position feedback. For MU-MIMO to work, the AP needs to calculate the location of the client devices, relative to itself, and thus select appropriate sets of clients with enough relative spatial separation in order to communicate with them. Unfortunately, aside from a few USB dongles and Samsung Galaxy smartphones, most client devices do not have chipsets that support MU-MIMO position feedback. Thus, while all AP vendors market their IEEE 802.11ac wave 2 products as handling a higher number of client devices with MU-MIMO, the lack of client device support makes the use of IEEE 802.11ac wave 2 for high capacity applications an empty marketing promise.

IEEE 802.11ax promises even more capabilities for handling large numbers of client devices and is actually specifically designed for high capacity applications on both the 2.4 GHz and 5 GHz bands. The primary technologies for high capacity include both support for bi-directional MU-MIMO as well as OFDMA, which slices up the channel into 2 MHz sub-channels that can each be used to communicate with separate client devices. To maintain backward compatibility, an IEEE 802.11ax AP shall use a transmit opportunity (TxOp) to take control of the channel and assign time and channel slots to each client device. While very promising in theory, the complexity introduced makes these APs extremely sensitive to co-channel interference during the TxOp. In addition, like IEEE 802.11ac wave 2, IEEE 802.11ax can only be effective if large numbers of client devices on the network support it. Otherwise, the AP will spend most of its time communicating with legacy client devices.

There is also interest in using IEEE 802.11ad (60 GHz) for high capacity applications, especially in educational environments and areas with a lot of walls or other structure, as this structure will serve to keep neighboring areas isolated from each other. Again, IEEE 802.11ad shall only be an effective solution if large numbers of client devices support the technology.

As a real-world reflection of the adage "you get what you measure," a client device like a smartphone or laptop is generally sold on its own potential bandwidth capabilities. IEEE 802.11ac wave 2 and IEEE 802.11ax do not actually increase the bandwidth of client devices but are intended solely to manage large numbers of devices simultaneously. For lower-end IoT devices, the design motivations are low cost and maximum compatibility, usually with very low amounts of data per device. Therefore, IoT devices are virtually always 2.4 GHz only, and some of them even use IEEE 802.11b or IEEE 802.11g chipsets. Accordingly, there is little motivation for client device vendors to deploy IEEE 802.11ac wave 2 or IEEE 802.11ax, as these technologies only serve to increase costs without increasing measurable performance.

A client device is always measured by its individual performance, not its performance in a crowd of other client devices. A capacity issue is, by definition, the network's problem, not the client device's problem. While IEEE 802.11ac wave 2 and IEEE 802.11ax may or may not provide valid technical solutions, they are certainly poor market solutions, because of mismatched business and marketing motivations between AP vendors and client device vendors. The dynamic will only change if the client device vendors have a strong market motivation to support these technologies, which seems quite unlikely in the foreseeable future.

With IEEE 802.11ad, the 2 GHz channel widths promise significant increases in bandwidth for individual client devices, so there is motivation for high-end devices like smartphones, tablets, and laptops to integrate IEEE 802.11ad capabilities into future products and sell those products with the usual market recipe of "faster wireless speeds."

6.5: Virtual Private Networks (VPNs)

Branch offices are geographically remote locations of an organization. While the organization usually has a primary location (i.e., headquarters), branch offices typically host a relatively small number of employees. Telecommuting employees may be considered a "branch office of one," and if they travel for work, their branch office is mobile by definition.

Nonetheless, the single employee or branch office likely still requires a secure WLAN solution. Many AP vendors offer branch office solutions, either built into their own platform or in partnership with vendors that specialize in routing and VPN appliances and applications. To implement a comprehensive and secure solution, several design issues must be considered.

First and foremost, some connection is needed to the primary location. Depending on the distance from the headquarters and whether the branch location is fixed or mobile, different options exist.

For short distances, such as connecting buildings on a campus, within the same office park, or even within the same town, wireless PTP / PTMP links might be practical, and would be the simplest and most cost-effective solution. This allows the remote building to connect to the main headquarters via a "wireless wire" and remain on the same Layer 2 network. However, larger organizations tend to be spread out across different cities or even different continents, making this solution impossible.

For fixed location branches, typically dedicated leased lines or Internet-based connections are used. When using *leased lines*, a monthly fee is generally required, and the costs can be rather high for even moderate data rates. However, the line will be dedicated to passing business traffic between the branch office and the headquarters. In some implementations, no local Internet is provided, and all user Internet activity must pass through the leased line. Leased lines come in many forms and speeds. They can be implemented as full or partial lines, meaning that the full potential data throughput of the line is made available or only part of it. Typically, each branch office is on its own Layer 3 subnet, and the leased lines form a *private network* between the headquarters and the branch offices. Leased lines have generally fallen out of favor as Internet-based connections have become both faster and cheaper, and as corporate workforces have become more mobile.

Hence, most modern organizations connect to their remote branch offices over Internet connections to each location, defining a virtual private network (VPN). Instead of using leased lines, the VPN builds a private tunnel through the public Internet to interconnect the branch offices and/or remote employees. VPN tunnels are encrypted to protect the traffic as it traverses the Internet.

VPNs come in a few different flavors. For individual / mobile employees, the VPN will run as a software application that the user launches after connecting their PC via Wi-Fi or wired Ethernet to a local Internet connection. Branch offices will either use dedicated VPN appliances or Layer 3 routers with VPN capabilities built in. Some AP vendors that cater to large enterprises have VPN tunneling built directly into their APs. This allows connecting the AP to the local Internet without needing additional controller or router hardware and allowing remote monitoring and management of the remote branch office APs from a controller housed at the corporate headquarters.

VPN solutions tend to add overhead to the connection, so the measured throughput is usually going to be less than the available bandwidth.

When implementing a branch office, it's important to understand your bandwidth needs to be based on where your servers are located and how you plan your wireless access. Most modern enterprise VPNs give you a choice between full tunnel mode and split tunnel mode. In full tunnel mode, all traffic is tunneled back from the branch office to the main VPN appliance at headquarters. This mode tends to consume a lot of bandwidth and computing resources at both the branch office and the corporate headquarters, since traffic between the branch office and the Internet need to pass over the Internet in an encrypted tunnel between the branch office and headquarters, and then again between headquarters and the Internet. In most cases, it would only take a few users at the branch office streaming Netflix to bring your entire corporate network to a grinding crawl! Most larger networks, therefore, operate in a split-tunnel mode, where only corporate traffic is transmitted via the VPN tunnel, and all Internet traffic from the branch office gets routed via the local Internet connection. The split tunnel architecture is illustrated in Figure 6.17.

Each VPN vendor seems to like to use their own terminology for these modes, and some vendors offer additional variations on the theme, such as combining VPNs with content filtering and establishing independent VPN tunnels for each VLAN as well as for remote AP management. It is strongly recommended that you work closely with your VPN vendor of choice to get everything configured properly, as VPNs tend to have a lot of settings to get "just right," and most tend not to be user-friendly.

Depending on the specific security requirements, an organization may opt to keep servers for applications like email, RADIUS authentication, content filtering, network monitoring, etc. either fully at headquarters or have locally replicated instances. For instance, if the branch office is utilizing 802.1X/EAP authentication for wireless users, but the RADIUS server is located at headquarters, then a user at the branch office cannot connect to the Wi-Fi at all without authenticating to a remote server. If the VPN were offline for any reason (e.g., an Internet disruption at headquarters), the user would not be able to connect to the branch network at all, even for Internet traffic.

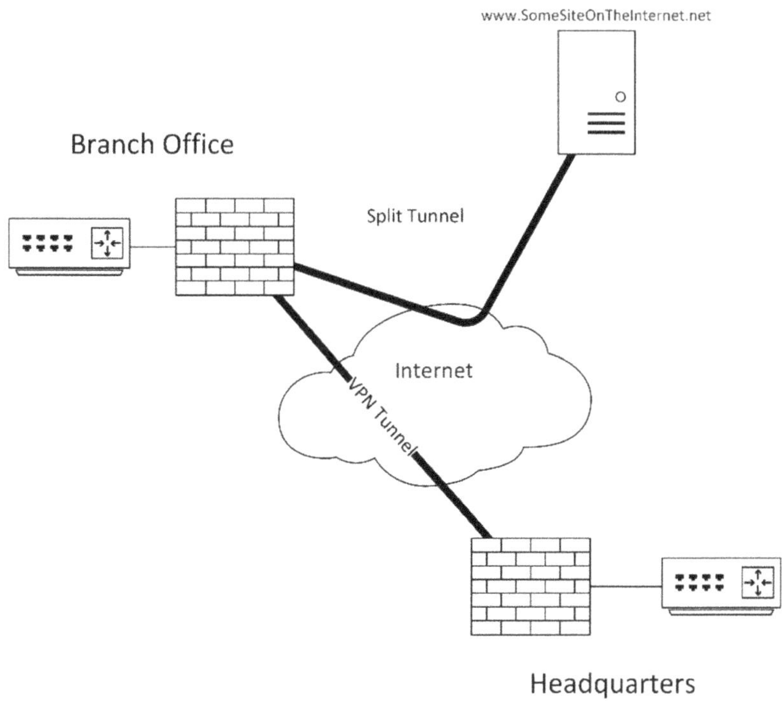

Figure 6.17: VPN example is utilizing a split-tunnel architecture

Many of these servers and services are becoming cloud-based, making them less dependent on VPN connectivity and dependent instead on Layer 7 app-based encryption. There is a distinction between private cloud and public cloud. Private cloud services are generally hosted by the organization in one of their private or leased data centers, whereas public cloud services are hosted by a third party, generally the firm providing the app or service. Public cloud is very useful for smaller organizations that have little if any internal IT support in terms of manpower or data center resources for such services. Private cloud usually requires dedicated internal resources, though the data stored in these applications is intrinsically "more secure," as it never leaves the organization's control[87].

Personal VPN

Since most guest networks do not utilize an encryption key, or utilize a WPA2 Personal key that is publicized, which defeats much of the effectiveness of that type of encryption, many end-users have looked to utilizing VPN technologies to protect their personal data traffic when using Wi-Fi at hotspots, hotels, cafes, and the like.

There are several services that have popped up in recent years to offer personal VPN services. Unlike their large enterprise counterparts, these services are designed to be both inexpensive and easy for users to install on their PCs and mobile devices.

When you utilize a personal VPN service, you are trusting your data to that personal VPN vendor, so you want to know that your privacy and your security are protected. Many "free" personal VPN services have cropped up, but most of these are actually malware, designed to access the very data that a user is trying to protect. Nothing in life is free, so the provider has to be making money somehow. There are independent reviews on the Internet of the best personal VPN services, and new ones keep cropping up as consumers become increasingly aware of their need for such a service.

6.6: Captive Portals

A *captive portal* is a Layer 3 mechanism used to authenticate users prior to allowing them full access to the Internet. Captive portals are typically deployed in network environments that don't utilize any Layer 2 authentication or encryption (i.e., open networks). Captive portals are mostly deployed for guest networks in hospitality and retail but can get deployed in any vertical where guest access to the Internet is required on an isolated VLAN. Thus, captive portals are common for guest networks across many different verticals, such as apartment buildings and student housing that offer Wi-Fi as

[87] In addition to public and private cloud is hybrid cloud. Hybrid cloud implements a solution using some public cloud resources and some private cloud resources.

an amenity, hospitals and healthcare facilities that provide Wi-Fi access for patients and visitors, government offices, schools, etc.

Since the captive portal is a Layer 3 application, it is usually deployed as a dedicated appliance, either as part of the site's gateway or as a bridged Ethernet appliance in-line with the gateway. Higher end captive portal appliances were generally cloud managed long before the term "cloud management" was even coined, though lower-end appliances operate as standalone devices.

When a user connects to the network and first attempts to reach the Internet, the captive portal will intercept outgoing HTTP traffic and redirect the user to a splash page. The user generally must enter some type of credential and/or accept Terms of Use to be allowed through the captive portal to the Internet. At a minimum, the splash page is usually branded with the logo of the property and/or the network operator and may include images relevant to the site.

Typical login methods for captive portals are as follows:

- Terms and Conditions (T&C): In virtually all captive portal networks, the user must acknowledge agreement to a set of terms and conditions to use the network. On most free guest networks, this is the only requirement. To log in, the user simply checks a box indicating agreement or clicks directly on the login button.
- Username and Password: This login method is common for MDU networks like apartment buildings and student housing, to ensure the user is authorized to use the network. This method is also what is used for paid access, where paid access is based on an hourly, daily, weekly, or monthly subscription. This way, the network owner can track who has paid for access and disable accounts automatically once the subscription expires. The captive portal can also be used to restrict upstream, and downstream bandwidth based on the user's paid subscription level, as well as limit the number of different client devices that can use the same login account.
- Code: This method is common in hospitality environments as well as conference centers, to ensure that only guests of the facility can use the network. The guest is given a code upon check-in to access the network and enters this code on the splash page. Most hospitality environments are setup to rotate the code on a

weekly or monthly basis, though some environments will have a fixed code for the life of the network, which ultimately proves insecure.

- Name and Room Number: This method is common in hospitality environments, and generally the captive portal appliance communicates with the hotel's property management system (PMS), so the user can only log into the network while they are checked into the hotel. This setup is required if the hotel is charging for the Wi-Fi service and having the charge included on the hotel bill, though even hotels that provide free access often use this method to ensure usage only while a user is checked in as a hotel guest.
- Social Media / Email: Many network operators use the splash page to collect social media or email addresses from users to populate their marketing lists. This method is common in retail outlets like shops and restaurants.
- Advertisements: To monetize their networks, some network operators use the splash page to show either image or even video advertisements from 3rd parties. By definition, the users are a captive audience.

In most scenarios, once a client device logs in, it will be MAC Authenticated for a certain amount of time, so the client device will bypass the splash page and automatically log in. In hospitality environments, this typically expires after 24 hours. In MDU environments, this assignment is usually permanent.

As captive portals generally only intercept HTTP traffic, any other traffic, such as HTTPS or POP3/IMAP from an email application will simply get blocked without explanation. This is especially problematic for devices that do not have web browsers built into them (e.g., gaming consoles, VoIP handsets, IoT devices, etc.). Such devices usually need to be manually MAC authenticated to be granted access to the network.

Most captive portal appliances are either ridiculously expensive and/or very difficult to program and manage. Usually, both are true. Accordingly, in recent years, many AP vendors have added captive portal capabilities to their controllers because of market demand. Because APs are Layer 2 devices and captive portals are fundamentally a Layer 3 mechanism, the AP is fundamentally the wrong place for this functionality. (Strictly speaking, it is an OSI Layer violation.) The captive portals offered by most AP vendors, therefore, tend to be very limited in their capabilities, and even that basic level of functionality requires convoluted liberties to be taken with the control architecture. Those AP vendors that also bundle Layer 3 solutions as part of their architecture have a better chance of doing this properly.

> Captive portals are generally detested by both users and network administrators alike. If used, they should cause as little intrusion as possible.

Most captive portal web pages are not designed and implemented very well. The login process often requires clicking on multiple screens to get access, in addition to having to provide email or social media contact information or, in some cases, having to watch an advertisement. Thus, most captive portals violate the usually unstated requirement to make access easy for users. To quote Keith Parsons, "users want Wi-Fi that is fast, free, and frictionless."

Captive portals are still widely used, however. They are necessary if the property is charging for the service, either via credit card or via integration with the hotel's property management system (PMS). While most hotel chains have stopped charging for Wi-Fi access, some higher-end hotel chains still charge for the service. Of these, most hotels have migrated to a hybrid model. In one common iteration, members of the hotel's frequent guest program can enter their membership number and get the service for free, whereas non-members have to pay. In another hybrid model, a fixed bandwidth constraint is applied for the free service, but guests can optionally buy higher speeds. In reality, hotels don't make money off of their Wi-Fi, despite years of trying to figure out ways to monetize their Wi-Fi infrastructure. Very few guests actually elect to pay for service, so the hybrid paid plans tend to get used as marketing tools. For example, hotels that specialize in hosting conferences will use captive portals to issue free login vouchers for conference attendees.

Even facilities that do not charge for service still utilize captive portals, based on the "requirement" to provide the T&C legalese warnings that every user agrees to but virtually no user ever actually reads. Such text is unique to each property and/or service provider, though it generally includes statements such as the user agrees that the service is provided as-is, the user won't download copyrighted material, the user won't hack anybody from this location, etc. It is not clear that stating such legalese and forcing users to acknowledge it is actually required to legally protect the network operator in case of an incident, though most network operators don't want to go through the legal battles and pay the expense to find out through litigation. In other words, it is mostly done to prevent litigation that may be decided on the part of the network owner in the end, but the network owner prefers to avoid the hassle.

> Hotel guests generally expect free and fast Wi-Fi. Bizarrely, many hotel reviews rate the quality of the Wi-Fi to be more important than the traditional hospitality fundamentals, like clean sheets and towels!

Network operators do need to have a mechanism to be compliant with the Communications and Law Enforcement Act (CALEA), which allows a Federal law enforcement agency to obtain a warrant to track the online connection activities of a specific subscriber. Most captive portal appliances will enable compliance by allowing a specific user to be tagged and have all of their metadata (e.g. IP addresses of external networks they are communicating with) to be logged to a syslog server. CALEA warrants are not very common, especially in the hospitality vertical, but there can be hefty fines levied against the network operator and property if a warrant is issued but the network equipment is not in place to comply.

6.7: Voice Over IP (VoIP)

Voice Over IP (VoIP) is a technology for making phone calls over the local network or Internet. It is typically found in large enterprise networks like corporations and hospitals, and usually involves a mix of both wired and wireless handsets. It is also not unusual to see VoIP installed as the guest phones in hotel rooms, though these are usually wired into a PoE switch.

When dealing with Voice over Wi-Fi (VoWiFi) applications, there are additional considerations that need to be made in the Wi-Fi network design.

- Latency: While phones are low bandwidth, the real-time two-way communication inherent in a phone call makes VoWiFi extremely sensitive to latency, so the use of multiple wireless backhaul links is not advisable.
- Signal: VoWiFi appliances tend not to be very well-behaved client devices. To maintain call quality, it is desirable to have an overly strong signal everywhere. The design guideline is usually an RSSI of -67 dBm throughout the network. By

comparison, typical Wi-Fi deployments usually only require an RSSI threshold in the range of -70 to -75 dBm.

- Roaming: It is important that a call not get dropped or appear to be interrupted in any way when a VoWiFi client device roams from one AP to another. Roaming must, therefore, occur very quickly, typically within 50 ms. Human perception is on the order of 90-100 ms, so any interruption that occurs faster than this will not be observed by the user. Many deployments, therefore, use WPA2 Personal security on a dedicated SSID / VLAN. For deployments that utilize WPA2 Enterprise, local RADIUS servers and implementation of IEEE 802.11r are usually required, with IEEE 802.11k and IEEE 802.11v being advisable.

In the past, these handsets used to always be dedicated client devices, making QoS easier, as a prioritized Voice VLAN could be established in most managed network switches to automatically prioritize all traffic on that VLAN. It is more common now to see SoftSIP applications running on smartphones, meaning that the smartphone is passing both voice and data and is connected to a single SSID.

It is a good idea to implement QoS on your APs and network switches. QoS comes as part of the standard for IEEE 802.11n/ac, so usually, this is handled automatically by the APs.

For QoS to work, the originating station needs to embed the appropriate QoS tag in the outgoing frame, so the network can recognize it. Unfortunately, since this is client device dependent, it isn't always supported by SoftSIP applications or even dedicated VoWiFi handsets. Thus, design for QoS, but don't simply trust it will be used in practice.

6.8: Internet of Things (IoT)

There has been much hype over the last several years about the Internet of Things (IoT). Broadly speaking, this is an array of Wi-Fi sensors, actuators and appliances to monitor and control physical entities over the Internet or across the local network. There are numerous smart home and smart building applications which include lighting and outlet controls, kitchen appliances, thermostats, water and pump monitors, doorbells, and an array of other devices. It is projected that the largest growth of Wi-Fi client devices will be in this area, and specialty IoT devices already are and will continue to have a larger presence in virtually every vertical market.

Aside from devices with integrated cameras, most IoT devices require little bandwidth and large coverage ranges. They typically utilize 2.4 GHz chipsets (sometimes IEEE 802.11b or IEEE 802.11g), both for low cost as well as for universal compatibility. Some IoT devices use other protocols on 2.4 GHz (e.g., Bluetooth, Zigbee), or even other frequencies (e.g., Z-Wave on 900 MHz). There are proposed standards like IEEE 802.11af (902-928 MHz) and IEEE 802.11ah (old TV white spaces between 54 MHz – 700 MHz) for use with IoT devices, though it is not currently clear whether these proposed standards will ever gain widespread adoption.

Zigbee operates on the 2.4 GHz band but uses a different channel numbering scheme for 5 MHz channels. If Zigbee devices are in use, the Zigbee channel is set to either 11 (below Wi-Fi channel 1) or 25 (above Wi-Fi channel 11), so as to keep it from interfering with the Wi-Fi.

To handle Wi-Fi IoT devices, the 2.4 GHz design is crucial. Most clients are designed to lesser standards than APs, so good 2.4 coverage is recommended with an alternating 1 / 6 / 11 channel pattern. While IoT devices are usually low bandwidth, there is the potential for capacity issues simply because of the number of IoT devices connected to each access point. If deploying many IoT devices, the design will likely require more APs at low transmit power settings to handle the volume of 2.4 GHz clients[88].

6.9: Oversubscription Ratio

Most service providers will sell service based on a service level agreement (SLA). Some providers may have the same SLA for all users, whereas others may offer tiered SLAs at different levels for different prices.

[88] The wireless IoT track (CWISA, CWICP, CWIDP, CWIIP, and CWISE), offered by CWNP, provides in-depth coverage of wireless IoT protocols.

Most service providers will specifically specify their SLA as providing speeds "up to" a particular data rate. Some commercial bandwidth vendors will offer a guaranteed SLA, though such bandwidth circuits tend to be significantly more expensive.

For the purposes of bandwidth planning, there is a useful concept from the early days of telephony known as the oversubscription ratio. This is based on the premise that not all users are going to be using the service simultaneously, so the total amount of bandwidth required is less than the sum of the SLA promised to each user. Thus, the total amount of bandwidth needed can be calculated as follows:

```
Total site bandwidth = # devices x SLA / oversubscription ratio
```

The correct oversubscription ratio to use will vary by vertical. For sites with users doing a lot of video streaming (e.g. Netflix), the oversubscription ratio will be lower. For a typical office deployment, an oversubscription ratio of 20:1 or 25:1 is usually appropriate, whereas for student properties the oversubscription ratio may be 10:1 or less.

6.10: Specific Use Cases and Vertical Markets

When implementing WLANs, it is important to consider specific use cases and the variances functional requirements between different vertical markets. A vertical market is a kind of industry or organization, such as healthcare, hospitality, or government deployments. While each facility will have its own layout and specific requirements, many functional requirements related to usage, coverage, and capacity are common within a specific vertical market. Understanding these requirements is critical for creating the right design solution for the application.

High-Density Design and Large Public Venues (LPV)

- **Usage:** Usage is entirely "event-based," so large numbers of clients will be using the network for brief periods of time. Almost all client devices will be portable, largely smartphones and tablets, with some laptops depending on the venue and event type. Primarily, the clear majority of users will need to be on a guest network, with a small number of users on a staff and/or point-of-sale (PoS) SSIDs / VLANs. There may be surges of simultaneous uploads of events (e.g. sports plays) hitting social media sites. Since most users are attending the event, there will generally not be too much video streaming from the Internet. However, the internal network may have wired, or wireless video surveillance, so internal wired and wireless capacity requirements also need to be assessed

- **Coverage:** Because of the large number of client devices, coverage is typically broken up into small sections. There will be a lot of APs deployed with minimum to low transmit power settings. In stadiums and arenas, placing APs under seats is not unusual, and especially directional antennas are common to carve up the coverage areas.
- **Capacity:** These networks are driven by having large numbers of client devices in a small area. Bandwidth limitations per device are critical, even with a large Internet bandwidth circuit available. Oversubscription ratios are likely to be in the range of 10:1 to 20:1.

There are AP vendors that specialize in high capacity applications. Such vendors usually have AP models that are specifically designed to handle large quantities of users.

Education

- **Usage:** The trend for years in education is to move towards online modules and videos to enhance the classroom experience. It is therefore not unusual for each student to have a tablet or laptop, and the classroom itself may have electronic whiteboards, wireless projectors, and other education-specific network appliances. There will generally be separate VLANs for students and teachers. Security will generally be WPA2 Enterprise, at least for school-owned assets. Guest access for visitors is also a typical requirement.
    - Content filtering is often a critical requirement in K-12 educational facilities to block inappropriate sites as well as social media. Content filtering may vary by VLAN and may also be adjusted at different times of the day, such as allowing social media during lunchtime or during after-school activities. This requirement can be satisfied either on a per-AP basis (depending on the AP vendor) or via a centralized content filtering appliance.
- **Coverage:** Each classroom will need coverage, as well as all public areas like cafeterias, auditoriums, libraries, gymnasiums, and outdoor courtyards and athletic fields. Administrative areas and offices generally also require coverage.

Older buildings often have cinder block or concrete walls between classrooms, which can make signal propagation difficult.

- **Capacity:** Education environments are a subset of high capacity deployments. Some vendors advocate "one AP per classroom", but this is bad when presented with a one-size-fits-all approach. The number of APs needs to be driven by the number of wireless clients per classroom and the building materials between classrooms. If many of the educational resources are Internet-based (vs. being coached on local servers), the oversubscription ratios are likely to be around 10:1.

There are particular AP vendors that specialize in education applications. Such vendors usually have AP models that are specifically designed to handle large quantities of users and have functionality like content filtering built into the platform.

Healthcare

- **Usage:** Hospitals and other large medical facilities incorporate a lot of different types of mobile devices, including pagers, push-to-talk phones, VoIP handsets, and health monitoring devices such as biometric sensors and heart monitors. There are also stringent security requirements for HIPAA compliance to maintain the privacy of patient data. Guest networks are also usually expected for patients and visitors. Multiple VLANs and SSIDs are commonplace. One common trick is to hide all but the guest SSID so as to prevent confusion for patients and visitors as to which network they should connect. Dynamic VLANs can also be used, with unrecognized devices being shunted automatically to the guest network. Note that many medical devices are notoriously poor when it comes to basic network security protocols, so strong VLAN isolation and properly configured firewalls are a necessity.
- **Coverage:** One of the challenges unique in these environments cab come with lead-lined walls for X-Ray rooms. If a facility has been renovated, some rooms that one wouldn't expect may be lead-lined, so on-site surveys, especially of older facilities, is generally advisable.
- **Capacity:** Capacity is generally relatively moderate. While such facilities are usually quite large, there generally aren't enough devices, staff, and patients in any area that capacity per AP is a problem. There are notable exceptions to this, such as university hospitals, but even those cases are generally limited to a few areas. For HIPAA compliance, most monitoring is done on site, so the traffic

medical device equipment traffic and surveillance will not be going to the Internet. An oversubscription ratio of around 20:1 is likely to prove adequate.

Retail

- <u>Usage:</u> Retail and restaurants are typically using the Wi-Fi network for guest access, internal operations such as order and inventory control, as well as *point-of-sale* (PoS). PoS systems GE really need to be PCI-DSS compliant for protecting financial credit card transactions. Strong isolation between VLANs and a decent firewall are usually necessary. Most client devices will be mobile, and barcode scanners for inventory management are commonplace in many retail sectors.
- <u>Coverage:</u> There is often a distinction between *front-of-house* where guests and customers are located and *back-of-house* which is purely for operations and facilities. SSIDs are often different in these areas; for instance, one generally does not want to provide guest access in the back-of-house areas.
- <u>Capacity:</u> Capacity is generally relatively moderate, though a banquet facility may require high capacity designs for its ballrooms. Guests tend to be transient, and operations usually do not take up significant bandwidth. An oversubscription ratio of around 20:1 is likely to prove adequate.

Warehouses

- <u>Usage:</u> The primary application is inventory control, so mobile laptops, tablets, and barcode scanners are common. Some newer warehouses are adding Wi-Fi controlled automated robots to pull inventory from shelves and to restock, which makes constant connectivity and roaming performance critical. There may be a corporate network as well in some environments, and guest networks may be deployed for employees to use their own personal devices.
- <u>Coverage:</u> Coverage can be challenging, as coverage is needed in the aisles, which are often 15 feet – 25 feet high and may be several hundred feet in length, to record when inventory is added or removed from shelving. For overhead AP deployments, the attenuation between aisles changes depending on the type of inventory and the varying inventory levels over time. Think of it as designing for an environment where the walls keep changing. Accordingly, it is common to use directional antennas to focus coverage in each aisle, which directs the signal where it is needed and is less impacted by inventory levels.

- **Capacity:** Generally, the usage on these networks is not high compared to most other verticals. An oversubscription ratio of around 30:1 is likely adequate in most environments.

Hospitality

- **Usage:** Hospitality is one of the more challenging environments, as the client device population of the guests literally changes on a nightly basis. It's also not uncommon for guests to have multiple client devices (i.e., laptops, tablets, and smartphones) connected simultaneously. Hotels will also use the wireless networks for staff and security. In New York City, there is an ordinance that all staff members must wear an emergency panic button that will indicate the approximate location of the staff member in the hotel when pressed and many of these systems work over Wi-Fi (usually on a separate SSID / VLAN).
- **Coverage:** There is often a distinction between *front-of-house* where guests and customers are located and *back-of-house* which is purely for operations and facilities. SSIDs are often different in these areas; for instance, one generally does not want to provide guest access in the back-of-house areas. Typical deployments will generally allow for one AP to cover a few guestrooms, depending on location and building materials, so the client density per AP is generally not high. An attached banquet facility or conference center may require a high capacity design for those areas.
- **Capacity:** Capacity is generally moderate to high in terms of bandwidth utilization by guests, as users are often streaming movies or television on their devices. Guests tend to be transient, and operations usually do not take up significant bandwidth. An oversubscription ratio of around 10:1 is generally required.

Student Housing

- **Usage:** Student housing tends to be one of the most challenging environments to manage. The students generally have a very large number of the most modern client devices and are generally utilizing large amounts of bandwidth. There may also be VLANs and SSIDs allocated for the property management staff and for security / surveillance appliances. While student housing can look like a normal apartment building, the units are smaller and thus generally there are more significantly more students living in a student housing property vs. an apartment complex of the same size.

- **Coverage:** Most modern specifications for student housing require both wired and wireless access. The wired access tends to be used by students to either (a) plug in their own Wi-Fi routers, and thus interfere with the building Wi-Fi, or (b) plug in servers for doing BitTorrent or similar content sharing services. If you do need to support wired ports in every student bedroom, it is recommended to disable them by default and only enable them upon explicit request. Such wall-plate Ethernet ports, however, can be convenient for cabling by using wall-plate APs in strategic locations. Student bedrooms are generally small, and construction is generally inexpensive, so generally, one AP will cover several student bedrooms and may be deployed one AP per suite or even one AP per 2-3 suites, depending on the size and layout of the suites and the building materials.
- **Capacity:** Capacity, in terms of both the number of client devices and the amount of bandwidth consumption, is going to the highest of any type of multi-dwelling unit (MDU) property. Limiting bandwidth per user is an absolute necessity in order to mitigate any abusive behavior (e.g., BitTorrent). An oversubscription ratio of around 10:1 or even 5:1 is generally required, whereas a normal apartment building may have an oversubscription ratio of 20:1.

Outdoor Environments (Parks, RV Parks, Campgrounds, etc.)

- **Usage:** Most people using devices are going to be on portable smartphones and tablets. Primarily, this will be used for guest Internet access, but there may be facilities and/or security and surveillance on dedicated VLANs.
- **Coverage:** Outdoor environments generally have significant cabling restrictions, so using mesh networks or point-to-(multi)point backhaul links to interconnect APs is typical. Note also that the APs need to be in weatherproof enclosures. IP65 is generally sufficient for most normal outdoor environments, but harsher environments (e.g., a saltwater beach) should be using IP67 or IP68. In urban or suburban environments, such as downtown parks, there may be significant Wi-Fi interference from surrounding residences and businesses. Directional antennas, especially sector and patch antennas, are generally useful in these environments for focusing Wi-Fi coverage over large distances with limited AP mounting options.

- **Capacity:** Capacity is generally fairly low to moderate. However, usage expectations are increasing over time. If the expected usage is not that high, a mesh network solution can be sufficient. However, for many RV parks and campgrounds, guests are streaming video and expect to have fast Internet service. In this case, it is better to use dedicated PTP / PTMP links for wireless backhaul that are interconnected to dual-band APs. An oversubscription ratio of around 20:1 is generally adequate.

Government

- **Usage:** Providing service in government facilities can be a long and arduous process, even for simple deployments for local government agencies. Security concerns are generally going to be paramount, so expect to use WPA2 Enterprise for government officials. Depending on the environment, there may be a guest network with a captive portal, as well as VLANs dedicated to security / surveillance / access control.
- **Coverage:** Coverage is generally about the same as any private sector environment, though expect physical security of the equipment to be a driving requirement. This may require both physically hiding and/or locking down the APs to prevent tampering, especially in prisons and courthouses. Many facilities are likely to be brick and concrete, so more APs may be required to provide adequate coverage to all desired areas.
- **Capacity:** Capacity is generally fairly low to moderate. However, usage expectations are increasing over time. An oversubscription ratio of around 20:1 is generally adequate.

6.11: Chapter Summary

In this chapter, you learned about advanced WLAN design scenarios and explored bridging, link budgeting, and specific use case scenarios. In the next and all remaining chapters, you will begin exploring what happens after the design: implementation, validation, and optimization.

6.12: Facts to Remember

Be sure to remember the following facts as you prepare for the CWDP certification, and be sure you can explain the details related to them:

- It is important to understand your requirements and constraints as completely as possible before undertaking any design effort.
- Typical Wi-Fi network design requirements can be categorized by usage, coverage areas, capacity, control, and integration.
- Typical Wi-Fi network design constraints include budget, aesthetics, cabling limitations, and power availability.
- Key design parameters that are controllable in a design are the selection of the AP and antenna, the AP locations, the transmit power settings, and the channel setting (center frequency and width). These parameters are not independent.
- AP vendors tend to specialize in particular applications. One size does not fit all.
- There are different types of wireless backhaul for different applications.
    - Client bridges are useful for getting a single wired client device connected to a Wi-Fi network.
    - Repeaters are only useful in small environments for fill-in coverage, losing 50% of their throughput per hop.
    - Mesh networks are only viable in scenarios where cabling cannot be run and performance is not critical.
    - Point-to-point / point-to-multipoint links are very useful in environments where cabling cannot be run, and can be used for both Wi-Fi and / or surveillance applications.
- Remember to take advantage of directional antennas in your design process. They are critical in point-to-(mutli)point applications and in certain types of verticals, such as warehouse aisles and outdoor coverage environments.
- Understand the difference between self-interference and external interference. Self-interference is from your own APs, and you have complete control to minimize these

effects in the design with an appropriate selection of AP location, AP transmit power and AP channel settings. External interference is from APs and non-Wi-Fi sources outside of your control, and you need to live with / work around this interference.

- When planning point-to-point and point-to-multipoint links, it is important to be well-versed in the different aspects of the link budget and understand how to adjust the height to ensure clearance of the Fresnel Zone and account for the curvature of the Earth.
- VLANs are useful in a wide array of Wi-Fi applications. A CWDP should be completely comfortable with configuring VLANs on all key network elements (i.e., APs, switches, and routers)
- Captive portals are in wide use for controlling access on guest networks, in lieu of encryption. There are several varieties / methods of restricting guest access.
- High-density applications generally require more APs with focused antennas and low power settings to carve up the coverage area into small regions with manageable numbers of client devices
- VPNs are useful for connecting multiple remote branches together via an encrypted Internet connection. It also allows for the wireless network and all security / access control to be centrally managed from the main headquarters
- Voice over Wi-Fi (VoWiFi) is generally extremely sensitive to latency and signal degradation, generally requiring stronger coverage areas.
- Usage, coverage, and capacity requirements vary widely by application and vertical market.

6.13: Review Questions

1. Why should you not hook up multiple wired client devices to an AP in client bridge mode?

    a. It violates the IEEE 802.11 protocol

    b. It loses Layer 2 information such as MAC addresses and VLAN tags

    c. It never works in any environment

    d. It loses 50% of its throughput per hop

2. What are the typical caveats with a mesh network?

    a. You generally lose 50% of throughput per hop

    b. There is a compromise with total potential data throughput of the AP

    c. There is a compromise with total potential user capacity of the AP

    d. All the above

3. What are the distinguishing characteristics of point-to-point or point-to-multipoint links?

    a. Wireless client devices do not directly connect to the AP

    b. Layer 2 information from both sides of the wireless link is preserved

    c. Wireless links use a different protocol than standard Wi-Fi

    d. All the above

4. Why should you minimize the number of hops in any type of wireless network?

    a. Each wireless hop adds latency (delay)

    b. Noise in the environment / retransmissions on each hop adds jitter

    c. Mesh and repeater networks can cause a 50% reduction in throughput per hop

    d. All the above.

5. True or False: The transmit power setting of the receiver AP is a factor in the link budget.

    a. True

    b. False

6. What is generally the largest single contributor to losses in the link budget?

    a. Attenuation

    b. Antenna cable losses

    c. Free space path loss

    d. Receive antenna gain

7. For a point-to-point link, at what point does the curvature of the earth start becoming an issue to be factored into the design?

    a. 1000 feet

    b. 1 mile

    c. 7 miles

    d. 30 miles

8. Where are captive portals most commonly deployed?

    a. Guest networks

    b. Corporate staff networks

    c. Point-of-Sale networks

    d. IoT / network appliance networks

9. True or False: The endpoints of a point-to-multipoint link should use directional antennas.

    a. True

    b. False

10. True or False: Student housing properties can be deployed just like regular apartment buildings or condominiums.

    a. True

    b. False

6.14: Review Answers

1. **The correct answer is B.** A client bridge turns a wired NIC (e.g., Ethernet port) into a wireless NIC for connecting to a Wi-Fi network. If multiple wired clients are connected, the Wi-Fi network will only see the MAC address of the client bridge, so the wired client MAC address information is lost. Furthermore, all connected devices will be on the same VLAN as specified by the AP's SSID, so different clients cannot be on different VLANs.

2. **The correct answer is D.** Mesh networks should only be used when wiring the APs is impossible or not cost-effective, AND the user capacity and data throughput requirements are not stringent. Since the radio spends part-time talking to clients and part-time talking to users, it loses 50% bandwidth per hop and reduces the total potential capacity of the AP in terms of both throughput and users.

3. **The correct answer is D.** Point-to-point / point-to-multipoint links do not allow direct connectivity, and either uses a variation of IEEE 802.11 known as WDS or proprietary protocols to interconnect the links. All wired Ethernet Layer 2 information (MAC addresses, VLANs, etc.) are preserved for each frame that traverses the link.

4. **The correct answer is D.** It is always good to minimize the number of hops for several reasons, including minimizing latency and jitter, and not incurring too much of a throughput penalty when using mesh networks or repeaters.

5. **The correct answer is B (False).** The transmit power setting of the transmitter AP factors into the link budget. On the receiving side, there is no concept of receiver power, as the signal is not amplified and is thus a "passive" activity.

6. **The correct answer is C.** The free-space path loss (FSPL) is the largest single contributor to the losses in a link budget. Attenuation occurs at each object in the environment and thus adds separately. Cable losses for antenna cabling tend to be zero or relatively minor. The antenna gain adds positively to the link budget, not negatively.

7. **The correct answer is C.** Above 7 miles, the curvature of the earth needs to be accounted for. Below that, the earth can be treated as flat.

8. **The correct answer is A.** Captive portals are used to control access of guest users to the network. These come in various forms, dependent on the vertical and the requirements to restrict access. Most corporate and point-of-sale networks will be implementing strong security to ensure only authorized users / client devices can connect. IoT / network appliances typically do not have user interfaces or integrated web browsers, making their use on networks with captive portals difficult.

9. **The correct answer is A (True).** When deploying a point-to-multipoint link, the endpoints should all have directional antennas pointed back to the base. The base station may use sector or even omnidirectional antennas to provide an appropriate field of view to all of the endpoints.

10. **The correct answer is B (False).** The rooms in a student housing property tend to be much smaller, meaning that there are on average many more residents in a student housing property than in a typical apartment building. Furthermore, students tend to utilize the network much more heavily than an average apartment building, so more bandwidth (i.e., a lower oversubscription ratio) is required.

Chapter 7: WLAN Deployment

Objectives Covered:

3.1 Ensure proper understanding and implementation of the design

3.2 Recommend or perform essential deployment tasks

3.3 Perform an installation audit for quality assurance

So far, we have explained how to produce a robust design and ensure you meet your client's objectives. A well-rounded wireless professional should also be able to guide his or her clients through the implementation of the wireless network. Explaining design decisions and providing clear instructions to all parties involved is essential to a well-executed deployment. Also, appropriate documentation of the process at all stages will allow for a smooth handoff when they complete the project.

An implementation meeting with all parties involved in the deployment of the WLAN infrastructure will allow you to explain the design, explain the installation steps, and avoid common pitfalls. We are fans of coming to such meetings with a good plan and a breakdown of responsibilities between each team and between team members. Your document should include:

- An executive summary: "A view from 20000 feet" of the project, materials, and costs.
- A timeframe of the project, including detailed implementation milestones.
- High-level design of the WLAN to tie the project's objectives to the solution you are implementing and introduce the general architecture, guidelines, and principles you have considered while designing the wireless network.
- Clear and well-documented floorplans of the WLAN deployment including pictures and any specific instructions.
- A low-level design of the WLAN including all the technical details:
    - WLAN architecture model
    - List of devices and FQDN
    - SSIDs
    - Management IPs
    - Security (AAA, RADIUS, etc.)
    - Monitoring information
    - Channels and power plans
    - Bill of materials (BOM)
- The physical topology of the related network infrastructure.
- The logical topology of the related network infrastructure.
- Any network-related requirements to put in place before WLAN installation.
- Any security-related requirements to put in place before WLAN installation.
- Any operational, monitoring, or support procedures for the IT staff and the WLAN clients, should you work in a phased project where the final handoff will only take place months after the first WLAN client connects to the network.

In short, this is an opportunity to showcase the content of your future final report and documents. By now, this report should be 95 percent complete. It is crucial that you keep this documentation updated and teach the staff to keep it updated. Maintaining the documentation can be the hardest part of any networking project because one can learn protocols and standards but the discipline of updating the documentation requires focus. If one or more APs are down and you do not know where they are, then any certifications or qualifications you may have are useless. If these APs are not visible, then you must know how to open and close ceiling tiles, because you will have fun for a little while trying to find your devices. Yes, that was sarcastic.

7.1: Implementation Meeting

Your presentation and explanations should go from the simplest to the most technical not only for logical reasons but because your non-technical audience may tune out because of the complexity of subjects, something which you would rather avoid.

Generally, the people who will be involved in the implementation of your design fall into the following categories:

- **Stakeholders or business party**: their interest is in knowing that the project's objectives are met technologically, politically, and financially.
- **Facilities, physical installation, and cabling party**: their interest is in understanding where and how to install your design.
- **Technical or engineering party**: their interest lies in the configuration, operation, and support for your design.

Stakeholders

Most business parties generally want to know how much and how long. They are concerned about the financial and political side of things and are usually not technical. In the previous chapters, a lot of information was provided to produce adequate design documentation. The part which will be crucial for this audience is an executive report, which should include:

- A suggested timeline with milestones
- A bill of material

- All costs (including support and training if necessary)
- A summary of the design and what it involves regarding efforts, HR needs and man/hour, contractors[89]

Installers and/or Cabling Contractors

The design document should include sufficiently clear instructions for the people who will install and connect the APs. We strongly recommend documenting any survey with pictures. Ideally, every location for an AP should be marked on a floorplan with the AP model and antenna orientation, mount type, and mount orientation. It is incredible to see how creative people can be if you leave any space to the imagination, so please try to be as precise as possible. Your name is on this deployment, taking the means to ensure that the resulting implementation is accurate is your responsibility.

If you are bringing new installers or technicians on board for this project, a few hours of mentorship and basic Wi-Fi operation and installation training is a sound investment. The Certified Wireless Technician (CWT) certification from CWNP is an excellent place for these individuals to start. These people do not have to have an in-depth understanding of the subject but here are some suggested topics to cover, if you choose to train your own team members:

- RF propagation basics
- Antenna patterns
- Building codes to respect, both regarding safety and aesthetics
- Security certifications and/or principles
- Any other norms your organization may have
- Any test you wish to perform to certify that their work is done well such as cable testing, AP association, or further picture documentation or inventory tasks

[89] Depending on the stakeholders involved, this meeting can vary significantly. Management stakeholders want to know that the solution will integrate with existing systems and meet their needs within budget and schedule. User stakeholders want to know that the system will perform well and allow the functions they require. Technical or administrative stakeholders will want to understand how the designed system will impact their routines and what will be required of them to maintain it. Additional stakeholders may be considered as well.

Should you need to hire a contractor, assessing their prior deployment experience, background and looking for someone who has training such as a the CWNP CWT, CWS or CWNA certifications are ways of ensuring that your design is cared for well.

Technical or Engineering Staff

Wi-Fi deployments come in all sizes. The people whom you may meet and explain your design to might be seasoned network personnel or newcomers to this world. Some may be tasked with infrastructure planning and architecture, others with operations, and others with customer support. In this perspective, your pre-implementation meeting should address these items:

- Roadmap of this deployment
- Location of WLAN devices in the logical topology of the network
- Any changes which must be implemented to accommodate the new infrastructure such as VLAN creation, firewall or routing changes, access-lists and the like
- List of installed devices
- Channel plan and frequency
- Deployment map
- Inventory and monitoring of devices
- Planned coverage and performance
- Antenna patterns
- Regulatory domains
- Building codes considered in the plan
- RF basics, if they are newcomers to Wi-Fi
- Network security aspects
- Any new WNMS, AAA, or DHCP servers included in the deployment, which they may have to support after completion of the project
- Any exception to the planned implementation: for example, if you are working on a hospital deployment, some rooms may be either out of scope or cannot have APs installed in them, such as an MRI room

In smaller implementations, a pre-implementation meeting may not be necessary. However, in large deployments, it would allow you to plan your work and resources appropriately. There may be preparatory work which must be completed for your installation to be functional, which should be discussed at this meeting to make sure any network equipment to which your APs will be connected is in place. The security and logical infrastructure should also be ready. If you request your installation personnel to confirm that the APs are functional, not only by testing their cables but also by ensuring that they can connect to the WLAN, delays could be incurred if the WLAN is not accessible. Remember that your installation personnel may need to document the deployment as they go as well. A smartphone and a methodology to record the implementation should be made available for those purposes.

7.2: Implementation Tasks and Phases

To make your implementation operations more efficient, we recommend that you perform some validation and optimization as you go. Here is a short list of the things to either keep in mind or perform during this phase (not always in the order presented):

- Inventory management and documentation
- Licensing model/support contracts
- Configuration of the network infrastructure that will host the WLAN
- Configuration of the wireless architecture
- Configuration of addressing schemes, DNS and DHCP
- Configuration of APs
- Staging, configuration templates and/or scripting
- Antennas, APs, and mounting
- Vendor Specifications
- Channel and power allocation plans, or radio resources management algorithms
- Cabling documentation, blueprint, or layout
- Monitoring
- Aesthetics validation
- Power availability
- Redundancy needs
- Defining the transition of responsibility from project to production
- Documentation (always)
- Documentation (did we mention it already?)

Inventory Management and Documentation

A method of managing the inventory of devices and ports which will be deployed is essential. It could come as an established database system or only from a spreadsheet, but it will help with the quality control of your project. Validate the BoM as devices are received and installed to ensure proper billing to the client and remove any confusion which may occur as installation is completed and support and licensing are processed.

Documentation at all steps of a wireless project is critical, and we cannot stress it enough. If something has specific instructions for installation, do not hesitate to emphasize it, as you may not be the person who will perform that step. Try to make documentation concise and adapted to your audience who may not be as technical as you. Some things that must be documented include:

- AP name, model, serial number, and antenna (if external)
- Wall plate number
- Cabling information
- Electrical information (in the event you are using power injectors)
- Location (room and building)
- Switch port, IP, FQDN name and VLAN
- Wireless topology and its related wired topology (which may or may not be available from your client)

Licensing and Support

As the APs are deployed, the licensing scheme must already be in place and ideally support contracts as well since some issues could occur in a phased deployment when a part is found to be defective. In such cases, one would need to replace some APs, antennas or even a controller, if your plan involves controllers. For example, it would be unfortunate to deploy hundreds, if not thousands, of APs to discover they cannot register to their controllers due to a lack of licenses. This problem could cause delays and bring up costs for the project. Figure 7.1 shows a controller licensing interface.

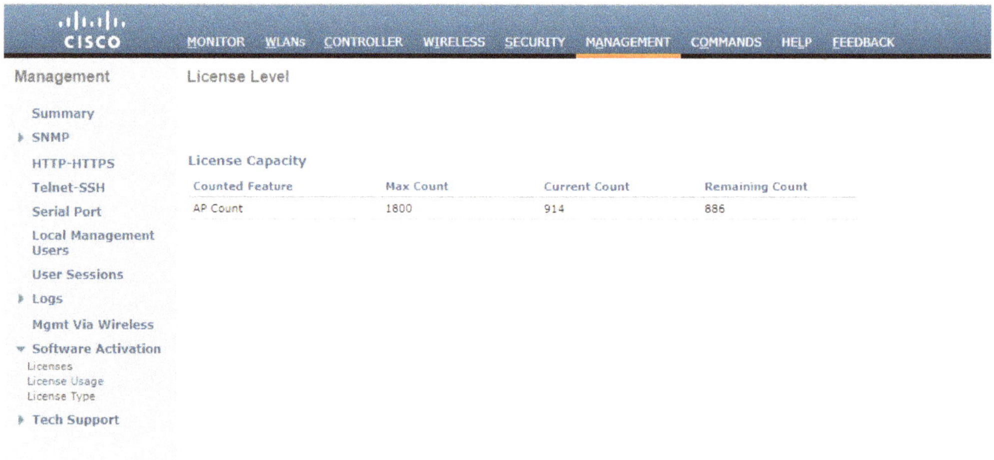

Figure 7.1: Controller Licensing Interface

Configuration of the Wired Network Environment

Depending on the scope of the project, you may have to make recommendations or configure the supporting network infrastructure. The ports, switches, VLANs, access-lists, routers, DHCP, DNS, and firewalls must be considered during implementation, especially if it a phased project.

Based on the wireless topology and network segmentation schemes, the switch ports will either be trunks or access ports. Trunks allow for multiple VLANs over a single link and are used in autonomous mode APs, or access ports in a management VLAN. This VLAN will tunnel back traffic to a controller and should be protected by an access-list, such as in Figure 7.2. Don't forget to allow traffic from CAPWAP ports through to allow APs to register to their controllers.

Small networks rarely include more than a multilayer switch or router. However, in large deployments, there will be several routers and, depending on the topology, some modifications to routes and/or routing protocols will need to be done to allow traffic to flow from the controllers to network resources and the outside world. It is not necessarily your responsibility to implement this aspect of the WLAN deployment, but we feel you should be aware that the wired network is essential to support your design.

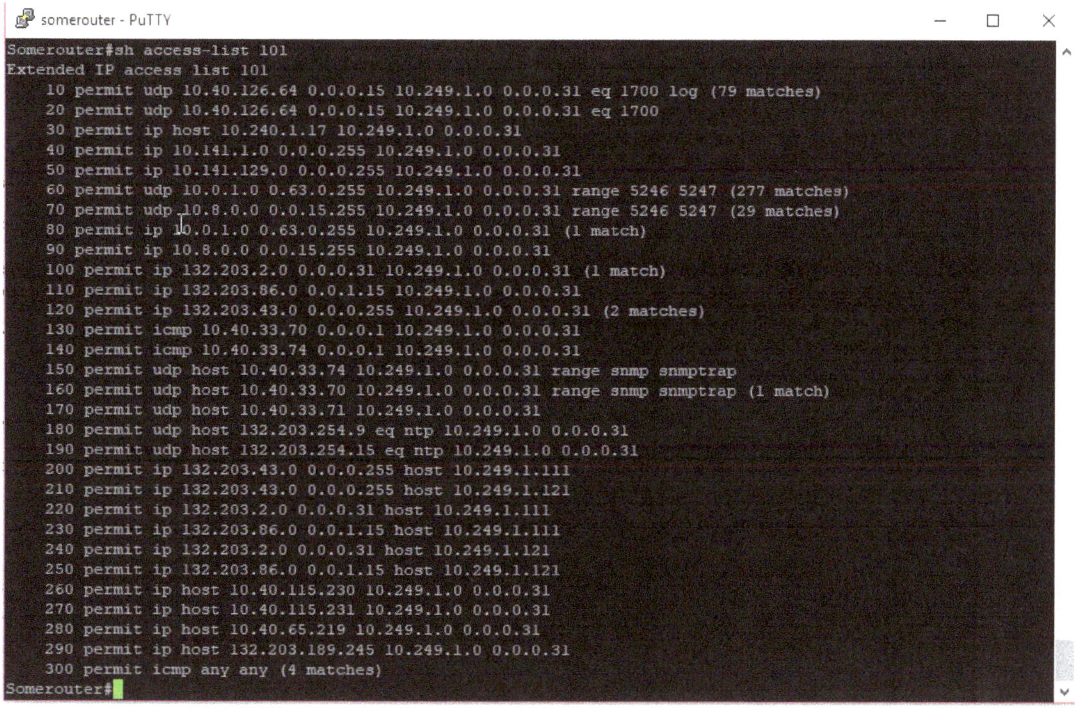

```
somerouter - PuTTY
Somerouter#sh access-list 101
Extended IP access list 101
    10 permit udp 10.40.126.64 0.0.0.15 10.249.1.0 0.0.0.31 eq 1700 log (79 matches)
    20 permit udp 10.40.126.64 0.0.0.15 10.249.1.0 0.0.0.31 eq 1700
    30 permit ip host 10.240.1.17 10.249.1.0 0.0.0.31
    40 permit ip 10.141.1.0 0.0.0.255 10.249.1.0 0.0.0.31
    50 permit ip 10.141.129.0 0.0.0.255 10.249.1.0 0.0.0.31
    60 permit udp 10.0.1.0 0.63.0.255 10.249.1.0 0.0.0.31 range 5246 5247 (277 matches)
    70 permit udp 10.8.0.0 0.0.15.255 10.249.1.0 0.0.0.31 range 5246 5247 (29 matches)
    80 permit ip 10.0.1.0 0.63.0.255 10.249.1.0 0.0.0.31 (1 match)
    90 permit ip 10.8.0.0 0.0.15.255 10.249.1.0 0.0.0.31
    100 permit ip 132.203.2.0 0.0.0.31 10.249.1.0 0.0.0.31 (1 match)
    110 permit ip 132.203.86.0 0.0.1.15 10.249.1.0 0.0.0.31
    120 permit ip 132.203.43.0 0.0.0.255 10.249.1.0 0.0.0.31 (2 matches)
    130 permit icmp 10.40.33.70 0.0.0.1 10.249.1.0 0.0.0.31
    140 permit icmp 10.40.33.74 0.0.0.1 10.249.1.0 0.0.0.31
    150 permit udp host 10.40.33.74 10.249.1.0 0.0.0.31 range snmp snmptrap
    160 permit udp host 10.40.33.70 10.249.1.0 0.0.0.31 range snmp snmptrap (1 match)
    170 permit udp host 10.40.33.71 10.249.1.0 0.0.0.31
    180 permit udp host 132.203.254.9 eq ntp 10.249.1.0 0.0.0.31
    190 permit udp host 132.203.254.15 eq ntp 10.249.1.0 0.0.0.31
    200 permit ip 132.203.43.0 0.0.0.255 host 10.249.1.111
    210 permit ip 132.203.43.0 0.0.0.255 host 10.249.1.121
    220 permit ip 132.203.2.0 0.0.0.31 host 10.249.1.111
    230 permit ip 132.203.86.0 0.0.1.15 host 10.249.1.111
    240 permit ip 132.203.2.0 0.0.0.31 host 10.249.1.121
    250 permit ip 132.203.86.0 0.0.1.15 host 10.249.1.121
    260 permit ip host 10.40.115.230 10.249.1.0 0.0.0.31
    270 permit ip host 10.40.115.231 10.249.1.0 0.0.0.31
    280 permit ip host 10.40.65.219 10.249.1.0 0.0.0.31
    290 permit ip host 132.203.189.245 10.249.1.0 0.0.0.31
    300 permit icmp any any (4 matches)
Somerouter#
```

Figure 7.2: Access Lists

Configuration of the Wireless Architecture

In a centralized or controller-based infrastructure, before performing deployment of the APs, it is necessary to setup and configure the controllers themselves. The AP configuration will be pushed to every AP upon registration to the WLC.

Remember, the AP may have some cached copy of the configuration, but most APs in use with controllers cannot be configured through their console or GUI unless they have completely lost their configuration and are in recovery mode. Should you happen to have a unit stuck in this mode, all is not lost, you should look at your vendor documentation to proceed with code recovery.

When the APs boot, they will register to a controller through various methods (depending on the vendor and preferred options such as DNS, broadcast, multicast, etc.).

The controller will validate the version of software running on the AP, make any required changes to that, and then push the configuration to the AP.

Centralized Deployment

Controllers come in many forms: network appliances, blades which can be inserted in multilayer switch chassis, integrated into smaller network switches for SMB deployments and even virtualized with VMWare. Additionally, some vendors offer virtual controllers within their APs so that one of the APs acts as the controller for the other APs in the network.

The designer's controller choice is based on the needs and infrastructure of the clients, but regardless of format, it will need to be primed before it can register APs onto the network. To prepare the controller configuration, ensure that you have the following information available to proceed:

- Access through the console, SSH, or GUI
- Administrator account and privileges
- Controller's FQDN
- IP addresses for in-band and out-of-band management
- Network topology and VLANs
- Any information related to mobility and its management, such as virtual IPs, MACs and mobility group name
- SSIDs and their full configuration including ACLs, DHCP, DNS, security parameters and additional options
- Security settings (management or CPU ACLs, RADIUS, TACACS)
- RF information per band (manual or automatic radio management, RF groups)
- NTP and any monitoring and backup facilities necessary (SNMP, TFTP, etc.)

Cloud-managed Deployment

If your design is managed through the cloud, an AP's boot process is somewhat like the centralized architecture. The AP will have to get its configuration from a central service, but the service is in the cloud instead of a local controller on your network. To complete the preparation of such an infrastructure, you will need the following information in addition to the items listed under the centralized controller section:

- Any protocol which must be allowed through a firewall
- Any NAT or PAT policies

- Any bandwidth reservation to ensure proper operation of the network should the backhaul be subject to congestion

Cooperative Control Deployment

With this variant, the control plane and management plane are kept locally on the APs. Thus, to deploy such an infrastructure, they must be individually configured or use an NMS server which will take over the management plane operations of the network. The server will push configurations to the AP on the hive to which it belongs. To deploy the manager, refer to the vendor's documentation and bear in mind that cooperative control is comprised of several protocols. If you implement ACLs and other security measures you should ensure that they are allowed access to network resources.

DHCP and DNS

To stage and deploy your APs, you will need IP addresses. If you have a few autonomous APs, you can elect to configure them with static addresses manually. In large deployments, and especially if you automate your configuration through TFTP and/or use a centralized management WLAN topology, it is recommended to use a DHCP server.

It will also be beneficial to assign IP addresses to your clients through DHCP. DHCP servers can push parameters to APs such as boot servers, boot file names, vendor-specific options or identification. A DHCP server is also convenient in a WLAN to assign addresses to clients associated with an SSID, DNS servers, and other parameters. In a routed network, remember to declare your DHCP servers in the networks where they will be used (or, if it suits your application, declare them globally in the network device configuration). Figure 7.3 shows an IP address management interface.

Centralized controllers often resort to DNS-based discovery. This behavior means that the APs can find their controllers based on a DNS lookup of the controller name. Typically, in the discovery process, the AP will query DNS servers for the controller's DNS name. There are other means through which your APs can find their WLCs such as broadcast or staging. For this information and the appropriate FQDN name for your WLC, refer to your vendor's documentation.

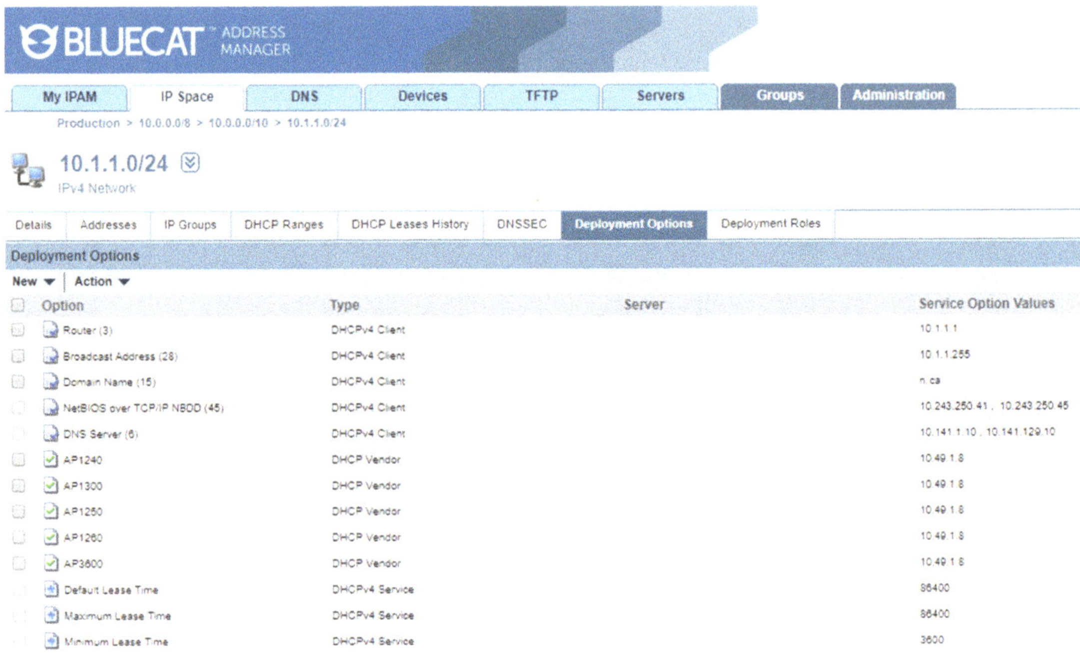

Figure 7.3: IP Address Management

NTP

To be able to get their configuration from the controller, lightweight APs require the use of an NTP server. You don't have to own an atomic clock to be able to use an NTP server: public servers are available such as *time.nrc.ca* or *time.nist.gov* which, if they are accessible from your network devices, will be able to give time to your network devices. Many organizations install an internal NTP server that synchronizes with the Internet-based atomic servers, and then the internal systems synchronize with the internal NTP server. The time given by NTP is in UTC, so you do not have to worry should your deployment be spread through several time zones. If APs and controllers agree on the time, they will communicate with one another. If the time is off, they may not allow communications.

For other architectures, using an NTP server also has benefits. If a problem should arise in the network, having all devices synchronized to the same clock, you can trust that the information in your syslog journals or other logs will be reliable in working out the play-by-play of the events which led to the situation.

Configuration of APs

As previously mentioned, centrally managed APs are configured through the parameters which you have already configured on the WLC. For cooperative control or autonomous APs, it is possible that you must configure each AP through a shell, the GUI, TFTP or some automated configuration deployment platform based on SSH or SNMP. Regardless of your deployment, you should have the following information on hand to proceed with the preparation of APs for deployment:

- FQDN of AP
- Type of deployment
- IPs for management interface or DHCP range available (ideally with reservations to assign a known IP to each AP)
- IPs for each VLAN interface
- Administrative credentials
- NTP, DNS, and DHCP server IPs
- Radio settings for all bands
- Security settings such as RADIUS server
- A configuration template with captive portal settings, disclaimers, ACLs, monitoring, SNMP and other such details

Should you need to configure only a few APs, it is entirely possible to proceed without a complete infrastructure. Some APs feature a console or management cable which will allow you to connect to a terminal interface. Others require the use of the network. Rest assured, most APs do not need a full DHCP deployment to work. They will send DHCP request packets to obtain an IP, and if there are no replies after a specific time, the AP will default to a non-routable IP documented in vendor literature. To connect, give your network card an IP in the same network and telnet into it or use the available web-based interface in the AP. Should you need to power it up and be short of a power supply, a midspan power injector will allow you to proceed and configure the device all at once.

Staging, Configuration Templates, and Scripting

Depending on the project requirements and available space, you can consider configuring APs before installation, something called staging. Staging requires setting up a lab which simulates the future environment and configuration. The advantage of this method is that fewer technical people will be required onsite when AP installation takes place. It could be a good idea particularly in the case of a deployment that includes hundreds of sites, such as retail stores and other multi-location organizations.

In the context of an autonomous mode deployment, consider using script-generated configurations and TFTP/SFTP. For this purpose, some will prefer scripting with PERL, Python, Go or Ansible. A CSV file with the parameters for each AP and a configuration template will make this script adaptable and easy to use. Templates can also be used to generate and deploy configuration of switch ports, VLANs, and any other repetitive network configuration tasks.

Of course, scripting is not in the scope of CWDP, but we strongly recommend to anyone involved in configuring, monitoring, or troubleshooting network devices to learn to code at least simple scripts. It will optimize the time you need to design and deploy WLANs, networks, assess infrastructures and further your career. Many learning possibilities adapted to the network engineer and even to the Wi-Fi engineer are available.

Antennas and Mounts

Many networks feature indoor APs with integrated antennas; however, external antennas are also available and very useful when covering specific or difficult areas. While you are planning your implementation with external antennas, you should be aware of their implications and installation requirements. We encourage you to be cautious in areas where they could be a hazard. Stored goods, mobile shelving, lifts, doors, conveyor belts, and other such mobile apparatus could cover or break them.

With regards to mounting, we strongly recommend that any AP or antenna be properly secured to prevent injury to persons or damage to equipment. Your manufacturer has recommendations for mounting, which we encourage you to follow. This practice is true for your customer's safety and your safety. On an AP-on-a-stick external survey years ago, a member of our team ended up with stitches on his forehead due to the mishandling of a tripod mast and an antenna with metal wings. With appropriate precautions, this incident could have been prevented.

Outdoor deployments are particularly well suited for external antennas. The reason is simple: you can keep your AP from the elements and avoid costly outdoor AP models. The cabling that is used for such installations is low loss cable which should be considered regarding loss, something that you've learned previously in the CWNA curriculum. Positioning external antennas outside must consider weather conditions: wind, snow, and ice could ruin your perfect placement. While you implement, confirm that the location will prevent snow accumulation and make sure to install with drip loops. They will prevent corrosion on the wire and water infiltrations. Drip loops are represented in Figure 7.4.

Figure 7.4: Antenna Installed with Drip Loops

In outdoor deployments, additional precautions should be taken to protect the network infrastructure from lightning. Some manufacturers have APs which are powered through coax cable and power injectors which feature an electrical ground and do not need arrestors: we leave it up to you, the designer, as to how to prevent such mishaps. Devices such as lightning arrestors can also prevent severe damage to your network equipment. They are fuse-like tubes placed between your antenna and AP: should lightning hit the antenna. The grounded lightning arrestor redirects the electrical surge to ground and protects the AP. Arrestors can induce some latency on the network, but it is rather insignificant compared to the impact of a barbequed telco closet with thousands of dollars' worth of smoked devices (and they don't even taste good). Figure 7.5 shows a lightning arrester from AccelTex.

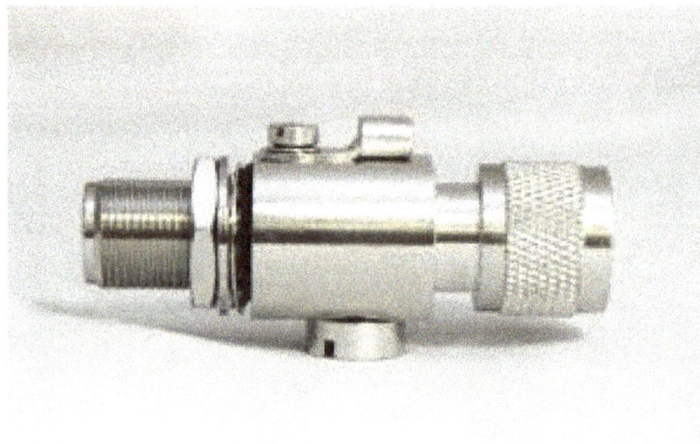

Figure 7.5: A lightning arrestor by AccelTex

A lightning arrestor's middle chamber is usually filled with gas to isolate both connectors. The spike in electrical current is redirected to the ground attached to it.

Did you think of documenting the staging, configuration template, and all network configuration? Did you update your layouts or blueprints, mounting procedures, and locations of APs, cable runs and antennas? This information is vital at every stage of a WLAN's lifecycle.

Vendor Documentation

Although the previous information is important to properly mount antennas and APs, you should always refer to your vendor documentation for instructions and safety information for installing network equipment. A long time ago, each vendor had different types of connectors to ensure EIRP compliance, but this practice has gradually faded, and a lot of antenna companies now offer antennas suitable for most manufacturers. Figure 7.6 shows documentation for an AP and antenna articulating mount, which includes mechanical and environmental specifications.

AccelTex SOLUTIONS

UNIVERSAL AP AND ANTENNA ARTICULATING MOUNT

The AccelTex Solutions Universal AP and Antenna Articulating Mount allows an AP and an antenna to be mounted in the same location. The mount can be fastened to a ceiling or wall, to an I-beam or to a piece of all thread. The mount allows both horizontal and vertical articulation of the antenna. This mount includes additional lengthening arms to allow for more antenna articulation and are removable for a more compact solution. This mount is compatible with all AccelTex Antennas. Additional antenna mounting plates are available for other vendor antennas.

Part Number: ATS-APANTARTMNT-UNIV1

> MECHANICAL SPECIFICATIONS

Dimensions: 10.7" L x 9.3" W x 11.0" H (271mm L x 237mm W x 281mm H)
Weight: 5.05 lb (2.29 kg)
Color: White
Installation: Wall, Ceiling, I-Beam, 3/8" All-Thread Mount, Metal Strapping
Max Vertical Articulation ±80°
Max Horizontal Articulation: ±80°

> ENVIRONMENTAL SPECIFICATIONS

Installation Environment: Indoor or Outdoor

> WARRANTY

1 Year Limited Warranty

> COMPATIBILITY

For use with AccelTex Solutions patch antennas. Additional antenna mounting plates are available for other vendor antennas.

> INCLUDED HARDWARE

Includes beam clamp hardware. Installer is required to provide any additional mounting/fastening hardware for other possible mounting situations.

Figure 7.6: Sample Vendor Documentation

Regardless of the external antenna which you may have connected, you vendor's support contract will not cover a third-party antenna. If they have any suspicion that the issue may have to do with a third-party device, they will require you to take it out of the equation to get support from them. Antennas, as passive elements, are generally not the part that breaks in a network but having to change one in a location that is not easily accessible could be costly. This must be kept in mind.

Channel and Power Allocation Plans, or Automatic Algorithms

Static versus dynamic (automatic) channel allocation has been the subject of many arguments in the WLAN community. Some are avid supporters of the manual configuration of channel and power, which allows for complete control of the available radio bands in a deployment. This allows for a robust and predictable design every time. Configuring every channel and power setting per AP can appear to be time-consuming but can be automated through NMS consoles or scripts. Going manual could save some troubleshooting time because it will not change over time, something that could impact your users' experiences.

The advantage of the dynamic power and channel assignments is that through some overlap, necessary for the algorithm to function properly, the network can easily "heal" itself should an AP go down or a coverage hole be detected. Depending on the vendor, the calculations performed by the controller will involve triangulation of the AP's location. This means each AP must be able to detect several APs in its surrounding area at a minimum power for a controller or mobility leader to be able to run the algorithm. Each AP will also go off channel for some time (60-100 msecs) to validate the data to feed the algorithm. Other parameters involved in the calculation also vary by vendor, but typically include:

- Channel Utilization
- Noise
- Interference (both Wi-Fi and RF)
- RSSI/RCPI

Any design should establish which type of channel assignment will be used before implementation. It would be a waste of time to design for one and switch to the other because it is the latest and greatest way of doing things. Unfortunately, a lot of people have been convinced they could play with this through aggressive and opportunistic marketing. There is no right or wrong choice here. They are tools which can work well in the hands of an informed professional who understands the algorithms, implications, and vendor documentation. A professional's design defines a set of channels and power which are best suited for a project. Will it be well suited for the next? As a wise engineer often says, it depends.

Cabling Documentation, Blueprints, and Layout

Cabling is essential to Wi-Fi. Proper cable installation and termination allows for a smooth AP deployment. Since the wires tend to have a longer lifespan, we suggest that you give some thought to provisioning the correct type of cable. At this point in the history of WLANs, some vendors will recommend CAT6a over CAT5e or CAT6. The current standard allows for speeds over 1Gig meaning that the backhaul, or wire onto which your AP bridges traffic to the network, may have to support multigigabit speeds. Table 7.1 provides a summary of wire categories and their basic specifications.

	Cat5	Cat5e	Cat6	Cat6a
Bandwidth	10/100/1000 Mbps	10/100/1000 Mbps	Up to 10 Gbps	10 Gbps
Length	328ft	328ft	180ft (10G) to 328ft	328ft
Cable type	UTP	UTP	UTP or STP	STP

Table 7.1: Wire Categories

Testing the cables should be part of the cabling installation workflow, and some people even include it in their installation reports. Installing APs and antennas sometimes requires specialized gear, lifts, and such devices. A simple cable test can prevent having to reschedule an installation job, and cabling firms have all the necessary equipment to perform it.

Care should be taken prior to sending anyone out to install any device to review the floorplan and positioning of APs. Personally, I (Mae) prefer to walk with my cable install manager and review the site and any space that could have been modified since the last time the floorplan was edited whenever it is possible. It allows us to prevent mishaps and confirm the exact location of every AP. Should one of us not be available, our team knows that the other person can give the appropriate information quickly to anyone in the field. Since I work for a university where a lot of buildings have older concrete or plaster walls, my partner can also show me where conduit will be necessary, something that must be done by electricians whose jobs are costlier than a cabler's. This allows a few savings along the way, something any project manager can appreciate. Figure 7.7 shows a floor plan with cable locations specified.

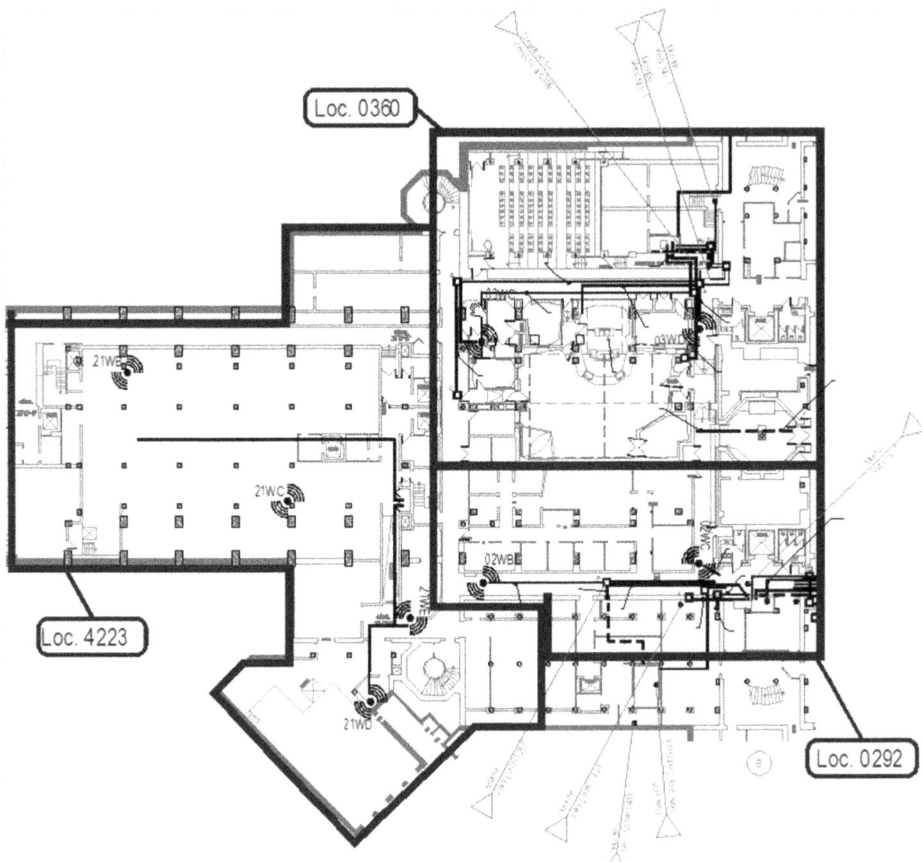

Figure 7.7: Floorplan with Cable Closet Locations

Of course, if you have the luxury of having very detailed 3-D CAD files of the location, this might not be necessary. You may also not have the required time, personnel, or budget to proceed as such. Regardless, bear in mind that blueprints are rarely exact and that the more details you can provide to your installation crew the easier and cleaner job you will get.

Some professionals like to use stickers to mark the location of APs on site, which will only work if nobody peels them off. To avoid a Little Poucet (*le petit Poucet*) adventure where the breadcrumb trail is lost, provide a detailed plan and picture of every AP location to your installation personnel. Report any change in your documentation and in your planning survey files. Some site survey software such as Ekahau Site Survey allows importing of survey files into a WNMS such as Cisco Prime. This would allow you to import the maps and the AP locations directly into WLAN management software, saving valuable time. Keeping your survey files up to date will save you time in delivering the project in this way and for final reports which you may have to provide upon completion of a project. Figure 7.8 shows a floor plan with AP locations specified.

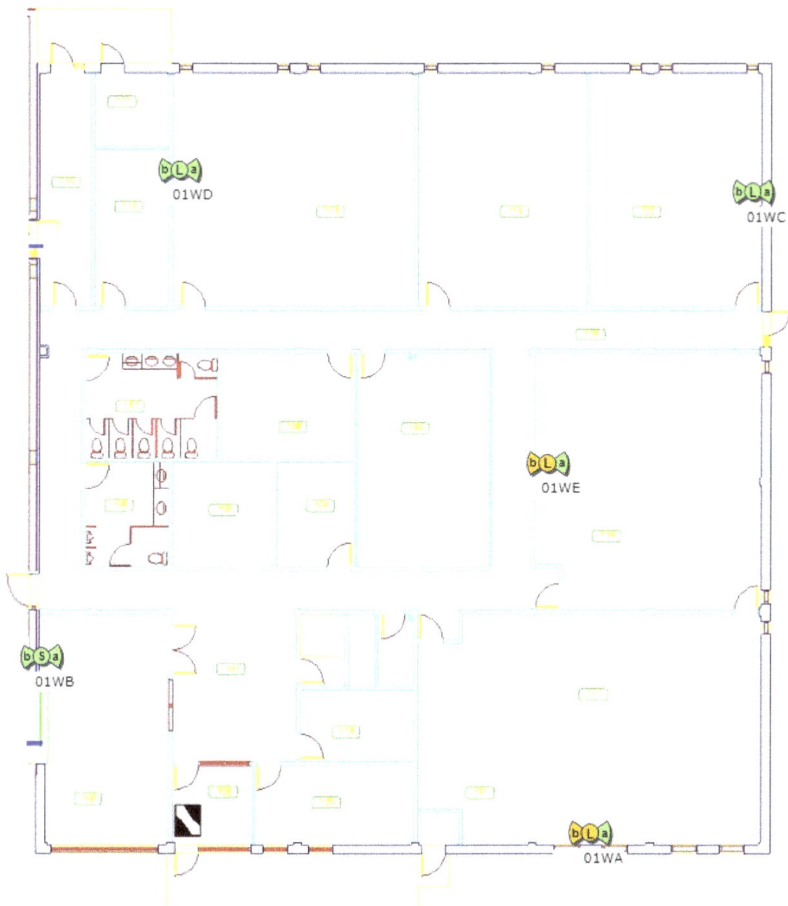

Figure 7.8: Floorplan with AP Locations

Monitoring

The concern of the current chapter is the implementation of the WLAN and APs. In a small project, the usual wireless project would involve gathering requirements, agreeing to the work that must be done, designing a WLAN, implementing, validating, and optimizing it. In larger projects, it is quite possible that the sheer size of the deployment requires you to work in phases. So, before validation is complete, some APs and parts of the WLAN will be used as a production network. Any part of a network that goes into production should be monitored. Should you have to work in such a project, discussions with the production/operations teams should be held to avoid any confusion on responsibilities, ensure that the required watches (ping, SNMP, or alarms on your controllers, APs or WNMS) are addressed. Figure 7.9 and Figure 7.10 shows two different monitoring interfaces.

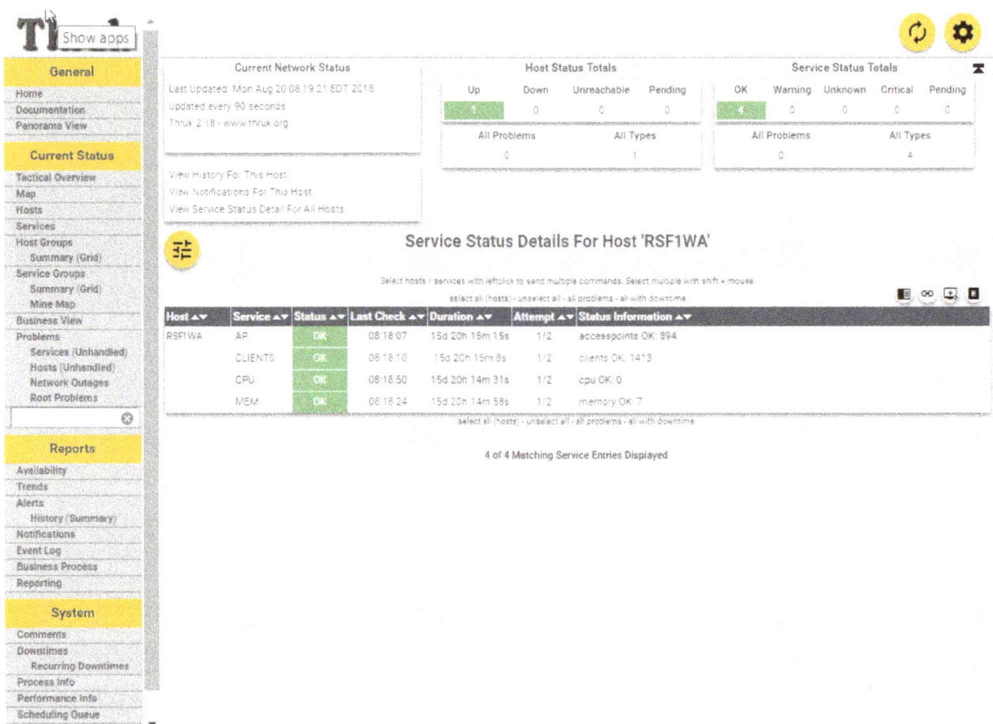

Figure 7.9: A Monitoring Interface

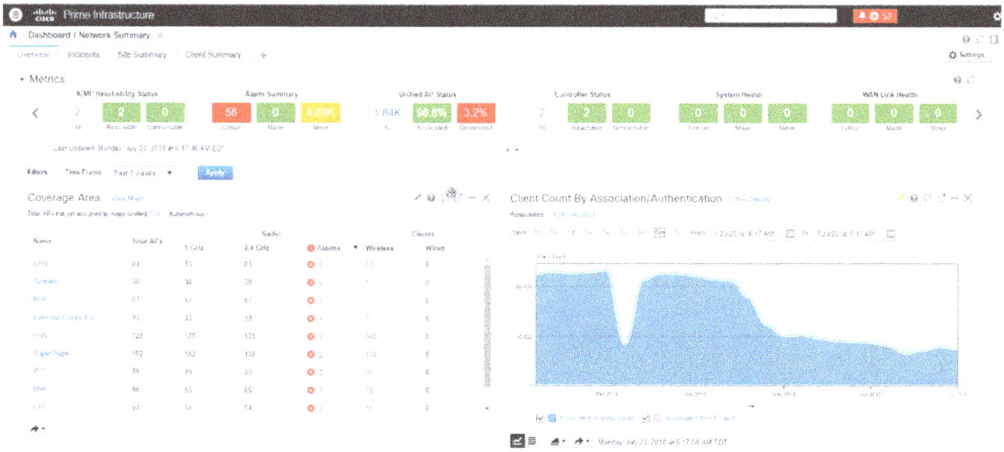

Figure 7.10: A Monitoring Interface

Aesthetics Validation

Some verticals are sensitive to aesthetics. The hospitality industry is particularly prone to trying to hide APs as much as possible, something that may be easier said than done. If you find yourself in the situation where aesthetics can be questionable, always try to check with your client prior to any cabling or installation to verify that he or she agrees with the look of the installation. Try to get this confirmation in writing to prevent any kind of misunderstandings. Some larger institutional deployments can be subject to an aesthetics and architecture review board, or in the case of older buildings, a patrimonial review board. This means that just choosing whichever installation you think is best in technical terms must be approved by this board. Here are a few things you can try to hide an AP or antenna without creating something our good friend Eddie Forrero (www.badfi.com) would call BadFi:

- Some manufacturers sell AP "skins" which doesn't void the AP warranty and only covers it with a camouflage film of plastic (Figure 7.11 and Figure 7.12 shows such skins)
- Some manufacturers sell enclosures and mounts of various types which can hide and/or secure APs
- Specialty antennas can hide as light bulbs, wall plates and in floor tiles
- At last, you can also see if your vendor will honor the AP warranty if you decide to channel your inner Picasso and paint the APs (be aware that lead or metallic

paint are not the first choices for that job due to their reflecting or absorbing qualities).

Figure 7.11: Example AP Skin

Figure 7.12: Tile-matching AP Skin

Since the design has already been done, should you run into an aesthetics problem, it should always imply that you must validate the impact a new antenna, enclosure or mount can have on the result. The designer (should you oversee implementation only for this project) must be informed, and documentation must be updated accordingly.

Power Availability

As the design is created, power should be considered. Depending on your selection of APs, 802.3af (PoE), 802.3at (PoE+), 802.3bt (PoE++) could be in use and require some additional configuration of the network devices or the use of a power injector. Be aware that though power has been standardized by the IEEE, some vendors may offer equipment which offers proprietary power only or require the configuration of additional protocols such as LLDP (neutral) or CDP (Cisco).

	802.3af "PoE"	802.3at "PoE+"	802.3bt "4PPoE"	802.3bt Type 4
Power available (PD)	12.95 W	25.50 W	51 W	71 W
Maximum power delivered by PSE	15.40 W	30.0 W	60 W	100 W
Power management	3 power classes	4 power classes: signature or LLDP based	3 power classes: signature or LLDP based	LLDP based only
Supported cabling	Cat5	Cat5	Cat5	Cat5

Table 7.2: PoE Specifications

While you are implementing a new WLAN, if the APs do not power up on a tested cable with proper port configuration, here are a few first steps:

- Double check that both the APs and the network equipment offer compatible PoE.

- Check the power load on the network equipment. APs initially boot with a higher power than their operational power. Should the switch have been filled up with other PoE device, sufficient power may not be available.
- Check that the cable length is not playing tricks on you: it isn't unheard of that faulty connections, or long cable runs could add loss which would mean that the AP cannot boot due to insufficient power.

  Redundancy Needs

Though some people may still think that WLANs are secondary networks and that critical resources are still using the wired infrastructure, we have seen many times that it is not the case anymore today. The design may include some redundancy features in the WLAN coverage, devices, interlinks and network services such as DNS/DHCP. When you implement them, take the time to test them. There is only one thing worse than having an untested backup, and that is to realize too late that your backup cannot be used during a failure because it was not configured appropriately. The next chapter will offer more insight into what should be tested, but we felt the need to mention redundancy here because of its importance to some deployments.

7.3: Chapter Summary

The implementation can be as simple as hanging one or two APs in a house or as complex as deploying hundreds of sites with thousands of APs each. Regardless, the key to a successful deployment is to be very well-prepared and organized. Getting all the parties involved in the installation to understand the result is your responsibility. Should you need support in doing so, implementing thorough documentation of all elements and procedures will prove very helpful. Make sure that any changes are reported in this documentation and that the team is made aware of them. Validation is not only the subject of the next chapter but also an ongoing process in a WLAN implementation. It's never too soon to check with your team to verify that the network dependencies have been met and the resources for which they are responsible (such as DNS, DHCP, and NTP) are in place so that you can install and configure the WLAN infrastructure. Quality control of APs or antennas starts with clear maps and indications before their installation. Should their aesthetics be a concern, don't hesitate to confirm with the concerned parties. There are various solutions available to address such situations and engaging the client is the first step to finding the right answer to any problem.

7.4: Facts to Remember

- The WLAN design plan should include an executive summary for managers to review and approve.
- The WLAN implementation meeting should include stakeholders, installers, and technical staff.
- You can break the implementation into phases and tasks to better manage it.
- While the WLAN design documents will include a BoM, the items purchased should make their way into an inventory management and documentation system.
- Ensure the implementers understand the need for proper licensing so that the APs will function when connected.
- Make sure implementers understand the VLAN requirements and have them configured appropriately on the switches.
- DHCP pools should exist and be large enough to handle the WLAN load.
- Any questions about AP configuration should be answered for the implementers.
- For outdoor antennas, drip loops help water to run off the cabling without damaging it.

7.5: Review Questions

1) A firm has requested that you deploy a centralized architecture which involves several APs in their new building in Tokyo. The controller will be located in their main offices in San Francisco. The APs can reach the controller but seem to experience issues getting their configuration from them. What could be the issue?
   a) The controller uses the same NTP but is not in the same timezone.
   b) The APs FQDN are not declared in DNS
   c) The APs cannot reach the RADIUS server
   d) The controller and APs' clocks are not sync'd to the same time

2) What should not be featured in an executive summary?
   a) BOM
   b) Timeline and milestones
   c) Costs
   d) Topology

3) What should always be featured in a low-level design?
   a) Costs
   b) Network topology
   c) WLAN addressing scheme
   d) Network management policy

4) In a centralized deployment, how is an AP configured?
   a) By prestaging it with an IP through its CLI or GUI
   b) By configuring the controller which will push the appropriate configuration
   c) By using an NMS and SNMP
   d) By using a script and templates with TFTP

5) You have an AP to configure but no DHCP server. This AP has no console port and a power supply. How can you log into it?
   a) Create a script with the appropriate parameters and a configuration template.
   b) Use a power injector
   c) Connect through the network port, waiting for the AP to use its default IP and configuring your laptop in the same range
   d) Use a console cable

6) You want to use an NTP server for a worldwide deployment with sites in more than 100 countries. Your staff has concerns that your logs and network devices won't have the right time and that it will be an issue for troubleshooting. Why are you not concerned by the time zones?
   a) NTP uses the local time zone on the network devices
   b) NTP uses UTC but network devices can display the local time if configured to do so.
   c) NTP has no impact on syslog
   d) NTP servers exist all over the world

7) You have been tasked with configuring a cloud managed deployment. You APs are powered and DHCP has given them IP addresses in the range 10.0.0.0/24. The DNS server they use resolves the name of the cloud controller. Yet no AP seems to be able to register to the cloud management control.

   What could be the issue?

   a) The code version on the AP is not matching the code version on the controller
   b) The access-list between each AP doesn't allow the management ports
   c) The firewall does not NAT the network range which in turn means that no AP can reach outside the network.
   d) The APs need more power

8) You are installing an external WLAN deployment in Houston, Texas. You have decided to use external antennas and are training your new cabling consultant to install them according to the vendor specification and best practices.

   Which of the following best practices can avoid corrosion on your wires and connectors outside?
   a) Using heat shrink over the connector
   b) Making a drip loop^
   c) Using a surge suppressor
   d) Using a low loss cable

9) Which industry vertical has the most aesthetic concerns?
   a) Retail
   b) Hospitality
   c) Healthcare
   d) Educational

10) Which of these methods will not allow you to stage or configure your APs effectively?
   a) DHCP and Configuration Templates with TFTP
   b) DHCP, DHCP options, script, and TFTP
   c) Centralized architecture
   d) AP configuration through SNMP

7.6: Review Answers

1) The correct answer is D. Controller and AP clocks must be sync'd. AP FQDNs and RADIUS don't prevent APs from registering to a controller
2) The correct answer is D. The executive's concerns are cost, material, and timelines. They do not need a topology for they are not always technically versed.
3) The correct answer is A. A low-level design is what the wireless professional needs to deploy the infrastructure. Costs are secondary at this point and have been covered by the high-level design.
4) The correct answer is B. By configuring the controller, which is the "brains" of the network.
5) The correct answer is C. The AP's DHCP requests will eventually time out, and you will be able to connect if your computer is in the same IP range. No matter how hard you try to push a configuration this AP has no means but the network port to get a configuration.
6) The correct answer is B. NTP uses UTC. The locale or time zone configured on the network equipment is of use to the network administrator for ease of reading logs.
7) The correct answer is C. Since your APs must reach the controller in the cloud, they need to be able to connect to the Internet. If you use any IP range that is non-routable as defined by RFC 1918, you will have to define a NAT scheme to allow for communication.
8) The correct answer is B. A drip loop will divert water away from connectors, something important to consider in a place such as Houston where rain is nothing short of torrential. Though heat shrink could help prevent corrosion, some water will eventually find its way to the connector.
9) The correct answer is B. Though I (Mae) sometimes wonder if the EDU space is not more concerned with aesthetics, the hospitality industry is the vertical best known for such concerns.
10) The correct answer is D. If your AP doesn't have an IP address, it cannot communicate, and SNMP bases itself on MIBs and not configuration templates.

Chapter 8: Design Validation

Objectives covered:

3.3 Perform an installation audit for quality assurance

4.1 Confirm the WLAN system is operational

4.2 Perform an RF validation survey

4.3 Perform client performance testing

After overseeing the deployment of the wireless network, a crucial yet often overlooked part of the wireless network's lifecycle is the validation of the installation. In fact, you may even be brought in to validate someone else's work for various reasons. Clients sometimes wonder why they should pay to prove the quality of your work: as a CWDP you should be able to explain why it is crucial.

At this point in your studies, you know that the wireless network is dynamic in nature: it will evolve and may behave differently from the RF model you designed to the best of your ability. Physical constraints, colocation with other businesses, a change in drivers, a new code version on the centralized controller: the slightest change can bring on both good and bad outcomes on the WLAN. In the process of a new deployment, the validation will provide a baseline which may require further tuning before the client signs the project off. When you are coming to a new site to troubleshoot, a validation will provide you with the first steps in identifying the issue the network is facing.

Let's call a cat a cat: a validation survey is not cheap, which is not enticing to a client. Having an individual walk around their facilities can be costly and have an impact on their operations. It can also add delays to putting the new infrastructure in production. To make your validation efforts effective, we have divided this chapter into three main sections: configuration validation of the deployment, site surveying tasks and last but not least, additional tests and tools.

Before you pop your laptop open and hit the record button on your site survey software, the following items should be reviewed:

- Physical installation of the APs
- Logical configuration of the APs and controllers
- Security aspects of the infrastructure
- Physical and logical installation of the supporting network infrastructure
- Configuration of the supporting services (DNS, DHCP)
- Captive portals

8.1: Physical Installation

Though you may have toured the facilities before the design of the network, it is always amazing how some installations can turn out to be bad. Whether it is a pipe for sprinklers which was not on the plan or an AP installer which felt creative in the way he was hanging the AP or orienting the antennas, a simple walkthrough of the site to check that the installation was performed to your standards will save you valuable time when you survey. Figure 8.1 illustrates this[90].

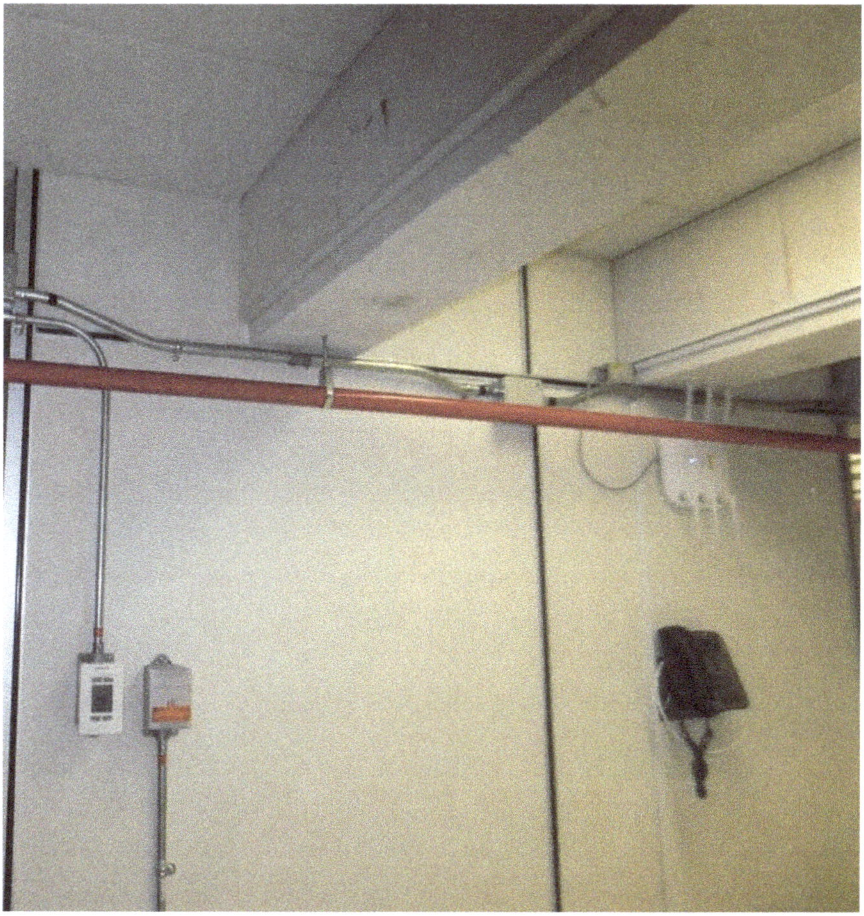

Figure 8.1: Physical Installation Challenges

Simple observation can sometimes give you a good indication of the checks you may need to do once this walk-through is done. If the LEDs on the APs were left visible, you

[90] To assist in avoiding such "dynamic" mounting solutions, be sure to take photos and mark specific mounting locations if you are onsite and able to do so during the survey.

could refer to your vendor's documentation to see if the APs are powered, have reached their controllers, and have sufficient power. Most vendors can even indicate through the led patterns whether clients are connected or if there are irregularities with the AP, such as power cycling, flash corruption and the like. Refer to your vendor's documentation for more information on device-specific patterns.

8.2: Logical Configuration

Sometimes the implementation is not done by a single individual. The client may have the personnel to configure the infrastructure but not the resources to design or deploy the APs. A quick assessment of the configuration will save you some valuable time as well. As an initial test, try to connect to the WLAN with a device of your choice. This might clue you into various issues such as:

- AP or controller misconfiguration
- Dynamic radio and power assignment issues

AP or WLAN misconfiguration issues

Not all logical AP configuration issues are impacting: they can be as simple as mislabeling the APs either physically or from the GUI. The fix is simple but elementary if you want to submit an accurate report to your client. You should also ensure that all your APs have their configuration: a simple ping test will not suffice. We've seen situations where quality control was done months later in add-on settings. Clients were informed that new APs were in place, and they could see them with pretty LEDs in a nice green color. However, they still had issues because the AP was up, but the radios were down. Sounds silly but it is a true story which we see more often than we'd like.

If you use a centralized controller design, make sure that APs have joined the controller they were designed to join. This could have an impact on redundancy strategies and roaming. In implementations such as wireless VoIP, delays in roaming can lead to dropped calls, jitter, and the like. The design document should be your reference as to how the APs were divided should you have multiple controllers.

In deployments where there are a lot of locations which may require a subset of SSIDs, vendors offer various means to advertise only the SSIDs one wants. Often this is done through logical "device groups." If you are using any kind of grouping scheme for your devices (a device group or hive for example), make sure that the APs are indeed in the right group, and that the SSIDs are bound to the right VLAN. There is nothing more frustrating than getting an IP for the next VLAN in the current VLAN only to discover you are the person who mapped it wrong in a specific group.

WLAN misconfigurations can affect both APs and user devices. Your physical inspection did not allow you to see if all bands used in the design are enabled: take the time now to review the configuration that was applied on your APs and/or centralized architecture. Ensure that the minimum data rate that was selected is indeed the one you intended based on the devices the network must support and your design document. Leaving lower rates enabled can have a significant impact on the size of the coverage cells, on roaming and on protection mechanisms. Hidden node problems and client misbehavior can also be caused by low data rates left enabled.

PoE checks for enough power to each AP can also save you time. A lot of manufacturers' devices will operate will a lower than prescribed amount of PoE: should it be the case, the AP will not be using all its features. Cutting our spatial streams, a band, or saving on other proprietary bells and whistles will allow the AP to be operational yet not to its full potential. Making sure the right amount of power is supplied before you conduct your site survey so that once you are walking the site with your laptop, you are really testing the deployment you intended to deliver to your client[91].

If you are using static channel and power configuration, a quick review should be enough to ensure that the channels and the power that was configured follow common best practices. By know you know that in 2.4GHz there are only 3 non-overlapping channels: 1,6 and 11. 5 GHz offers more: the number will depend on the design and the devices which you are using. Keep in mind that some channels in 5GHz are subject to DFS regulations. The local regulatory body's policy will vary: FCC-affiliated countries will not impose DFS on channels 36-48,149 – 165 whereas ETSI and APAC countries will each have their own preferences. Check that power is at most as strong as that of your

---

[91] As you're beginning to see, many of the "problems" encountered in validation only occur if someone disregards the design or if the requirements process and the survey (a real survey, not a design-as-you-go effort, but one intended to gather the right information for a proper design) are not completed properly. Defining the requirements and then ensuring the environment can meet them with a designed WLAN avoids many validation issues.

least capable device and respects the regulatory body's EIRP/power limits. It will not do any good to blast out signal.

Dynamic power and channel, aka RRM, will require that you do a quick review. Among things to check: available channels as well as the various settings which may have been tuned. RRM needs to settle before you can do your validation survey. Try to avoid doing a survey validation right after installation; some time is required for the RRM algorithm to operate. Some vendors will require it to run for 100 minutes, others for longer than that. Refer to your vendor documentation for details: in the absence of any specific parameters allow at least 12 hours for a large deployment to run the algorithms several times and stabilize. Your validation survey data will be closer to the real day to day operations of the network once it is put in production.

8.3: Security Aspects of the Infrastructure

Security parameters are an important part of the configuration as well. The goal may be to protect clients, access to the infrastructure and/or other networks. Access-lists to protect unauthorized access to the CLI/GUI of your network components should always allow the following ports (depending on your topology)

- BootP/BootClient (UDP 67/68)
- NTP (UDP 123)
- LWAPP (TCP 12222/12223)
- CAPWAP (TCP 5246 5247)
- RADIUS (old: TCP 1645/1646 and new 1812/1813)

Access-lists which protect the WLAN should allow for the following ports:

- DNS (TCP/UDP 53)
- BootP/BootClient (UDP 67/68)
- PPTP/ESP/AH/GRE/IPSEC (if you are using a VPN)

The simple test we suggested earlier of logging onto the WLAN may have lead you to discover that the infrastructure does not meet its authentication goals. Enterprise authentication requires a RADIUS server, whether embedded in the infrastructure or a standalone server. This should be in place when you are ready to validate the installation. Table 8.1 summarizes the main characteristics for each flavor of EAP.

	Server-side certificate	Client-side certificate	Credentials	Vendor support
PEAP	YES	no	MS-CHAPv2 username and password	Most
EAP-FAST	NO:PAC	NO:PAC	MS-CHAPv2 username and password	Cisco
EAP-TLS	YES	YES	Certificate	Most
EAP-TTLS	YES	NO	MS-CHAPv2 username and password, token	Juniper/Funk

Table 8.1: EAP Method Characteristics

Should you run into an issue with 802.11n rates, don't forget that the 802.11n standards call for WPA2 and AES to achieve MCS rates.

VPN connections are sometimes required. Validations for this type of connection should include the type of tunnels and routes offered by the VPN client. Make sure that the traffic is routed to the right instances.

As previously mentioned, most WLANs do not make use of public IP addressing space: a NAT or PAT policy will have to be configured on the edge of the network. A quick check to make sure that there are sufficient addresses for all your clients will prevent future mishaps. In the event you have enough resources for more than a PAT pool, we suggest that you plan for overflow: it is surprising how fast this kind of resource can be depleted.

8.4: Physical and Logical Aspects of the Supporting Infrastructure

As a CWDP it is not expected of you either in the test or in your day-to-day job to be an expert at wired networking. However, your understanding of the surrounding infrastructure is important because of the impact it can have on the operation of the WLAN.

Most APs are connected to switchports in large deployments. APs configured as thick APs (or Autonomous mode APs/distributed mode APs) will require the use of trunks to allow for multiple VLANs to be carried over the wire. Be aware that it is possible to either allow all VLANs or restrict the list to a few explicitly declared VLANs on the port. If any of them is not listed on the port's configuration, then the related SSID will not work properly on the related AP. Enhanced security features such as Mac address or DHCP limitations on the switchports could also create issues. Keep them in mind to prevent problems.

Depending on the manufacturer of the network equipment and of the wireless components, PoE issues can also occur. Low PoE may not prevent the AP from being up, but it may be operating with a reduced set of features. The APs' datasheet should mention the required PoE standard: some network gear may not be compatible with this standard. For instance, Cisco once had their own version of PoE. When they purchased Aironet and rolled out the AP 1010, a model they had acquired, we faced an issue: all our switches back then were using Cisco PoE. Trying to use an IEEE 802.3af device with a switch which provided Cisco PoE did not work: midspan power injectors had to be used to provide compatible power to the APs. This was in the early days of PoE, but it could happen again with the coming standards in WLANs. Make sure to refer to the manufacturer's documentation to ensure older network devices will be able to provide power. It has been our experience that some network gear requires the use of additional protocols such as LLDP to power more power-hungry devices adequately.

In the weeks and months prior to the launch of 802.11ax devices by most manufacturers, and although the design of the WLAN should have taken this into consideration, it is possible that APs will require either more than one cable or a single cable that runs at 2.5, 5 or 10 Gigs. When validation comes around the corner, ensure that the uplink to the network runs the adequate speed to support your wireless infrastructure.

Deploying a fast and efficient WLAN includes considerations on the wired infrastructure in logical terms too. Some architectures rely on routing policies, security mechanisms and such which could have an impact on wireless devices. Let's just say for example that the wired infrastructure does not implement QoS or that QoS is not carried throughout the entire network. The impact on a wireless voice WLAN could be huge: even if voice packets are prioritized over the air to the AP and controller, once they are out of the wireless realm no priority will be given to the voice packets anymore. QoS, should it be implemented on the wireless side should always be supported on the wired side. It must be implemented end to end to make a difference. A quick validation now could save you a couple of hours of troubleshooting down the line.

Routing and traffic management strategies should also be reviewed. Traffic shaping, queueing, or policing the input or output of networks can distort your performance measurements. If it is in use and you are required to meet specific performance targets, document a baseline of the manipulations which are done if it is not already part of the design document: further performance testing may require either that the policy be reviewed or temporarily removed so that you can test your infrastructure free of those constraints.

A lot of architectures rely on equipment and link redundancy. Validation of these mechanisms can be quite exhaustive: we suggest you stick the environment close to the wireless infrastructure. For example, port aggregation (or EtherChannel) between members of a mobility group can prevent a redundancy and failover scenario from working properly. Depending on the topology selected in your implementation, protocols such as HSRP, VRRP or GLBP may have been implemented. They are not part of the CWDP exam, of course, but make a network resilient. Test redundant network devices, line cards, HA controllers and any other redundant part of the network so that the WLAN can sustain some degradation if this is required by your client.

As we've previously discussed in chapter 7, DHCP and DNS are important parts of the wireless infrastructure for the services they provide to both APs and clients. There should be a sufficient number of available IPs for your clients. The lease time must be long enough to allow smooth operations yet short enough to prevent wasting resources on transient traffic.

8.5: Captive Portals

We strongly recommend you test the captive portal, should that be the solution you had to implement for an SSID. Its use should respect the parameters required by your customers or security team yet be simple to use. A multitude of devices will use the captive portal: testing with the most common types present in this implementation could save you a few trouble tickets when the WLAN is put in production. Test the onboarding process with a selection of common devices as well as your client devices to ensure a smooth registration process if registration is required. Captive portals are often just a gateway to the Internet: make sure that they remain as secure as possible for their providers. It is easy to assume that the Captive Portal's traffic will be isolated from regular traffic, especially if you are only taking care of the wireless side of the house. Just as easy in fact as forgetting to create or apply an ACL or bridging a VRF.

All these items cannot be validated through a site survey. The list we've supplied may look daunting, but all of it can be validated in a few hours. We believe any site survey should be representative of the infrastructure it validates. It would be unfortunate to spend a day's work testing an infrastructure to discover the reason why you are not seeing all SSID's as you are done with your day.

As a CWDP, it is expected that you can explain, mentor personnel who will perform or perform a validation survey yourself. The term survey is sometimes subject of Twitter brawls in the wireless community as it may sound imprecise. As a matter of fact, I (Mae) remember in the very early days of my career in wireless networking being really confused by the term. What was a "site survey" versus an "exit survey" and a "validation survey," terms I would see used by VARs, telcos and other consultants? After all, I only wanted to know where my APs should go!

An exit survey is a validation survey post-installation. Its purpose is to be able to confirm to your client that the WLAN you have deployed is performing as expected. It is also a way for you to make sure your design was correct and deployed adequately. This exercise will also allow you to test the WLAN's security parameters, applications, roaming, connectivity and client performance. Since you may have been called onsite following someone else's installation, you can also use the term "remediation survey." They are really one and the same. Only the angle differs: exit or post-installation survey is done after the project is complete, and remediation is because you suspect something is wrong.

There are many ways to perform validation. Back in the very the early days of Wi-Fi software solutions didn't really exist: surveys were often done with software provided by manufacturers and taking readings according to the map, recording the data themselves on the spot on a map or in a spreadsheet. This method is like doing a spot check with a specialized tool such as a NetScout G2 or software such as InSSIDer or Wi-Fi Explorer. For a small area, it could save you time over a site survey. Be aware however that applications on your mobile phone are using its wireless adapter to test. If you are deploying for a client with different capacities, this might not be the most accurate way to test.

Figures 8.2, 8.3 and 8.4 show various parts of a report from AirCheck G2.

Site survey software such as iBwave Wi-Fi, Ekahau, and AirMagnet allow you to record data points on a map to analyze the situation later. In larger deployment, these solutions allow a better view of the whole situation. If you cannot use the targeted client device as an adapter to perform your validation, it is possible to test it against the values of the site survey device to calculate an offset. Using this measurement in your software will allow you to visualize the results from the client's perspective and, in the case of a remediation, be able to further plan according to a target device.

Be aware though that client devices are not calibrated, so there will always be variations in the measurements for two identical devices. Should you be interested in learning more about client variation and even participate in a crowdsourced effort to illustrate and better use the offset capacity of the site survey software, you can visit rssicompared.com. This project's aim is to illustrate and better model various wireless clients. Figure 8.5 shows you the variations for 12 identical MacBooks in a given setting. Each dot on the map is the reading of the RSSI for one device at 3 meters. The calculated variation is done by Blake Krone, creator of this project.

Most people use site survey software to conduct validation. Using such software, you will be able to test the following items:

- Coverage
- RSSI, RCPI, SNR
- Channel and power
- Interference (ACI/CCI)
- Roaming
- QoS
- Security
- Network infrastructures & DNS and DHCP
- Applications: Throughput, loss, jitter, data rate
- Wired and wireless redundancy.

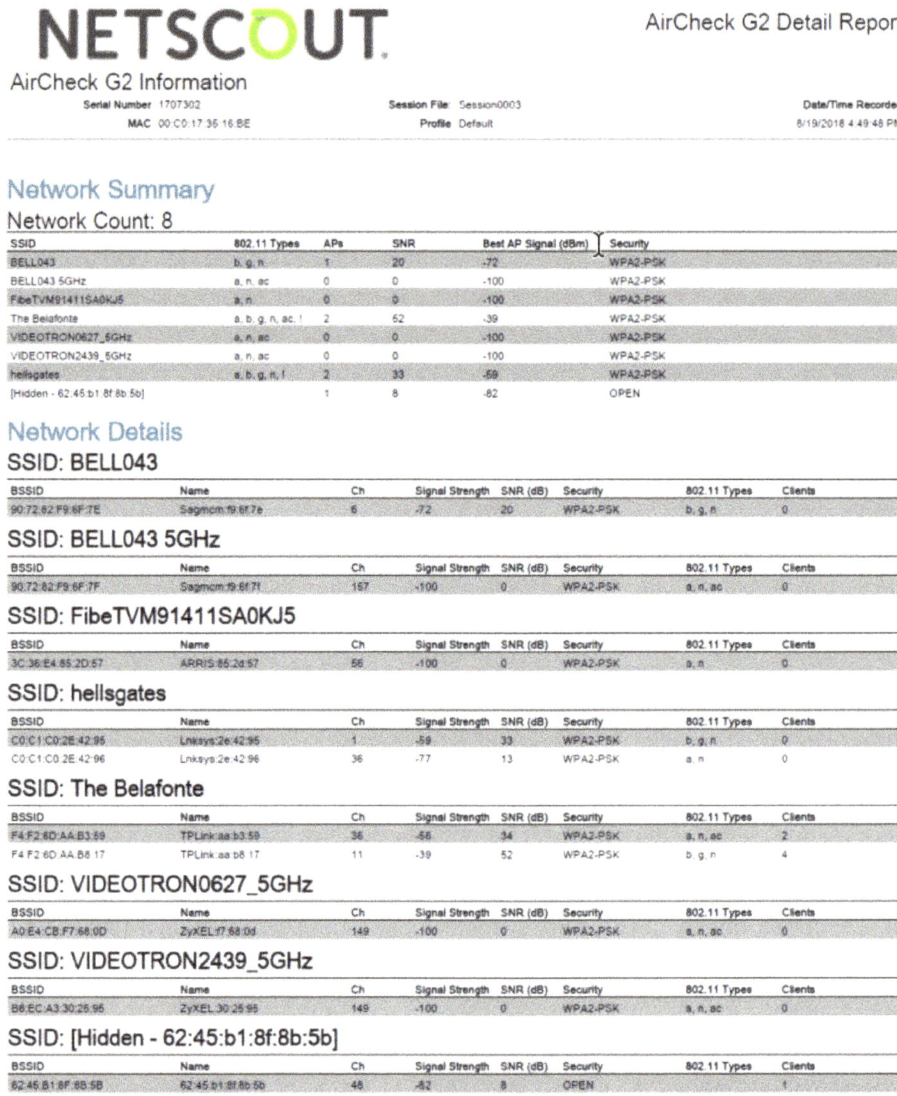

NETSCOUT

AirCheck G2 Detail Report

AirCheck G2 Information

Serial Number 1707302 **Session File:** Session0003 **Date/Time Recorded**
MAC 00:C0:17:36:16:BE **Profile** Default 8/19/2018 4:49:48 PM

Network Summary

Network Count: 8

SSID	802.11 Types	APs	SNR	Best AP Signal (dBm)	Security
BELL043	b, g, n	1	20	-72	WPA2-PSK
BELL043 5GHz	a, n, ac	0	0	-100	WPA2-PSK
FibeTVM91411SA0KJ5	a, n	0	0	-100	WPA2-PSK
The Belafonte	a, b, g, n, ac, !	2	52	-39	WPA2-PSK
VIDEOTRON0627_5GHz	a, n, ac	0	0	-100	WPA2-PSK
VIDEOTRON2439_5GHz	a, n, ac	0	0	-100	WPA2-PSK
hellsgates	a, b, g, n, !	2	33	-59	WPA2-PSK
[Hidden - 62:45:b1:8f:8b:5b]		1	8	-82	OPEN

Network Details

SSID: BELL043

BSSID	Name	Ch	Signal Strength	SNR (dB)	Security	802.11 Types	Clients
90:72:82:F9:6F:7E	Sagmcm:f9:6f:7e	6	-72	20	WPA2-PSK	b, g, n	0

SSID: BELL043 5GHz

BSSID	Name	Ch	Signal Strength	SNR (dB)	Security	802.11 Types	Clients
90:72:82:F9:6F:7F	Sagmcm:f9:6f:7f	157	-100	0	WPA2-PSK	a, n, ac	0

SSID: FibeTVM91411SA0KJ5

BSSID	Name	Ch	Signal Strength	SNR (dB)	Security	802.11 Types	Clients
3C:36:E4:85:2D:57	ARRIS:85:2d:57	56	-100	0	WPA2-PSK	a, n	0

SSID: hellsgates

BSSID	Name	Ch	Signal Strength	SNR (dB)	Security	802.11 Types	Clients
C0:C1:C0:2E:42:95	Lnksys:2e:42:95	1	-59	33	WPA2-PSK	b, g, n	0
C0:C1:C0:2E:42:96	Lnksys:2e:42:96	36	-77	13	WPA2-PSK	a, n	0

SSID: The Belafonte

BSSID	Name	Ch	Signal Strength	SNR (dB)	Security	802.11 Types	Clients
F4:F2:6D:AA:B3:59	TPLink:aa:b3:59	36	-56	34	WPA2-PSK	a, n, ac	2
F4:F2:6D:AA:B8:17	TPLink:aa:b8:17	11	-39	52	WPA2-PSK	b, g, n	4

SSID: VIDEOTRON0627_5GHz

BSSID	Name	Ch	Signal Strength	SNR (dB)	Security	802.11 Types	Clients
A0:E4:CB:F7:68:0D	ZyXEL:f7:68:0d	149	-100	0	WPA2-PSK	a, n, ac	0

SSID: VIDEOTRON2439_5GHz

BSSID	Name	Ch	Signal Strength	SNR (dB)	Security	802.11 Types	Clients
B8:EC:A3:30:25:95	ZyXEL:30:25:95	149	-100	0	WPA2-PSK	a, n, ac	0

SSID: [Hidden - 62:45:b1:8f:8b:5b]

BSSID	Name	Ch	Signal Strength	SNR (dB)	Security	802.11 Types	Clients
62:45:B1:8F:8B:5B	62:45:b1:8f:8b:5b	48	-82	8	OPEN		1

Figure 8.2: Report from AirCheck G2

Access Points

Access Point Count: 10

Access Point	802.11 Types	Ch	SSID	Best AP Signal (dBm)	SNR (dB)	Security	Auth Class	Clients	Supported Rates	Basic Rates
ARRIS:85:2d:57	a, n	56	FibeTVM91411SA0KJ5	-100	0	WPA2-PSK	None	0	6, 9, 12, 18, 24, 36, 48, 54	6, 12, 24
ZyXEL:30:25:95	a, n, ac	149	VIDEOTRON2439_5GHz	-100	0	WPA2-PSK	None	0	6, 9, 12, 18, 24, 36, 48, 54	6, 12, 24
Sagmcm:f9:8f:71	a, n, ac	157	BELL043 5GHz	-100	0	WPA2-PSK	None	0	6, 9, 12, 18, 24, 36, 48, 54	6, 12, 24
ZyXEL:f7:68:0d	a, n, ac	149	VIDEOTRON0627_5GHz	-100	0	WPA2-PSK	None	0	6, 9, 12, 18, 24, 36, 48, 54	6, 12, 24
62:45:b1:8f:8b:5b		48	[Hidden - 62:45:b1:8f:8b:5b]	-82	8	OPEN	None	1		
Sagmcm:f9:6f:7e	b, g, n	6	BELL043	-72	20	WPA2-PSK	None	0	1, 2, 5.5, 6, 9, 11, 12, 18, 24, 36, 48, 54	1, 2, 5.5, 11
Lnksys:2e:42:95	b, g, n	1	hellsgates	-69	33	WPA2-PSK	None	0	1, 2, 5.5, 6, 9, 11, 12, 18, 24, 36, 48, 54	1, 2, 5.5, 11
Lnksys:2e:42:96	a, n	36	hellsgates	-77	13	WPA2-PSK	None	0	6, 9, 12, 18, 24, 36, 48, 54	6, 12, 24

Figure 8.3: Second Part of AirCheck G2 Report

Interferers

Interferer Count: 5

Type	Id	Event Count	Peak Power (dBm)	Average Power (dBm)	Affected Channels	Duty Cycle (%)	First Seen	Last Seen
Possible Interferer	0	3	-35	-35	6-12	74.7	8/19/2018 4:45:06 PM	8/19/2018 4:45:16 PM
Bluetooth Device	1	1	-45	-45	1-13	0	8/19/2018 4:45:21 PM	8/19/2018 4:45:21 PM
Possible Interferer	2	1	-24	-24	011	77.4	8/19/2018 4:46:47 PM	8/19/2018 4:46:47 PM
Possible Interferer	3	1	-24	-24	011	77.4	8/19/2018 4:46:49 PM	8/19/2018 4:46:49 PM
Possible Interferer	4	1	-24	-24	011	77.4	8/19/2018 4:46:50 PM	8/19/2018 4:46:50 PM

Channel Summary Table

2.4GHz

Ch	Freq (GHz)	APs	Clients	Interferer Events	Avg Total Util(%)	Last Total Util(%)	Avg 802.11 Util(%)	Last 802.11 Util(%)	Avg Non-802.11 Util(%)	Last Non-802.11 Util(%)
1	2.412	1	0	0	13.23	7	3.45	1	9.77	6
2	2.417	0	0	0	12.23	2	0.91	0	11.32	2
3	2.422	0	0	0	13.73	4	0.18	0	13.55	4
4	2.427	0	0	0	13.91	3	0.14	0	13.77	3
5	2.432	0	0	0	16.14	3	0.23	0	15.91	3
6	2.437	1	0	0	18.91	9	0.36	0	18.55	9
7	2.442	0	0	0	17.62	19	1.05	0	16.57	19
8	2.447	0	0	0	17.48	13	0.67	1	16.81	12
9	2.452	0	0	0	17.43	13	0.29	0	17.14	13
10	2.457	0	0	0	18.81	21	0.52	6	18.29	16
11	2.462	1	4	0	16.1	19	1.62	5	14.48	14
12	2.467	0	0	0	13.24	10	0.95	0	12.29	10
13	2.472	0	0	0	11.14	11	0.29	0	10.86	11
14	2.484	0	0	0	1.9	0	0.19	0	1.71	0

Figure 8.4: Final Part of AirCheck G2 Report

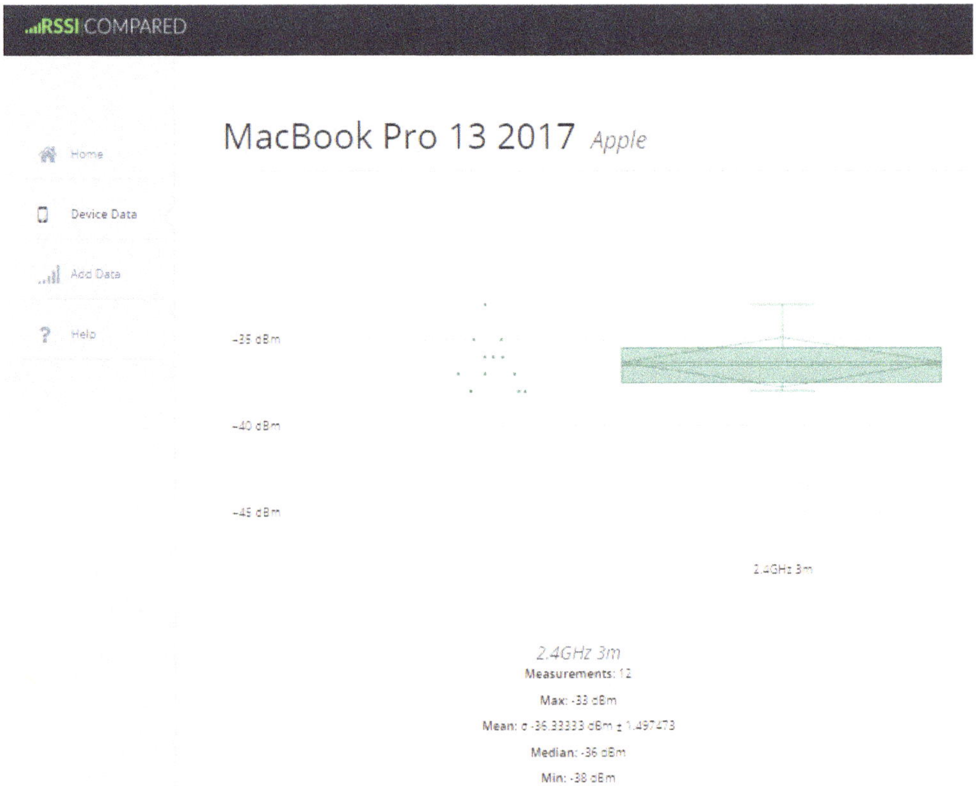

Figure 8.5 RSSIcompared.com

Depending on the type of validation you wish to perform, the surveying device may or may not be associated. The terms passive or active site survey are often used at this point in the process. Most software nowadays allows you to perform both the passive and the active survey all at once using a USB hub with multiple adapters. Using multiple adapters will also allow you to be able to perform your validation quicker because the time required to scan all the channels available in all bands will be split across the adapters. Figure 8.6 shows a multi-adapter confuration option using a hub. Alternatively, a special device, like Ekahau Sidekick, which contains multiple adapters, could be used.

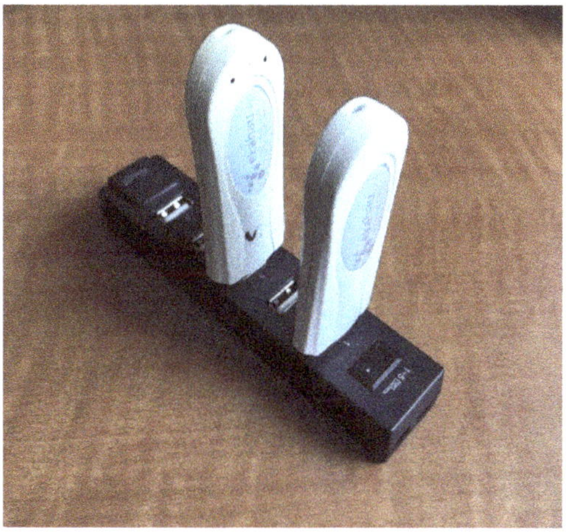

Figure 8.6: A Multi-Adapter Configuration

Usually, a validation survey is a passive survey. The reason for it is quite simple: no device will ever behave 100% the same, so the value of an active survey is, to most people, questionable as a validation tool. There are things that the active will test over the passive survey, which is why we decided to present both types. However, we advise you to opt for a combination of passive survey and real-world device tests to present the most accurate view to your client of the state of his infrastructure.

Figure 8.7 : Ekahau Sidekick

8.6: Passive Survey

This validation will allow you to test the following items[92]:

- AP positioning/presence
- Signal Strength
- RSSI/RCPI
- SNR
- Channel and power
- Interference (both ACI and CCI, depending on how you test)

The passive survey is based on management frames. It does not exchange data frames with the network. It is a view of the current state of the network which will let the designer know that his work, as well as the installation and configuration, was done correctly. The advantage of the site survey software is that you can define walking paths and follow the exact behavior of the wireless network in time. Some site survey software includes spectrum analyzers and various ways to collect data.

In the resulting file, most will list the APs that are detected, either by MAC address or AP name. They may recombine the various mac addresses which have been attributed to a specific AP, and some can even auto-place APs where the algorithms calculate they are located. This localization is based on triangulation and the accuracy of your data. Therefore, a slow, steady walking pace and regular data acquisition ("clicks") are recommended. Be careful when conducting a site survey: it's easy to get either lost in the map or click in the wrong spot. The software will calculate a result regardless.

If your site survey results look inaccurate versus your model, you should either question the staff who performed the validation or deployed the network or both. Since we assume you have been careful in validating that the APs were configured properly according to the design document, the reason might be that there are obstacles in the RF path which were not considered. Should you notice such an obstacle, make sure you are

[92] The term survey now reveals its post implementation purpose. You are now surveying to "see" what the network looks like after installation. The purpose of this survey is to gather information to validate requirements and, where requirements are not met, to make recommendations for optimization so that requirements are met.

surveying as thoroughly around it as possible to be able to assess the RF behavior and possibly address it as effectively as possible.

Signal Strength

The very first metric that we all look at is the signal strength. It is the most basic assessment of WLAN quality: the RF power which is emitted from its AP's management frames. The visualization of the signal strength will allow you to view what the E-plane and H-plane chart were trying to illustrate in an antenna or APs' spec sheet. For voice deployment, the recommended maximal signal strength is between -65 and -67 dB. For data the devices tend to be a little more forgiving and use better wireless components but still we prefer to keep maximum signal strength at about the same level to ensure that we have a margin, should we need it. Don't forget that the data rate the client gets is dependent on the signal strength. The AP manufacturers often provide information regarding the maximal signal for each data rate in AP data sheets: we suggest you refer to this documentation to make sure you achieve the signal strength you need for your client's application.

When you are reviewing your results, don't forget to look at secondary coverage if it was requested: most site survey software will only show you primary (aka the strongest) AP data. Various visualizations are available depending on the software you are using, so make sure to be comfortable with the display whether it is color-coded, presented in dB or in dBm. Also bear in mind that if it is in dBm, you may be looking at RSSI or RCPI in the labels, but know if it is truly RSSI (per the 802.11 definition) or colloquially RSSI as used in the WLAN industry, which is far more common in tools today.

RSSI and RCPI

They are a device's view of the network. The RSSI is a proprietary way of displaying signal quality, but, as previously stated, most tools use it synonymously with dBm. Be aware though that it bases itself only on the header of the packets. Depending on the packets' type, they could be sent at the lowest mandatory speed, something that would mislead you. Slower speeds can cover more ground: if your goal is to support a set speed, this should be kept in mind to avoid disappointing your client. RCPI is a normalized view of the signal quality in dBm. It is calculated over the entire packet, including header and trailer, which makes it a more accurate metric. It is also not a secret sauce recipe that varies from vendor to vendor. Enterprise-grade Wi-Fi tools do not use either RSSI or RCPI[93] for signal strength, but dBm is used instead; however a CWDP should remember the difference between the two and the proper definitions.

SNR

If you remember from your CWNA studies, the SNR is the ratio of the level of the signal to the level of the noise in an environment. This metric will be measured in a passive site survey and can allow you to know if a set device will be able to reach its performance target. You may have -45dB in signal strength, but if the area's noise is -60, your SNR will be too low to reach any decent kind of throughput on the WLAN.

If you do not have a set target per your client's request, bear in mind that most devices require at least 20 dB (data) to 25 dB (voice) and ideally more than that (28dB is the minimal recommended SNR for high-speed MCS rate with 802.11n/ac).

Channel, power, ACI, and CCI

If you have implemented a static channel and power plan, the passive survey will allow you to validate that the deployment has been done exactly like you were expecting. It is now time to check that there is no strange propagation pattern causing APs on the same channel to be too close to one another or too many APs on top of each other. Figure 8.8 shows the APs in the Frequency Monitor within Ekahau Site Survey.

As you may recall from CWNA, ACI is the interference created by the presence of RF in channels which are overlapping one another. The reason why Wi-Fi professionals are preoccupied with ACI is simple. Since 802.11 uses CSMA/CA and can understand the headers of all other 802.11 devices on the same channel, all 802.11 devices can be polite and take their turn at using a channel. However, when a device is on another overlapping channel, its capacity to hear the initial channel is affected. Since it cannot say whether another device is using the channel or not, it may try to send. If two stations on adjacent channels send at the same time, you can get collisions and retransmissions. Severe ACI can have a snowball effect which will make the WLAN unusable. By

---

[93] Sadly, RCPI, which is a much better metric that RSSI as defined in the 802.11 standard, it not used by hardly any tools. For a detailed description of RSSI, see *RSSI – A Changing Definition* located at the CWNP blog: www.cwnp.com/rssi-changing-definition/

eliminating the sources of interference on overlapping channels and implementing a sensible channel plan, you can mitigate ACI. Figure 8.9 shows an AP causing ACI for channels 1 and 6.

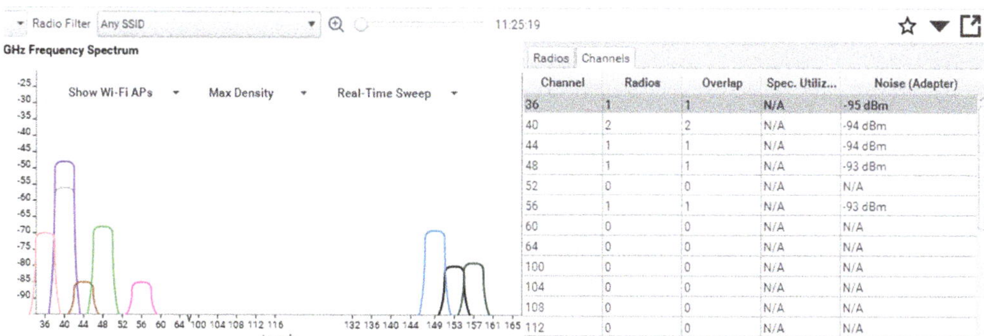

Figure 8.8: Ekahau Site Survey Frequency Monitor

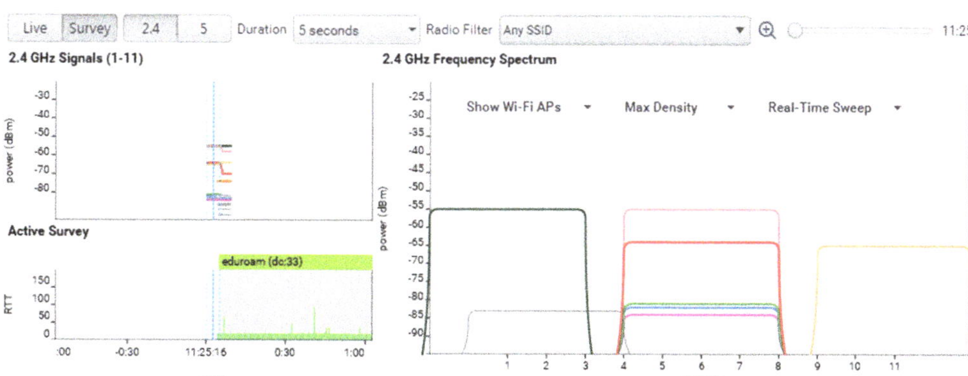

Figure 8.9: Frequency Monitor showing ACI

CCI is basically the competition of devices on a channel where there is a finite amount of space available. As an analogy, think of CCI like a race event at the Olympics. On a race track, there is only so much space for competitors, just like there is so much airtime available on a band. Trying to stick too many racers on the track will not make the competitors any faster: it will slow them down because they don't want to hurt one another. The more space between them will allow them to focus on their one mission: racing. Just like an AP's mission: exchanging data.

The effect of high CCI is contention: the WLAN will be subject to a higher load of management traffic and retransmissions. High contention may trigger protection mechanisms, meaning even more traffic. If your passive survey shows heavy CCI after an APoS survey, pre-deployment, check if the person in charge of the APoS walked to the cell's edge. It might explain why, considering the power of the APs, you see so much CCI. This is particularly crucial if your network is using 40MHz or larger channel plans. For example, using a 40MHz channel, CSMA-CA mechanisms will try to detect a carrier sense at -82dBm up on a primary channel, and energy at -62dBm and above on a secondary channel (802.11n) or -72dBm (802.11ax).

Your site survey software will identify the most common source of CCI: the APs. Keep it mind that even though they are not mapped by the software, the clients themselves are also a source of CCI: making cells smaller will help in mitigating their part of the CCI issue. An adequate channel and power plan can help you achieve the perfect balance.

In the case of dynamic channel and power assignment, the passive site survey will allow you to validate the behavior of the algorithm **at that point in time**. It is possible to refer to it later as some form of a baseline of things upon putting them in production. Misbehavior of RRM could be detected in this way. You could also use the results to tune the algorithm further if the data you have gathered is different from your RF model. Radio resource management can work well, but it is not all magic. You should know how to tune it so that it can do what you want it to do. Some manufacturers have complete guides to help you achieve the results you hope for, so look them up to further understand how you can make RRM the best it can be.

8.7: Active Survey

The active survey will allow you to test your device's behavior on the real network. Since it is not really an accurate representation of the reality of the network with your clients' devices fully active, most Wi-Fi engineers do not use active surveys as a validation tool. You will gather data most likely from an adapter which is different from the "least capable most important device" that the infrastructure must support. Unless there is a requirement from your client to get an active validation survey, we suggest you stick to a passive and opt for more meaningful tests with the actual client devices which will be in use on the network, something which we will cover later in this chapter. For now, a few words about active surveys.

Active surveys generally require the entire WLAN and related LAN infrastructure and services to be in place. These are the elements it can test:

- Network infrastructures & DNS/DHCP
- Security
- Throughput
- Retransmissions
- Packet loss,
- Jitter
- Delay.
- Roaming
- QoS

Network and Related Infrastructures

Most wireless deployment today rely on DHCP to distribute IP addresses to clients. Not only should you validate that the infrastructure gives you an IP address, but it must also provide the right DNS, gateway, and mask. Site survey software allows you to ping a common destination (such as a central public resource on your network or on the Internet) which will allow you to test both WLAN performance and configuration of the wired side including DHCP, NAT, firewalls, DNS, AAA, routing, and security.

Security

Since an active survey requires a connection to the WLAN, it will test the implementation of the security mechanisms used to authenticate, whether using a passphrase or 802.1X. Clients should be able to associate and authenticate on the WLAN with minimal delay. Some devices are particularly sensitive to long association times: additional tuning of the AAA methods in use or the supporting servers may be required for the network to perform adequately.

Throughput

Throughput is about quantifying the upper layer data that you are transferring over the air. It is not about marketing numbers but real usable information transfer. Rule of thumb is the advertised speed divided by two for lower data rates: so, for example, 802.11a and 802.11g were up to 54 Mbps with throughput between 23-25 Mbps. 802.11n was around 150 Mbps for common implementations with between 75-100 Mbps throughput, and 802.11ac is around 600 Mbps for common implementation with between 300-450 Mbps. As the data rate goes up, the percentage available for throughput does as well in many cases. The exception is a saturated BSS with lots of devices contending for the medium and extra overhead from protection mechanisms, which can bring the throughput back down again to closer to 50 percent of the data rate. This is due to many factors such as the bandwidth in use, the security and protection mechanisms, usage level of the AP, RF environment, and WLAN configuration.

If you've ever had to support or use a congested link, you may know that there are lots of tools on the Internet available that "test your bandwidth." Aside from not being part of any site survey software, most wireless professionals will try to avoid taking metrics from such tools. The reason is simple: you are not only testing the link from your device to the AP, you are also testing the network infrastructure the WLAN rests on, the Internet pipe that it is providing you access to the web, the meanders by which your packets will make its way to a defined server including the server's infra, Internet pipe, and even its hardware. This is in no way as precise as an IPerf server simply because you have no control over any part of the path aside from the WLAN, LAN and, in some way, the uplink to the Internet. If you choose to test the throughput against an IPerf server, your site survey software may require additional configuration: refer to the product documentation for more details.

How can you test throughput while conducting an active site survey? Ekahau and AirMagnet support IPerf and iBwave offers a proprietary solution called NetScan. Essentially these tools allow you to exchange data packets with a server and then compiles the statistics for the exchange. Keep in mind that if you are testing throughput, you must be careful where you are putting the server because it will have an impact on the result. You should keep the server within the network and ideally in a centralized location.

Retransmissions

As you know, Wi-Fi uses ACKs (or block ACKS) to confirm the reception of packets. If the corresponding ACK is not received, the transmitter will try to resend the packet. This attempt is what we call a retransmission. If you detect through an active survey's Iperf or through a real-life device test with some form of packet analysis utility that there are more than 10% retries on the WLAN, this is a major concern which should require your attention. 5-9% means that you should be careful and pay attention to your cell size, data rates, hidden nodes, ACI, and noise. Less than 5% can be accepted: it would be abnormal not to see any retransmissions on a WLAN. Retransmissions are one of those things which will create a snowball effect if not controlled, creating more traffic, more contention and more problems until you address it.

Packet loss

Packet loss occurs when data packets do not make it across the medium. This is often due to congestion on the network. An active survey can give you a sense of packet loss if you are running either a ping test or an Iperf test. Some types of communications are more sensitive to packet loss than others: for example, Voice over WLAN will not fare well with more than 1% packet loss. Data networks will tolerate more than that but more than 3%.

Jitter

A concern which is typical of voice and video application, Jitter is the variation in the reception of a stream of packets. It is often caused by either misconfigurations, contention or queuing issues. For voice networks, the acceptable jitter is ideally below 20msec. More jitter will create echo on audio transmissions. An active site survey will give you information on jitter though it might be minimal: other tools which we will cover later will be chattier if ever jitter is a big issue.

Roaming

There are generally two types of users on a wireless network: mobile users and nomadic users. Mobile users must remain connected all the time. This would be the phones, cameras, hospital IoT devices and such. Nomadic users are the laptops, barcode scanners and other devices which need to access the network almost anywhere but do not need to be connected all the time and can afford to lose the connection for a few seconds.

It is not easy to test roaming as rigorously as most would hope to be able to test it. Client devices are not calibrated, and roaming algorithms are usually proprietary and dependent on OS and drivers. This is to say that your test will be good for the device which you are testing at that moment: an identical device could behave differently. Since an active survey may not be the best way to test roaming for your clients, we suggest you use several actual devices for a definitive test. The active survey will only allow you to get a general idea of where the deployment stands in terms of roaming ability and behavior, mobility domain or group configuration and the like.

Capacity

Initially, wireless networks were there to complement wired networks. However, since then, the tendency has been reversed: we tend to think in terms of Wi-Fi first and wired second. It is safe to say that demand has exploded while at the same time the number of connected devices has grown exponentially. Those of us who come from the wired side will be tempted to think that capacity is a simple thing to address. After all, if you are missing switch ports, all you need to do is add another line card or another switch. With wireless it's different, there is only so much airtime that is available. Once you've reached the physical limit, throwing another AP to the mix to add capacity isn't going to help, it will make things worse!

Site survey software will be able to give you a general idea of the capacity of your network if it is equipped with an embedded spectrum analyzer. However, this is not in any way a guarantee that your client devices and applications will be able to support the requested capacity. Even emulation of clients cannot do that. The reality is that capacity testing can only be done by…**observing the actual WLAN live in action with the requested capacity**[94]. It is one thing to create a theoretical model of what a WLAN should

[94] This is not to say that capacity cannot be planned. Testing versus planning is key. You can certainly plan capacity and if you considered every single possible variable, the implementation should function as planned. But can you consider every possible variable? No, you can't. Things will change from the time of the requirements engineering and survey tasks to the time when the design is completed, and the implementation is performed. Therefore, testing the network in production is an important step, which may lead to recommendations for optimization to achieve the capacity required in the beginning.

be, and to that effect we suggest you look into Andrew von Nagy's Capacity Planner referenced elsewhere in this book as well and shown in Figure 8.10.

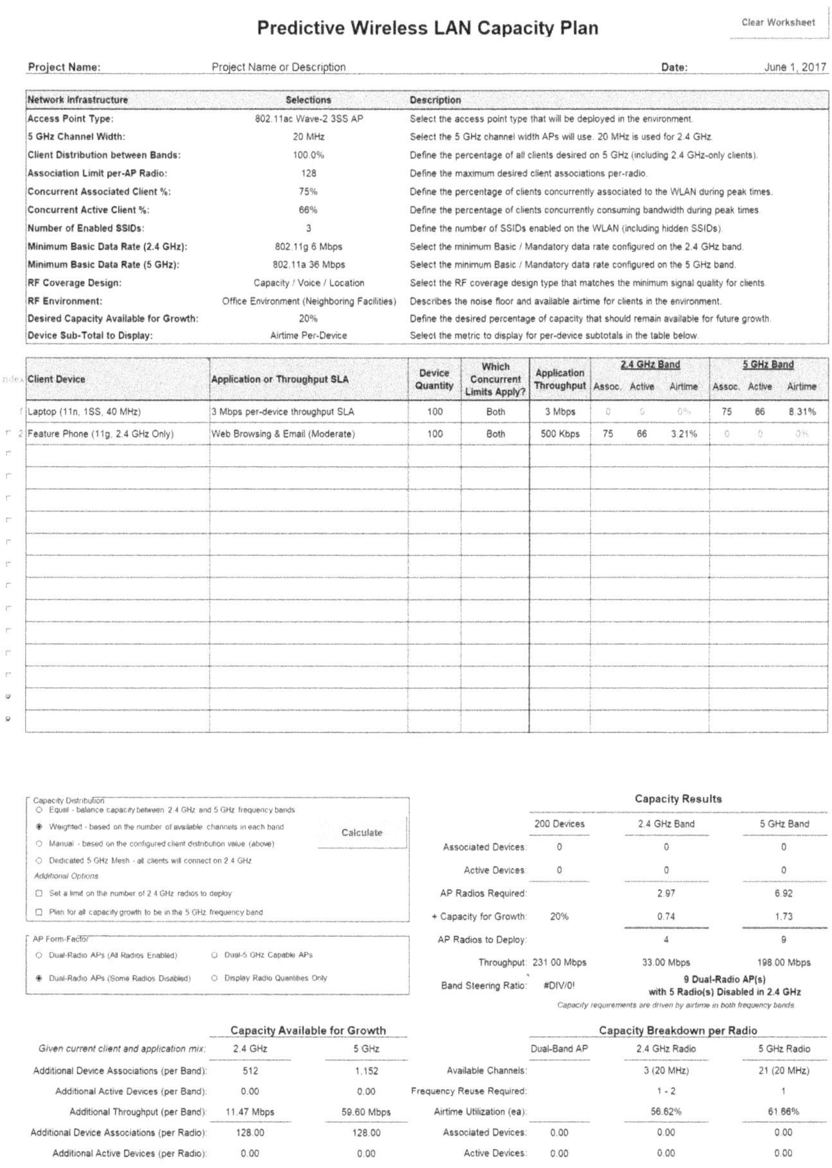

Predictive Wireless LAN Capacity Plan

Clear Worksheet

Project Name: Project Name or Description | Date: June 1, 2017

Network Infrastructure	Selections	Description
Access Point Type:	802.11ac Wave-2 3SS AP	Select the access point type that will be deployed in the environment.
5 GHz Channel Width:	20 MHz	Select the 5 GHz channel width APs will use. 20 MHz is used for 2.4 GHz.
Client Distribution between Bands:	100.0%	Define the percentage of all clients desired on 5 GHz (including 2.4 GHz-only clients).
Association Limit per-AP Radio:	128	Define the maximum desired client associations per-radio.
Concurrent Associated Client %:	75%	Define the percentage of clients concurrently associated to the WLAN during peak times.
Concurrent Active Client %:	66%	Define the percentage of clients concurrently consuming bandwidth during peak times.
Number of Enabled SSIDs:	3	Define the number of SSIDs enabled on the WLAN (including hidden SSIDs).
Minimum Basic Data Rate (2.4 GHz):	802.11g 6 Mbps	Select the minimum Basic / Mandatory data rate configured on the 2.4 GHz band.
Minimum Basic Data Rate (5 GHz):	802.11a 36 Mbps	Select the minimum Basic / Mandatory data rate configured on the 5 GHz band.
RF Coverage Design:	Capacity / Voice / Location	Select the RF coverage design type that matches the minimum signal quality for clients.
RF Environment:	Office Environment (Neighboring Facilities)	Describes the noise floor and available airtime for clients in the environment.
Desired Capacity Available for Growth:	20%	Define the desired percentage of capacity that should remain available for future growth.
Device Sub-Total to Display:	Airtime Per-Device	Select the metric to display for per-device subtotals in the table below.

Index	Client Device	Application or Throughput SLA	Device Quantity	Which Concurrent Limits Apply?	Application Throughput	2.4 GHz Band Assoc.	2.4 GHz Band Active	2.4 GHz Band Airtime	5 GHz Band Assoc.	5 GHz Band Active	5 GHz Band Airtime
1	Laptop (11n, 1SS, 40 MHz)	3 Mbps per-device throughput SLA	100	Both	3 Mbps	0	0	0%	75	66	8.31%
2	Feature Phone (11g, 2.4 GHz Only)	Web Browsing & Email (Moderate)	100	Both	500 Kbps	75	66	3.21%	0	0	0%

Capacity Distribution
- ○ Equal - balance capacity between 2.4 GHz and 5 GHz frequency bands
- ● Weighted - based on the number of available channels in each band
- ○ Manual - based on the configured client distribution value (above)
- ○ Dedicated 5 GHz Mesh - all clients will connect on 2.4 GHz

Calculate

Additional Options
- ☐ Set a limit on the number of 2.4 GHz radios to deploy
- ☐ Plan for all capacity growth to be in the 5 GHz frequency band

AP Form-Factor
- ○ Dual-Radio APs (All Radios Enabled)
- ○ Dual-5 GHz Capable APs
- ● Dual-Radio APs (Some Radios Disabled)
- ○ Display Radio Quantities Only

Capacity Results

	200 Devices	2.4 GHz Band	5 GHz Band
Associated Devices:	0	0	0
Active Devices:	0	0	0
AP Radios Required:		2.97	6.92
+ Capacity for Growth:	20%	0.74	1.73
AP Radios to Deploy:		4	9
Throughput:	231.00 Mbps	33.00 Mbps	198.00 Mbps
Band Steering Ratio:	#DIV/0!	9 Dual-Radio AP(s) with 5 Radio(s) Disabled in 2.4 GHz	

Capacity requirements are driven by airtime in both frequency bands

Capacity Available for Growth

Given current client and application mix:	2.4 GHz	5 GHz
Additional Device Associations (per Band):	512	1,152
Additional Active Devices (per Band):	0.00	0.00
Additional Throughput (per Band):	11.47 Mbps	59.60 Mbps
Additional Device Associations (per Radio):	128.00	128.00
Additional Active Devices (per Radio):	0.00	0.00
Additional Throughput (per Radio):	2.87 Mbps	6.62 Mbps

Capacity Breakdown per Radio

	Dual-Band AP	2.4 GHz Radio	5 GHz Radio
Available Channels:		3 (20 MHz)	21 (20 MHz)
Frequency Reuse Required:		1 - 2	1
Airtime Utilization (ea):		56.62%	61.66%
Associated Devices:	0.00	0.00	0.00
Active Devices:	0.00	0.00	0.00
Avg. Throughput:	30.25 Mbps	8.25 Mbps	22.00 Mbps

Figure 8.10: Predictive Capacity Planner

However, bear in mind that cramming 10 APs in a single room to support the clients and their applications may not be feasible because of the laws of physics. There is only so much airtime available and adding more APs to support more clients on an already overloaded network will only make matters worse[95].

Site survey software is an excellent tool to validate and confirm the state of any deployment, whether it was just completed or about to be redesigned. It does have its limitations though and is only as precise as the type and precision of data you are collecting. There are also other types of tests which you may elect to do on a deployment to ensure it is in good working order. Should you choose to use a site survey and those tests, your client would get a very precise 360-degree view of the health of his WLAN.

8.8: Additional Forms of Validation Testing

Most Wireless engineers, when thinking of validation will think of maps and site surveys and such when the simplest form of validation is to use the WLAN. Using the newly built infrastructure for some time with a typical client, or even better, your client's target least capable most important device can provide you with a good sense of the quality of the network.

Capacity and roaming can be tested most accurately by using the target devices in a live environment. Though he may have struck the common memories as the "Verizon guy," the add which features a man with his phone asking the famous "can you hear me now" question is not far off: this is the best way to test roaming. While conducting such a test, data on the quality of the voice call can be obtained on the devices themselves through their GUI and logs. Some companies such as NetScout offer testing software to get an assessment of voice readiness, just like some manufacturers offer tools to assess a

[95] Cell size reduction has its limits. As a general rule, you cannot reduce a cell size, as to its potential for interference with other cells, below that of the clients. In other words, turning down the AP transmit power does NOT reduce the cell size to infinity. The reason is simple, clients transmit at the power they transmit at and some of them have no algorithm to lower the transmit power at all. Of course, an exception to this exists in OFDMA in 802.11ax because the clients must adjust their transmit power in an uplink OFDMA transmission according to the rules of the standard and these adjustments are based on the received signal from the AP. Therefore, in some distant future, we will be able to have very small cells measured in a few meters, but for now most cells are measured in tens of meters in the real world.

wireless deployment for voice or location, sometimes even embedded in their wireless network monitoring solution such as in the following Figure 8.11 and 8.12.

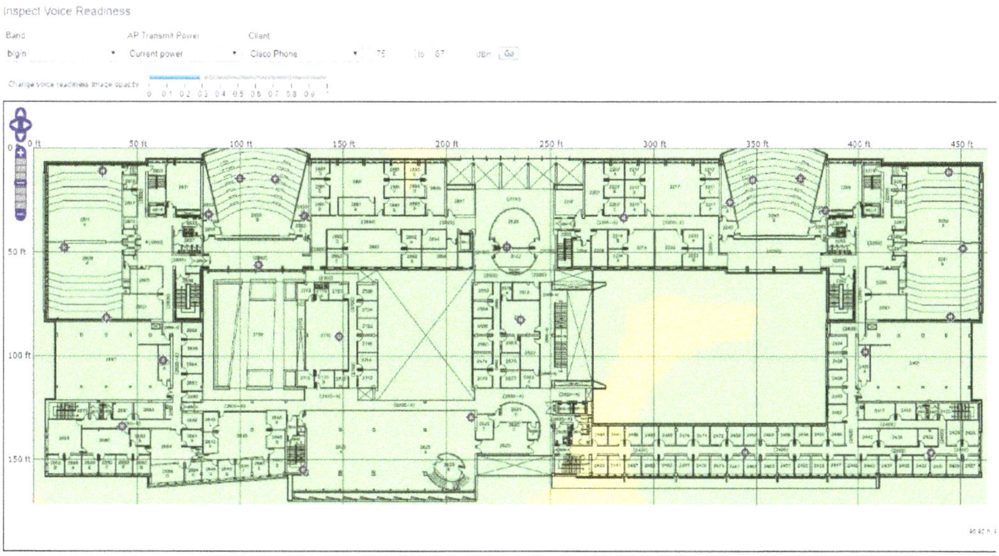

Figure 8.11: Inspecting Voice Readiness

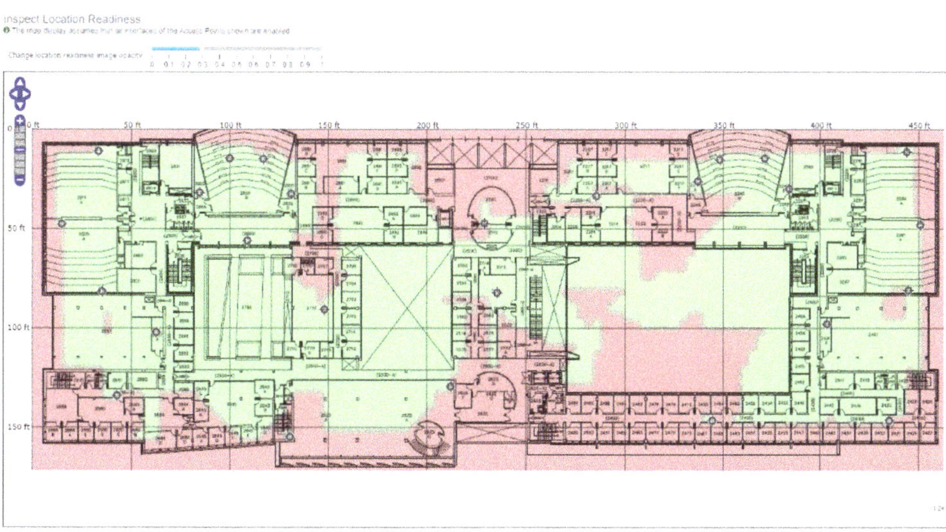

Figure 8.12: Inspecting Location Readiness

If the goal of testing roaming behavior does not include supporting voice deployments, there many other methods available. Tools such as the Netscout Aircheck G2 (shown in Figure 8.13), AirMagnet WiFi Analyzer and various scripts such as the awesome UbuntuRoamTest by Nigel Bowden, CWNE #135, available on GitHub, can help you in your efforts to measure roaming accurately. Though there are people who argue over what "total roaming time" should be calculated over, it is safe to say that as a CWDP your concern should be about sticking to being as close as possible to a real-life scenario, including testing with your client's target device. Consistency is key and remember: ***Perfect is the enemy of good.***

8.9: Testing throughput and performance

As we said, you should try to stick to the target device in all the measurements you take. To test throughput and capacity accurately, using tools such as IPerf3[96] and clients based on it such as Wi-Fi Perf Pro would be the best way to assess your deployment. IPerf 3, when used for Wi-Fi testing can test the following:

- Backhaul capacity
- Throughput per AP, per client
- QoS information
- Jitter
- Loss

Various ways of using IPerf exist. The NetScout Aircheck G2, a beloved tool among Wi-Fi circles, implements an IPerf test which either requires a test tool or an IPerf server. Ekahau and AirMagnet both implement a version of an IPerf client in their site survey software. Other site survey software manufacturers offer proprietary software which may be the equivalent. iBwave's Netscan and TamoSoft's Throughput tester are among those which come to mind.

[96] You can download iPerf for various platforms at iperf.fr/iperf-download.php. It is available in binary form for Windows, Android, iPad, iPhone, macOS, and Linux. You can also download C++ source code for the tool, which allows for customization to meet your needs.

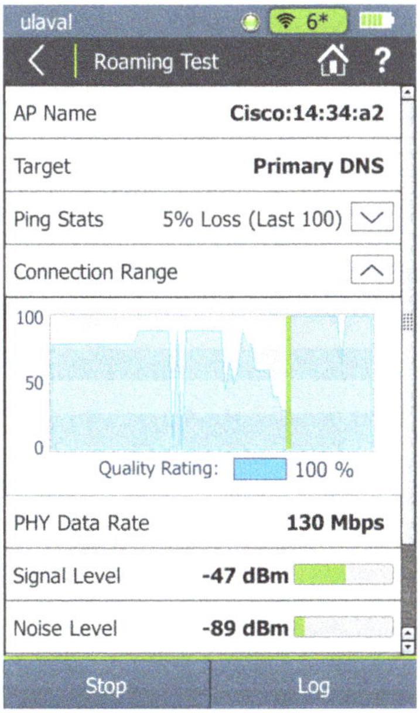

Figure 8.13: Testing Roaming

8.10: All-in-One Tools

There are a few additional tools which we would like to introduce because of their value in doing validation testing. These tools range from the open source free license to several thousand dollars. We strongly believe that a knowledgeable Wi-Fi engineer can get his information from any tool, regardless of budget considerations. This list is in no way exhaustive: should you have a favorite which is not listed here, it doesn't mean that it is not good. It only means you should blog, podcast, present at a conference to pay it forward to your fellow wireless professional who may benefit from learning something new.

NetScout AirCheck G2

One of the most popular tools among Wi-Fi pros is the NetScout AirCheck G2. A handheld device, it can give you a 360-degree view of the state of the WLAN. From testing the ethernet cable into which your AP will be connected to roaming analysis, this tool does it all. During validation, you can use it to test the following items:

- Available networks, APs including their security settings, data rates,DNS, DHCP
- Information about clients: security, associated AP, Data Rate, supported PHYs
- Interferers and channels, occupation, nature (802.11 or not)
- Roam testing
- Iperf testing (to their proprietary tool or to an Iperf Client)

WLAN PI

A standalone platform based on the Nano Pi, the WLAN Pi is the brainchild of Jerry Olla, CWNE #238. This tool can be built on a budget and run open source software such as:

- H.O.R.S.T (Provide a brief assessment including RSSI, data-rate, and info on traffic)
- Kismet (A classic open source software whose graphical interface can help view traffic types, data rates, and other information)
- Iperf (most flavors)
- HTML5 Speed Test: This local flavor of the famous speed test allows you to test uplink, downlink, loss, RTT, and jitter.

The WLAN PI can even act as a remote sensor to Adrian Granados' Wi-Fi Explorer Pro, sending back to the software data which was gathered in the field. Its small form factor makes it easy to carry around, and for the price, you cannot beat all it can do. It is shown in Figure 8.14.

Figure 8.14: WLAN PI

Wi-Fi Explorer Pro

A complete solution to assess your wireless network, Wi-Fi Explorer (Figure 8.15) allows the use of various remote clients to gather the most relevant information about a WLAN. It provides a 360-degree view of all broadcasted SSIDs, channel utilization, security parameters and spectrum analysis. A lot of bells and whistles come with it such as filtering, network coloring and more. This tool is the go-to tool for Wi-Fi engineers who run MacOS. Unfortunately, you cannot run it on a Windows platform.

As an aside, since you can capture wireless traffic on MacOS and not on Windows which hides the juicy Wi-Fi stuff you are looking for when troubleshooting, we think the OS your test gear runs may be the subject of intense questioning, especially if you are aiming at going down the CWNE path. CWAP is all about the analysis of packets, which can be performed natively on Apple's platform.

As we've mentioned previously, validation is all too often overlooked. A thorough validation survey to ensure your coverage objectives are met is essential to a successful deployment. Though they will always be present, ACI and CCI can be mitigated by ensuring that the power, channel plan, and overall WLAN configuration is correct. To address most application benchmarking, network capacity and connectivity, the tests which will provide you with the most accurate assessment of the situation are functional tests. Work on the wireless network, observe the behavior of the clients while roaming. Use the client devices as test tools and do not shy away from capturing the traffic live. After all, this is what your client wants: a usable WLAN infrastructure that is functional, stable, robust and on which his business can depend.

Wi-Fi Explorer is a macOS-only tool. To get similar functionality on Windows, consider the Acrylic software line. These tools provide similar functionality, and they offer an entire suite of tools useful for heatmap generation, design, scanning, and protocol analysis. Their tools are available at www.acrylicwifi.com. For frame capture, they support 7 different adapters at the time of writing with 802.11ac. They also support many 802.11n adapters. Additionally, they support BLE analysis. Figure 8.16 shows the Acrylic Wi-Fi Professional tool, which provides exceptional WLAN scanning abilities on Windows. The Acrylic Sniffer tool provides enhanced frame capture for Windows environments and integrated well with Wireshark for full frame decoding.

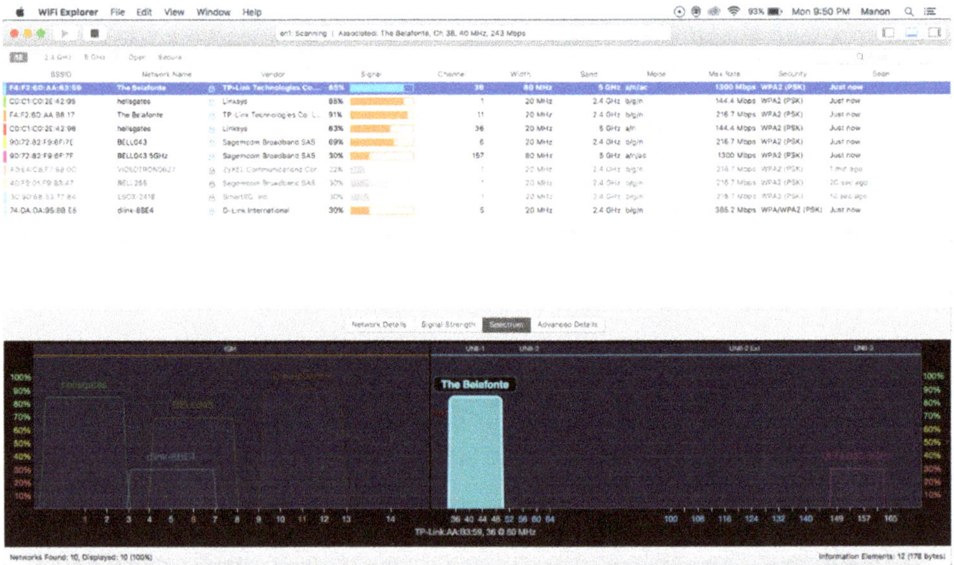

Figure 8.15: Wi-Fi Explorer Pro

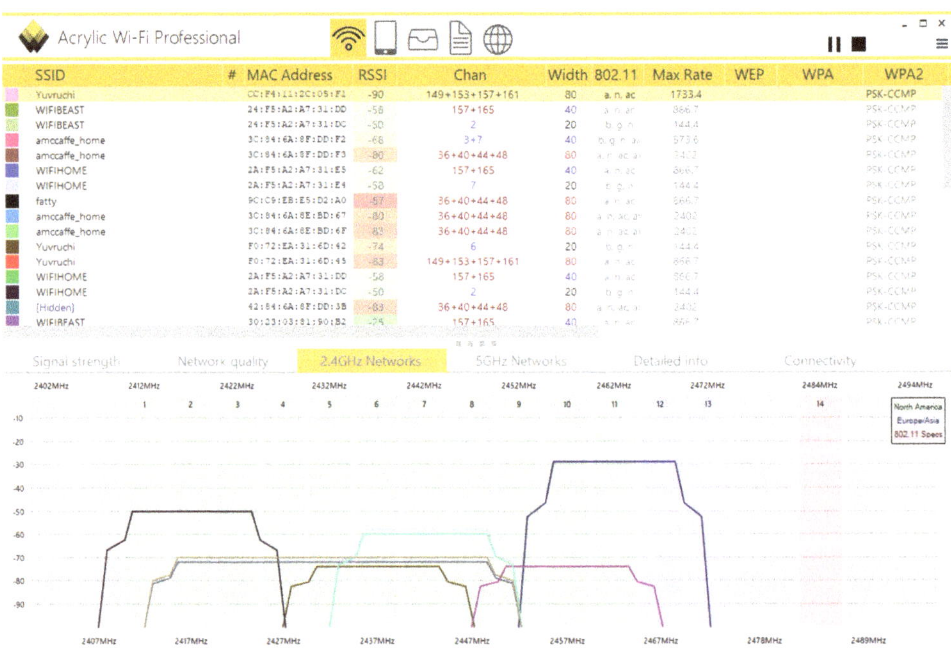

Figure 8.16: Acrylic Wi-Fi Professional

8.11: Chapter Summary

In this chapter, you learned about post-validation surveys. You explored the various tools and processes used in this task. In the next and final chapter, you will see how to make or recommend optimizations based on discoveries from the post-validation processes.

8.12: Facts to Remember

- Always validate the locations and configurations of APs before concerning yourself with surveys. If they are not where they're supposed to be, you already know the WLAN design was not implemented correctly.
- Validate that security configurations are correct using scanning tools like Wi-Fi Explorer Pro or Acrylic Wi-Fi Pro.
- Test captive portals, if in use, to verify proper functionality.
- Perform a post-validation survey using your WLAN survey/design software and compare/contrast the real network with the designed network. Always learn from this for future designs!
- If your mostly interested in AP configuration and signal strengths, a passive survey will suffice.
- If you require more detailed data, such as throughput and actual data rates achieved, an active survey will be required.
- Document any problems that can be resolved for the optimization phase.

8.13: Review Questions

1. You are performing a passive site survey to validate coverage of a new WLAN. Which of the following items cannot be tested in this way:

   A. Signal Strength
   B. Coverage overlap
   C. Interference
   D. Data Rate

2. Your client has identified his target device to be an iPhone 8. Lucky for you, you carry one in your pocket and the client provided a sample unit. You install client software on both units and test roaming, throughput, RSSI and SNR. What cannot be the reason to explain this behavior?

   A. Clients are not calibrated
   B. One device runs more applications than the other
   C. Drivers are not identical
   D. OS version is different

3. You would like to assess the capacity of a new WLAN which supports VoIP. A coworker suggests you use a traffic generator to simulate a load on your network. You would like to explain to him why it will not achieve your objective of testing the network capacity for VoIP. Pick the best explanation to offer to him

   A. Traffic generator are costly
   B. Traffic generator can overload the network
   C. Traffic generators cannot simulate realistically load, traffic and roaming in VoIP
   D. Traffic generators cannot test packet loss

4. CCI or co-channel contention is a common phenomenon in WLANS. Which of these devices do not cause CCI on your WLAN?

   A. WLAN Pi
   B. 802.11 client
   C. Wireless Router
   D. Cordless DECT 6 phone

5. You have connected a client device to the WLAN using 802.11ac with WPA2/AES/802.1X authentication. You ping your default gateway and can reach the local server mail.mynet.com in the 192.168.1.0/24 subnet to retrieve your email. Your IP is in the 192.168.10.0/24 subnet. When you try to update your Facebook status to let the world know your deployment works, it seems that you cannot reach the website. What could be the cause?

A. DHCP issue: you don't have the right gateway
B. ACL issue: traffic is not allowed
C. DNS resolution: DHCP did not have them configured in your subnet
D. NAT issue: your subnet is using an RFC 1918 private ip: you need NAT to be allowed outside the LAN.

6. A WLAN requirement for your client states that you should support VoIP. Which of these parameters doesn't meet VoIP commonly accepted requirements for WLANs?

A. Roaming time of less than 20 msec
B. SNR=20dB
C. Signal Strength=-67dB
D. Less than 1% packet loss

7. Which of the following tests is the most reliable throughput test available?

A. Site survey software using a generic adapter in an active survey
B. Aircheck G2 software testing to a DHCP client
C. Packet capture using the target client device
D. Running an iPerf client or throughput tester on the target client device

8. Delay can be catastrophic for WLANs, but some devices are more sensitive to it than others. Your target device is a laptop running a YouTube video. What is the maximal delay this application can stand?

A. No more than 20 msec
B. No more than 150 msec
C. YouTube is buffered so a few seconds is fine
D. YouTube is overly sensitive, so no delay is possible

9. You just completed the deployment of a brand-new dual band WLAN. During your walk around, you saw that the APs were positioned correctly and powered. Yet using site survey software you cannot see any AP in the 2.4GHz band, even 2 feet away from them. What could be the cause?

A. APs are single-band 5GHz
B. APs are using RRM
C. The 2.4 band is enabled on the controller but not on the AP
D. The PoE in use on the switches is not standard vs the 802.11af AP.

10. You have decided to do an active survey only to test a few parameters. Your co-worker points out to you a brand-new medical device which was installed in the last week which features a nice 802.11ac W2 router. He asks you on which band this wireless monster is configured. How can you tell him from your current test?

A. You can't: active surveys associate you to the WLAN: you are only gathering data about the deployed WLAN
B. You can tell him by looking at the list of detected devices
C. You can gather data points around the device to acquire data
D. You cannot because your current deployment is 802.11ac W1.

8.14: Review Answers

1. The correct answer is D. Data Rate is measured using active surveys.

2. The correct answer is B. Drivers, OS versions and the fact that client devices are not calibrated can all explain your results.

3. The correct answer is C. The best way to assess capacity is by observing it in real time. Client devices, drivers, OSes and such can influence the behavior which a traffic generator cannot simulate.

4. The correct answer is D. A cordless DECT 6 phone cannot generate CCI on your network because it is not using the same band

5. The correct answer is D. You can reach the email server on another subnet of the LAN. You can resolve its name. So, DNS, ACLs and DHCP are not the cause. NAT is.

6. The correct answer is B. SNR should be 25dB or better.

7. The correct answer is C. The best test to validate a deployment is done using the target device, often called least capable most important device.

8. The correct answer is C. YouTube is buffered making it more tolerant of delay (except with children but they come with an embedded delay detector)

9. The correct answer is C. The deployment is dual band, APs are powered so there is no incompatibility in PoE and since even 2 feet away from the AP there is no signal, the most likely culprit is the 2.4 interface directly on the AP.

10. The correct answer is A. Passive surveys scan all channels and all bands: active surveys require a connection to a specific AP or set of APs.

Chapter 9: Design Optimization

Objectives Covered:

4.4 Recommend appropriate physical adjustments

4.5 Recommend appropriate RF adjustments

4.6 Recommend remediation for application issues

4.7 Implement knowledge transfer and hand-off

Now that you have successfully defined, designed, deployed, and validated a WLAN, you are ready to fine tune it and hand it off to your customers. Wait, maybe your customer had a Wi-Fi problem and brought you in to solve it. You have not performed any of the "3 Ds" (Define, Design and Deploy) but validation and optimization are often the bread and butter of a Wi-Fi engineer. In this chapter we will address the following topics:

- Physical adjustments
- Configuration adjustments
- Remediation strategies for common issues
- Reports, training, and project handoff

9.1: Physical Adjustments

As we've mentioned previously, it is very common to find APs installed in the wrong location, inappropriately mounted, or simply in the wrong spot, though it appeared as the perfect location on a 2D plan. Relocating an AP to a different location should involve reviewing your design, if you are the implementer or, at a minimum, modeling a subset of the deployment to assess the impact a move could have on the WLAN. Just for the sake of review, here's a list of things the WLAN pro should try to avoid when repositioning an AP:

- Moving shelves
- Moving racks
- Areas with reduced clearance (ex: lots of lifts roaming through a factory or warehouse)
- Rooms with specialized medical equipment (MRI machines)
- Hidden in a ceiling (bonus points if it is between concrete or steel beams, or other metallic pipes and ducts)
- Anywhere it can be obstructed by goods or equipment
- Hallways (especially if they are far from your clients)
- Atriums
- Behind any type of metallic mesh (including those found in gyms protecting clocks)
- Behind any lead or concrete structures.

- APs within fifteen feet or less of each other (if on the same band and different channels)

The list could go on for a long time, but for the sake of brevity, we will stop here. Remember that your APs should be in a location convenient to the user and not convenient for aesthetics[97]. You want to provide a wireless connection to the users, keep it close to where they gather. Try to use the sources of attenuation to your advantage to create smaller, more manageable coverage cells which are less subject to CCI.

Antenna orientation is something that was long debated when MIMO came along with 802.11n. Some people, hoping to create more multipath, which could finally be used to a Wi-Fi engineer's advantage, argued that external antennas should be placed at an angle from one another. Others were sticking to the manufacturer's alignment recommendation[98]. Unless you are working on a point-to-point link, you can even stray a bit from observing antenna polarization: a mismatch will cause at most reduced signal strength for the device. For more information on antenna propagation patterns, we suggest that you look at your antenna's data sheets; they should provide an E-plane and H-plane view of the signal propagation pattern for your antenna.

Should you change AP models or antenna models, one thing that you should always keep up to date is the design documentation including RF model, design and any facility layouts you may have. Switching an antenna or an AP for another is an exercise which must be modeled in order to prevent further problems from occurring. If you have ever heard a wireless professional say that an AP is an AP, remember that he was only trying to tell you that platforms will not change the function of an AP. He is absolutely not saying that APs or antennas can be exchanged at will.

---

[97] Of course, exceptions always exist. For example, in a zoo, the deployment absolutely must meet aesthetic requirements. The environment should look as authentic as possible. The good news is that skinning can blend the APs into the environment so that they are barely noticed.

[98] The truth is that antennas are best in similarly polarized positions in most cases. Research, and extensive research at that, has shown this to be true. However, smart antenna arrays, which use intelligence behind multiple different antennas, often with different polarizations, are the optimal solution, but they are also exorbitantly expensive for most WLAN deployments.

9.2: Configuration Adjustments

Power, channel, and channel widths can all be friend or foe. If your validation survey has determined that the channel plan was not respected, then the first thing you should do is to make sure this is corrected. Additionally, if interferers or changes in the environment have made the channel plan ineffective, it should be adjusted accordingly. Sadly, such adjustments often have a rolling impact requiring that changes be made throughout the network[99].

Symptoms of a WLAN where AP power is too strong are:

- High CCI
- Hidden nodes
- Asymmetric power
- Very large cells
- Sticky clients and odd roaming behavior

Since APs have both more powerful and more sensitive radios, they will hear one another very well. If you are lucky, they will be polite with one another and "only" cause CCI, pushing the channel utilization to a higher level. With more utilization, the network performance will be impacted because everyone on the channel will get their turn to talk less often.

Another consequence of high power is the hidden node problem. In a large cell, it is possible that two clients can hear the AP and exchange with it; however, due to the clients' transmit power capacity and receiver sensitivity, they may not be able to hear one another and, thus, attempt transmission at the same time, resulting in corruption and retransmission.

Often another symptom of power that is too high is asymmetric power. APs come with antenna gear which is more sensitive and more capable than the average commercial client device. So, if an AP is blasting its signal to 100 mW, a client device may associate with it: after all, the signal is really strong. This is not representative of the client device's

[99] This reality is one of the strongest arguments for automatic channel selection. Not only does it help to resolve such immediate change requirements, but it helps when new interferers are introduced in the future. In very large networks, a rolling static channel plan change can easily become a multi-day project.

capacity to return its answer. Though individual devices may differ, the typical client cannot emit at a power higher than 40-50 mW and may not be able to reply as loud and clear to the AP as the AP can send to it. As it was mentioned previously, one should always aim at using a maximum TX power on APs which is the maximum power of the least capable, most important device (LCMID) that must be supported on the WLAN. This power level is typically 40 mW and sometimes less.

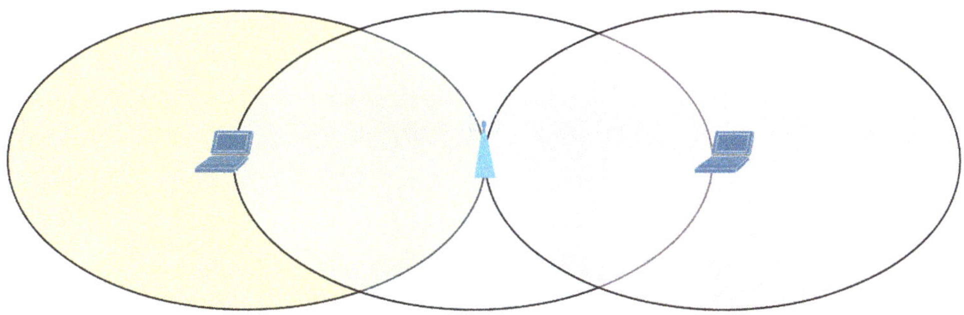

Hidden node: The laptop on the left can hear the transmission of the AP but not that of the laptop on the right, and vice-versa, while the AP can hear both laptops.

Figure 9.1: Hidden Node

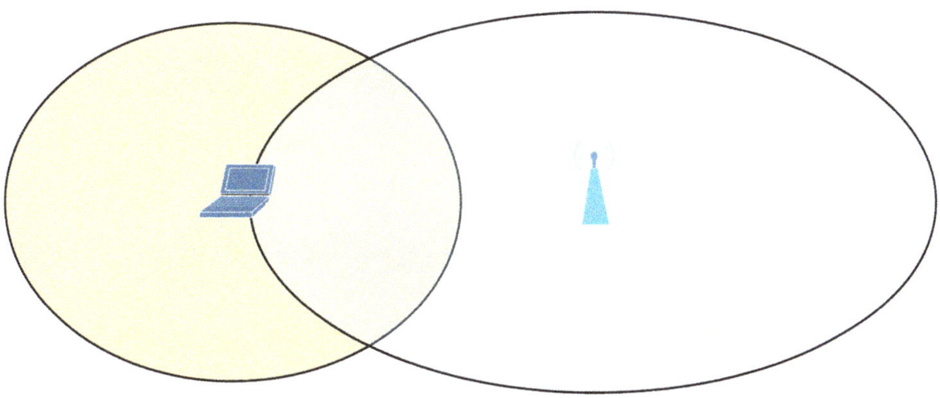

Asymetric power: The laptop on the left can hear the transmission of the AP but it cannot reach it because its own transmission power isn't as strong.

Figure 9.2: Asymmetric Power

Fifteen to twenty years ago, when the modus operandi of WLANs was to achieve coverage, care was not always taken to keep cells to a manageable size. If we had a nice, strong signal, it gave the impression that the network was reaching every nook and cranny and coverage was achieved. Devices back then were not as fast as devices today and certainly not as abundant, so the impact on performance was not as critical back then. In today's environment, where devices are hungry for speed, capacity, and performance, large cells mean that there are often too many devices per AP, extending the rates in use to much lower data rates and making a mess of performance. Lowering power on APs as well as cutting out data rates may help to make smaller cells, which results in grouping fewer clients and using more effective transmissions speeds. It may not be possible to make every cell use a basic rate of 24 Mbps, and we suggest you assess your client's capacities before you cut down on the enabled rates, but common wisdom recommends disabling 802.11b rates (1, 2.2, 5, and 11 Mbps in 2.4 GHz). Considerations should be extended to 5 GHz low rates as well, such as disabling 6 and 9 Mbps.

2.4 GHz is also subject to a lot of CCI and ACI. Depending on the situation it is possible that you will have to turn off this band on dual-band APs to maintain 5 GHz service while trying to achieve some quality in 2.4 GHz. Should it be the case, your validation survey will show significant CCI even if the power in use on the band is very low. Some manufacturers offer APs where you can opt to use dual 5 GHz instead of dual-band service: should it be your case, try to remember to keep both channels non-adjacent as well.

Sticky clients and odd roaming behavior can also be symptoms of power issues, although they may not be the very first thing one will notice. Remember, the roaming algorithm of a device is a proprietary feature. So, if power is very high, cells very large and the magic roaming sauce is based on AP power, your devices may fail to roam and stick to an AP, even if it is two floors above where the device is located and offers a lower rate. Alternatively, the device may roam to an AP which is further away than the one in front of you.

To address all these symptoms of the high-power problem, make cells smaller and use APs set to the highest power of the least capable device. Lowering the power will create coverage holes which may have to be addressed by either planning an addition in your RF planning software or testing it on site, depending on your practices. Once the additional AP is in place, don't forget to update the project's documentation accordingly.

Symptoms of a WLAN where AP Power is Insufficient

Several problems can occur when the AP power is too low for the client devices, at least, based on the number of APs deployed. The general rule of thumb is that the more you lower the AP transmit power, the greater number of APs required. The following are symptoms of AP transmit power set to low for the number of deployed APs:

- Coverage holes
- Asymmetric power
- Roaming doesn't work very well: voice call quality is poor

A deployment where the power of APs isn't strong enough for the number of APs in the space will be plagued by coverage holes. In addition, the asymmetric power issue that exists with APs which have the power set too high can exist the other way around: clients can get their frames to the APs, but the APs can't get their frames to the clients (this is an extreme scenario). One symptom of insufficient power is voice quality: if the signal is not strong enough, delay, jitter, packet loss will impact call quality and make it clear that there is a lack of coverage. If this doesn't do it, your CEO's call to the board or the university's rector call with the education minister's drop will make it clear that you have a problem.

If you are using static power assignment, the first thing you may try to do is to check what is the maximum power possible. (Remember, the maximum power possible is the maximum power of LCMID (least capable most important device). Your RF model should allow you to visualize the impact of pushing the power higher. Try to find the middle ground: you should aim at closing the gap and creating some overlap but keep your cell as small as possible. You may require additional APs. Just like power changes, try to visualize the addition in your RF model before, and don't forget to add this AP to a BOM and to all documentation.

You may have seen or read about cell overlap: most vendors recommend 15-20% for voice and 10-15% for data. A lot of people wonder how you can measure that, and though there are formulas to calculate exactly the distance, size, and a whole collection of other physical aspects of the cell in an ideal setting, we would rather have you understand what it means in practical terms. This is especially important to understand if you are trying to fix the power of your APs.

The idea behind the x-percent overlap is to allow for roaming and a safety margin. If your LCMID requires that the signal be -67 dB or better on AP A and that you want it then to roam to AP B, its maximum signal strength requirement has not changed. It's still -67dB. So, when you walk all the way to the edge of the cell for AP A, you should, at least, see AP B at -67dB as well. Site survey software allows you to model or survey it and check secondary coverage.

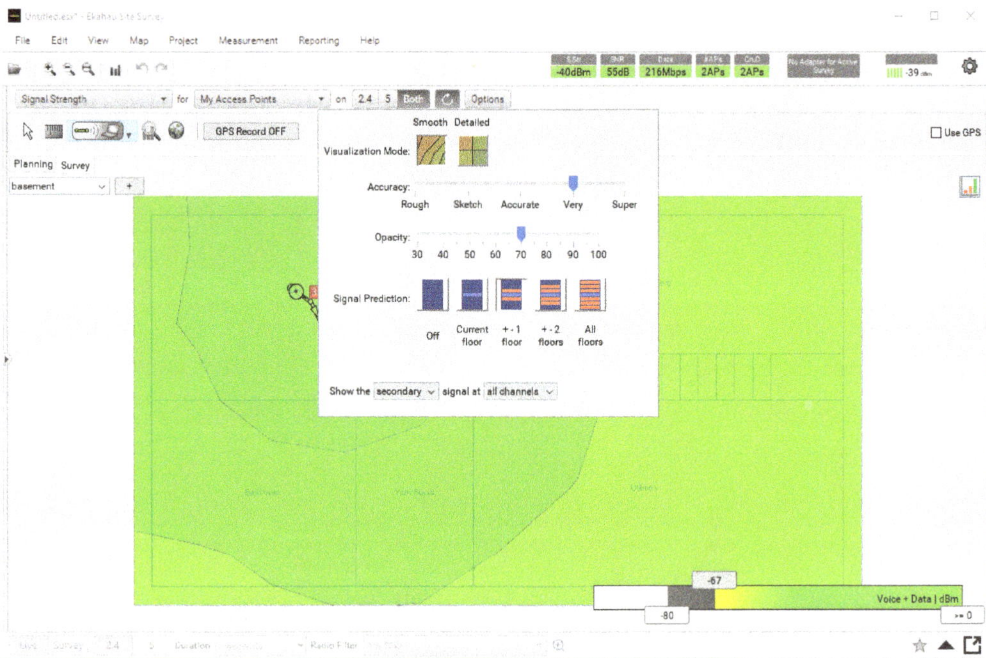

Figure 9.3:View Secondary Coverage

Depending on where your intervention comes within the WLAN's lifecycle, you may have a full channel plan complete with relevant design information, Tx power, and channel widths. It is also possible that the deployment doesn't have any of those things, in which case we recommend you produce as much documentation as possible given the situation.

The first thing to do is to establish whether you are coming in on a network which was designed for static or dynamic channel assignment. Though it might sound strange, there are plenty of networks which have been built years ago which have been switched to automatic channels and power because of marketing misinformation and bad recommendation of overzealous professionals who were thinking that the idea was good because it was featured in a list of best practices. This is not an urban legend, we have seen it often. If the APs are not deployed to accommodate automatic/dynamic channel assignment and the parameters for the algorithm are not set correctly, simply enabling it will not resolve problems.

Symptoms of a WLAN with inadequate use of dynamic channel and power mechanisms

- Inconsistent power settings throughout
- Inconsistent channel plan throughout
- Inconsistent cell size
- Symptoms of both insufficient and high power

In dynamic WLANs, while an initial post validation survey cannot be used as an accurate guide to what the channels and power should be, it could give you some clues as to the quality of the AP placement and initial channel and power plan. If the dynamic channel and power algorithm is set adequately, the resulting values should be like the RF model of the network and like the initial validation survey (in terms of logic).

Sometimes, though, good design tends to prevent it, you will walk into a deployment where APs have been placed with an unobstructed view of other APs. That could be in a hallway or a warehouse. If the minimum power and maximum power have not been set prior to the execution of the dynamic power selection, the required power may have been set too low. Calculations of the required powers are also time-based. If the calculation happens in a large high-density lecture hall prior during off hours, power may be set too low when the hall fills up with students whose bodies are great sources of natural attenuation.

The point is that it is always preferable to tune dynamic channel and power so that it meets the coverage objectives at a minimum and the LCMID's maximum TX power at a maximum. Most manufacturers allow you to do so on a per-band basis. The timers for the algorithm can also be manipulated: see your manufacturer's documentation for details.

When the channel plan seems inconsistent, there might be sources of attenuations that were not considered in the design or added after the deployment's completion. APs require at least 3 other APs to triangulate their position in the WLAN accurately and thus assess channel and power. From the AP's CLI or GUI, it is possible to check which APs it sees. However, depending on the manufacturer, the RSSI indicated for every neighbor may be what this neighbor see as TX power for the AP you are logged on. Site survey software could clue you into what's going on by looking at the spectrum analyzer and point-to-point data, but the CLI may be the fastest way to go[100].

What can you do to control dynamic channel plans? You can often configure the following parameters:

- **Transmit power range:** This setting allows you to configure that minimum and maximum transmit power levels within which they can be dynamically adjusted.
- **Channels:** This setting allows you to select the channels to be used by the algorithm. You can disable channels you do not want used, such as 2-5, 7-10, and 12-14 in 2.4 GHz, or the DFS channels in 5 GHz.
- **RF groups:** This setting allows you to group APs together into collections under a management policy. For example, you can indicate that the APs on one end of the building participate in a group that cannot use particular channels due to known interferers in that space.
- **Coverage hole detection threshold:** This setting allows you to specify the signal strength that should be considered the minimum received from a client. If a frame is received below this threshold, the AP/controller assumes a coverage hole must exist (otherwise, the station would have roamed to a different AP).

A Word on Channel Widths

With 802.11n and subsequently with 802.11ac it is possible to bundle channels. Of course, in 2.4 GHz bonded channels cannot be used, but using them in 5 GHz may require that you consider channel separation and how CSMA/CA and CCA work.

Be aware that doing so may cause ACI if the channel plan doesn't have neighboring APs on non-adjacent channels. OFDM, the modulation technique in use with 802.11a/g/n/ac/ax is known to be subject to high spectral regrowth on sidebands. In plain French, that means that the sidebands with OFDM are larger and their power stronger than other modulation techniques. For example, with a 22 MHz channel, DSSS and

[100] To be clear, RRM and other dynamic channel management methods from various vendors are not simply about channel assignment. They deal with interference detection, traffic load, and other factors as well, which result in channel assignments, transmit power adjustments, and potentially other settings, such as channel bandwidth.

HR/DSSS have up to -50 dBr at ± 22 MHz of the center frequency. With a 20 MHz OFDM channel, at ± 30 MHz of the center frequency, the signals can still have energy at -40 dBr.

Additionally, using larger channels may allow for higher throughput and better performance. However, be aware that management and control traffic, as well as legacy devices, are still using the primary channel. Should the primary channel be filled up with that traffic, having a secondary channel will not be useful. 802.11ac (as 802.11n) requires both the primary and secondary channels to be clear in order to clear CCA. This means that the primary channel requires enough SNR to detect signal (SD at least 4 dB) and energy (at least SD+20 dB) to pass CCA. On top of that, the secondary channel needs to detect a signal at -72 dB or better and energy at -62 dB. All of this data makes any WLAN professional think twice about using bonded channels, especially in dense environments. As a matter of fact, most manufacturers tend to recommend 20 MHz channels in high-density deployments to avoid the side effects of bonded channels. The strategy in high-performance WLANs is simple: get the client on the medium as fast as possible to get it off the medium as fast as possible so that it is free for the next client who wants its turn.

Finally, wider channels also increase sideband spectral regrowth. Figures 9.4 and 9.5 illustrate this phenomenon thanks to Andrew von Nagy's historic blog that was at RevolutionWiFi.com.

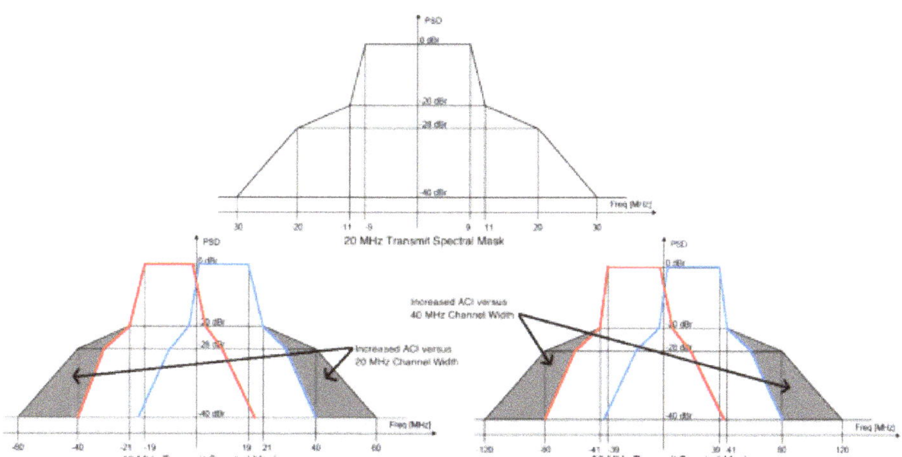

Figure 9.4: Spectral regrowth in bonded channels. Source: www.revolutionwifi.com by Andrew von Nagy

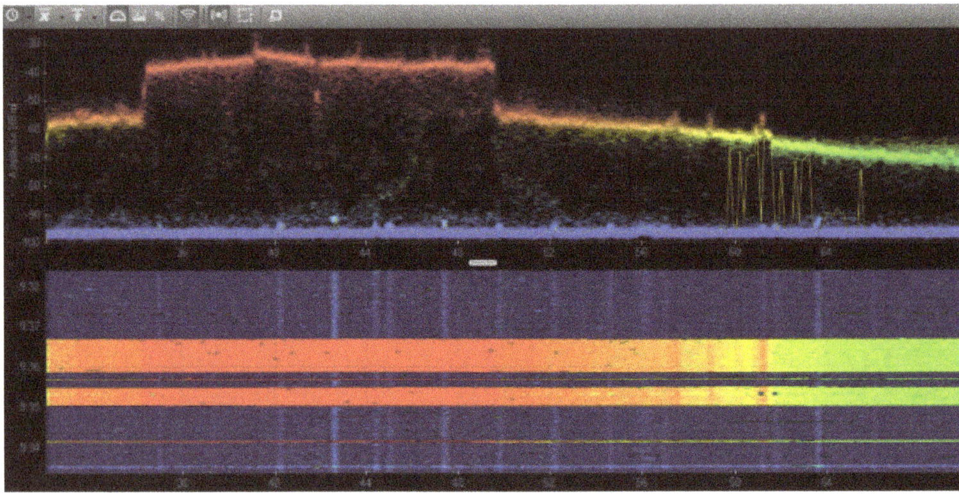

Figure 9.5: View of spectral regrowth, 80MHz channel, spectrum analyzer Source: www.revolutionwifi.com by Andrew von Nagy

For an in-depth technical overview of the bonded-channel spectral growth and filtering problem, see the paper: www.ti.com/pdfs/bcg/channel_bonding_wp.pdf

9.3: Remediation Strategies for Common Issues

The WLAN may appear functional from a user's standpoint. Most devices can log on and use the network, but there seem to be performance issues with applications which require to be transmitted within a specific window of time. This is typical of QoS problems. Often encountered in voice deployments, the symptoms will include increased latency, delays, packet drops and jitter. QoS, if the client requires it to be implemented, should always be implemented end to end to serve its purpose. To troubleshoot these issues, we suggest trying a differential approach. Do all clients see this kind of behavior? Is the behavior tied to an application in particular?

Bear in mind that QoS is based on 802.11e but that manufacturers are implementing as they please. Cisco and Aruba, for example, use WMM which is based on the standard. However, Cisco also uses CCX which proprietary.

Once you are certain that the issue is QoS, then you should try to look at the markings to ensure that there are no issues with them or their conversion once they cross network boundaries. To do so, you may require a packet analyzer or specialized tools such as Tamosoft's Throughput test or Iperf. The later will allow you, depending on implementation, to assess the performance based on specific markings while the first bases itself on WMM categories.

Figure 9.6: Tamosoft Throughput tester: note the QoS Traffic types are WMM categories

Interference

Though sometimes wireless engineers wish that 2.4 GHz was dead, it is not dead just yet. It is plagued by interference from neighboring networks, non-802.11 sources and all sorts of IoT devices from the pregnant manikin woman and her wireless baby to the dishwasher control system.

To diagnose an interference problem, you will need a spectrum analyzer. Some site survey software has it as a feature while others allow for integration into site survey software though are separate products. It is also possible to find spectrum analyzers as standalone software such as Metageek's Channelizer, which is used with a specialized data acquisition card or integrated into NetScout's Aircheck G2 v2.

To mitigate interference, you should first try to locate it. Using your spectrum analyzer, note the signal strength which can currently be detected. Then proceed to turn 90 degrees clockwise. Note the signal strength now. Repeat this step until you know you have found the strongest signal strength. At that point, you should walk in the direction of the strongest signal: if it becomes weaker, come back on your path. Repeat the clockwise detection to validate in which direction the strongest signal is now detected. Eventually, you should come to locate the source.

Another method is provided with the NetScout Aircheck G2 which features a nifty "locate" function. The tool can be used with a directional antenna (the previous method's directionality was mostly due to your body's natural attenuation) or with the embedded omnidirectional antenna. The G2 will beep and display a simple signal meter and graph to show signal strength. The beeps become faster as the signal strength goes up. Eventually, you should come to find the source of interference, it should be very close to the area where the signal is the strongest.

Now that you have found the source, there are two ways to deal with interference: either you remove the device or turn off the Wi-Fi source embedded in it or you accept to make changes to your channel plan to accommodate the signal on this channel. It is not always easy to handle these kinds of issues, especially when there are no official policies in place which may forbid the use of such devices.

If the device is a non-802.11 device such as a microwave, consider having it replaced by a unit which doesn't leak as much signal or relocating your APs to make the impact of the oven less significant on the WLAN. Alternatively, if you determine that the microwave radiates energy mostly in the low portion of the 2.4 GHz band, you can configure the nearest AP to use channel 11 and vice versa if the microwave radiates energy mostly in the high portion of the band.

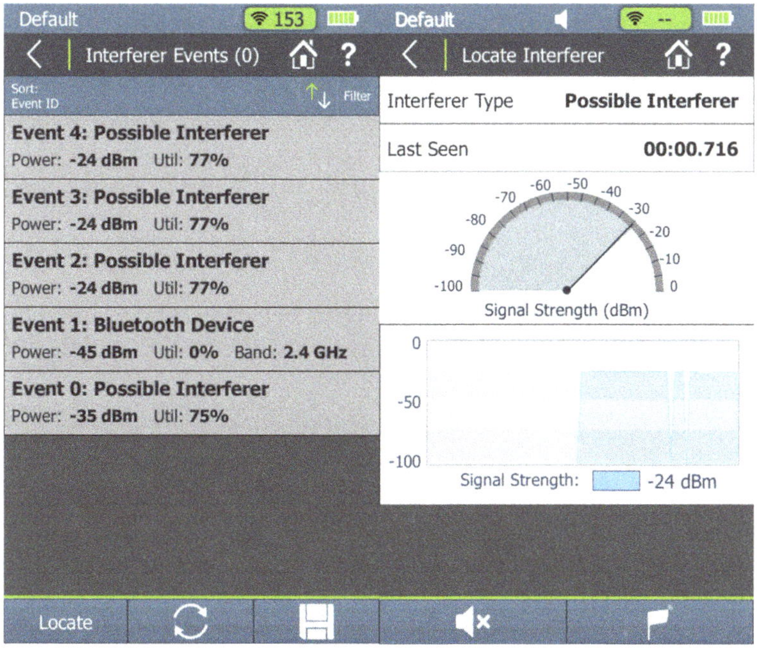

Figure 9.7: Searching for an Interferer Using the NetScout Aircheck G2

Quick Reference to Resolving Post-Validation Issues

RF coverage issues

- Tuning power.
- Validating dynamic power and channel settings.
- Tuning dynamic power and channel settings.
- Adding, removing, shutting, or reassigning flexible band antennas to 5 GHz.

Capacity problems

- Review the applications which must be supported and assess its needs in airtime.
- Review the number of SSIDs to free up airtime which is currently used by unnecessary beacons
- Assess if additional APs can solve the issue based on your RF model.
- Consider using directional antennas in high-density environments.
- Consider using flexible band APs and dual 5 GHz settings on non-adjacent channels.

- Test capacity in a real live environment: trying to test capacity by emulation is inaccurate and time-consuming. A spectrum analyzer may help assess if there's airtime left.

QoS problems

- Review QoS markings on both RX and TX traffic: a packet capture tool will help you in that task.
- Validate conversion of markings.
- Ensure that QoS marking follow throughout the path to the destination.
- Validate uplinks in the path from source to destination, adding links to channels if necessary.
- Remove low rates to free up RF airtime if RF congestion is suspected.

Security configuration problems

- Review security requirements: ACL, firewall, and authentication policies.
- Validate credentials and user rights.
- Validate the reachability of RADIUS or LDAP server.
- Validate the certificates and expiration dates.
- Validate the root and any intermediate certificates which might not be enabled on the RADIUS server or accepted by the client.

Client connectivity issues

- Start with the client, unless the issue is reported by a lot of clients at once.
- Validate all credentials, access rights, symptoms, location, date and time at which the event occurred.
- Test in an area where coverage and performance have been validated.
- Use debug functionalities on your infrastructure to see where the problem may be: is it really a Wi-Fi problem or a DHCP, DNS, certificate or network infrastructure issue.
- Work your way up from the client to the network.

Interference issues

- Validate with a spectrum analyzer the nature of the suspected interference.
- Using the spectrum analyzer or a handheld tool, try to locate and remove the source of interference.
- If you cannot remove the source, then try to review your RF model to accommodate the channels rendered unusable by it.

- If there are no rogue policies in place, it could be a good idea to suggest such a policy to your client.

9.4: Reports, Training and Project Hand-Off

When deployment has been completed and validated, a fair amount of documentation should have been generated. It is now time to transfer the project to your client. Depending on the nature of the work that was done, you may have to train the following categories of clients:

- End users
- Support staff
- Network staff

The primary clients of a WLAN are, of course, the end users. They are going to be on the WLAN, and without documentation, it is often surprising to hear how they manage their way to a functional connection (or lack thereof). A readily available guide and proper communication efforts will save the support personnel a tremendous amount of time. Some deployments offer an open SSID which leads users to download pre-configured profiles directly onto their devices: if required by your client don't forget to document its code, the versions that were tested as well as the way it works for the average user.

Support staff generally do not need to know the inner workings of the network. They will be the frontline to the client. Coverage maps, DHCP ranges, DNS and NAT are more than likely their concerns. If a VPN or supplicant is needed to connect to the wireless network, ensure that its usage and limitations are documented as well. Installation procedures and uninstallation procedure, as well as a cheat sheet of troubleshooting steps and information required to relay a ticket higher up in the chain, are time savers. The chances are that your client already has a process in place for incident and tickets, but should you need to establish one, it should include the following information at a minimum.

- Date/Time:
- Phone Number
- Email
- Name

- Wireless username
- Device Type
    - Operating system
    - Specific model (if known)
    - MAC address of device experiencing issues (if known)
    - The signal level of device experiencing issues (if known)
- Location
- Detailed problem description:
    - Does the service work elsewhere?
    - When did you notice the issue?
    - Is it constant?

The network operations and architecture staff will require the entire project document which will provide them with all the details necessary to operate the network and integrate it into the current and future architectures. It may be a good idea to send them a copy for review before the handoff meeting to allow them to read it and think of any questions they may have, something you can address at the final meeting.

Please note that for all categories of clients, the hardest thing is to keep documentation up to date. Should any modification be made by you or the client in the network or on the WLAN, the document will remain relevant only if it is maintained regularly.

Documentation and files provided to the client will vary with the client's requests and needs as well as your company's policy. We suggest the following structure for your report: should you need to add or remove some sections or use a corporate template, feel free to do so. As your experience grows, you will eventually create your own personalized template which will suit your needs. We decided to take the perspective of the entire project though we understand the deliverables in the final report will be based on the agreement between you and your client. The goal here is to get you writing the things which, as technical professionals, we prefer doing.

- Abstract
- Executive summary
- Engineering summary
- Project background, requirements, justification, SOW, and constraints
- Maps of the facilities
- Logical Topologies

- Physical Topologies
- RF modeling files
- RF modeling report
- Device inventory or BOM
- Deployment flowchart, timeline
- All WLAN configuration
- All additional network configuration
- Any server or service configuration
- Validation
- Procedures
- Final signoff

A short introduction which doesn't give any specific details to explain the rough lines of the project, the expected outcomes, benefits, deployment and report structure is useful. This is typically called an abstract.

It shouldn't come as a surprise that executives are generally not technical people. They prefer to get a bird's eye view of the work that was done to report up the chain of command. A good executive summary should include

- The project's goals: the what and why of the solution. Ex: Client X would like to support wireless barcode scanners on its wireless infrastructure and update its database in real time to offer up to the minute information to its online customers.
- A brief description of the work that was required and completed.
- Any important highlights which could outline limitations in the project, design choices or any changes that were required and agreed upon by the client.
- How this solution meets the initial goals.

An engineering summary is also useful. The target audience for this document is network engineers, technicians, and architects. What this category of clients wants to know is why things were done the way they were done and what's new in the infrastructure. Add to it a high-level summary of how it works, and they have all they need. They may go through the entire document that you will leave behind, but if they only need the information for an occasional support shift, this summary will be all they go through until they really need it. It should include:

- Design choices regarding devices and their physical installation for routers, switches, APs, antennas.
- Design choices regarding any additional network configuration.
- Information on services, IP ranges, DHCP and DNS configuration, NAT, etc.
- Any special tweaks, details or configurations and related explanations.

Background: Justification, SOW, and Constraints

The initial meeting, as well as any subsequent meeting prior to the deployment, should have generated a paper trail: customer questionnaires, maps, current information on the infrastructure. These items, as well as the statement of work, should be included in the main report. They are the starting point, the basis upon which the project was created. Highlighting the SOW as your goal and the LCMID or any other specific devices which should be supported is a good idea. Don't forget to list the limitations or any constraints which were mentioned or discovered through the project. This document and the WLAN are what is left once you are gone onto another project and as such, should be able to prove that you have done your due diligence in delivering a project that was not only complete but on point to address your client's needs.

Maps of the facilities

Some customers have maps of the facilities, whether they are to scale is something you have assessed by visiting the premises. Since the report is something that will be used by you as the designer and by the person who will either conduct the APoS, validation and/or optimization, we suggest you include them in this report. Should any notes be required as to physical constraints or access constraints, they can also be provided so to save time to your coworkers and client down the line.

RF modeling file and RF report

This one is sometimes debated. Some people will prefer using a report generated by the site survey software to provide the information required about channel plan, power, resulting coverage, SNR, channel overlap and a variety of other parameters. It should also provide information on the names and locations of every AP, antenna, or bridge. Others will also provide the RF modeling file itself in addition to the various report. Some may even elect to only generate this report as documentation. Coming from a background where there is no such thing as too much documentation as long as it is relevant to the project, we suggest that you provide as much information as possible.

Logical topologies

Often in Visio format, these maps will allow you to understand the data path and interactions between the various devices in the network. VRFs, policies, routing, QoS notes, NAT, firewalls, ACLS, anything that can affect the data path should be included there.

Physical topology

This is a diagram which can be used by someone in charge of the installation to proceed with the physical wiring and installation of all WLAN components except for APs. It will feature port numbers, device model, name, IP addresses and interconnections as well as wire types and length (the later if necessary).

Both the logical and physical topologies are used as cheat sheets by network personnel to prepare for maintenances and understand quickly where a failure could have its source. Make sure that your topologies are clear. Use agreed upon icons and clear information. Both topologies can help you in writing up a list of required devices, also known as a bill of materials.

Bill of materials

This list should be exhaustive and include inventory information (such as serial number, model):

- Interconnection patch cords
- Wired devices (to support the infrastructure)
- Wireless devices
- Servers (NMS, DHCP, DNS)
- APs and bridges
- Power injectors
- Antennas
- Licenses for APs, controllers, and other network devices
- Support agreements (with manufacturer)

- Any other billable physical items

Deployment flowchart

The idea behind this flowchart is to give an idea of the timeline. Some customers have very strict change policies and providing this flowchart as well as information on the changes may help with traceability. It will also detail the various steps in the deployment, something which could be valuable to you and your team.

WLAN configuration

The entire configuration of the controller should be provided to your client if you oversee setting it up. Details should include

- SSIDs
- IP ranges
- Security
- VLANs
- ACLS
- QoS markings
- Device groups
- Dynamic channel and power adjustments to the controller

If you are using AP in autonomous mode, your WLAN documentation should also include a sample template of the AP's configuration to allow your customer to be able to replace a defective unit. Moreover, providing the WLAN configuration could also help you train the customer's personnel if necessary.

Additional configuration: wired configuration

The wired network is elemental in deploying a WLAN. If any changes are required on the wired side of things, providing all the additional configuration and/or complete configuration for new network devices may help your client's network staff in troubleshooting issues which could arise from time to time. Don't forget the security configurations and NAT: they are equally as important.

Additional server or service configuration

All additional network configuration such as DHCP ranges, options, DNS settings, NTP should also be documented. Should you need to set up a captive portal in the deployment, it should be documented as well. Don't forget any configuration backup schedules and location of files, all administrative passwords you may have set for servers, services and network devices of all kinds. For passwords, it is best to use an encrypted file: tools such as LastPass will allow you to create one which you can leave along with the documentation.

Validation

Validation is the final seal of quality of your deployment. You should provide at a minimum a report from the site survey or another tool which you have used for validation. A list of other tests and related files should also be provided. Document the devices, OS, driver version and specific parameters which have been used in your test as well.

Procedures

These documents could be as simple as how to get support through your firm's help desk and a list of phone numbers or emails of various support entities. They could also be written for onsite support personnel as a cheat sheet such as the one provided earlier on in this chapter. Your client's network personnel may also appreciate procedures for routine maintenance operations such as adding or removing an SSID, enabling or disabling a WLAN, band or changing RADIUS servers or updating the code on their WLAN devices. The list of procedures you can include will depend on your project, your client, and your own firm.

Final handoff

As the WLAN deployment moves into production and your obligations have been fulfilled, we suggest you have one last meeting with your customer to hand in your report and possibly sign an official handoff, should it be your company's policy to do so. This meeting will allow you to go through the major points of the report and answer any questions that remain before handing your customer the keys to their WLAN.

Whether you are brought into a WLAN which needs to be optimized or completing a brand-new deployment, documentation is essential to all WLAN implementations. Not only can it help you optimize channel plans, power settings, and dynamic radio management, but it can also serve as a baseline for the future. There will always be sources of interference, rogues, ACI and CCI. As a CWDP it is now your turn to try to walk the thin line between a good wireless network and a bad one. While trying to observe as many best practices as you can within your client's requirements, remember that "The best is the enemy of the good." (Voltaire)

9.5: Chapter Summary

In this chapter, you learned how to address issues discovered in a post-validation survey. This is an important final step for the WLAN designer. You have now defined, designed, deployed and validated a WLAN from start to finish. It's time to relax and wait until that first stubborn device is introduced to the network and you can troubleshoot all over again.

9.6: Facts to Remember

- You can reduce CCI, in many cases, by reducing AP output power as long as the APs still provide the required coverage as defined by signal strength.
- You cannot remove all CCI through AP configuration because much of it is caused by client devices.
- When optimizing, keep your least capable, most important device (LCMID) in mind.
- Instead of worrying about percentage of cell overlap, ensure some number (typically 2-3 depending on the design, or more in some cases) of APs can be seen from each location.
- Inadequate dynamic channel management configuration can result in inconsistent power settings, channel usage, cell sizes and data rates throughout the network.
- Using wider channels is not typically the solution to insufficient capacity; instead, use better channel planning, more APs with lower output power, and facility materials to attenuate signals and achieve more capacity.
- Throughput testers can be useful, but only during times of actual user activity to simulate a true possible user experience.
- If QoS is not operating as expected, check the QoS settings on each switch and router where applicable as well as the WLAN profiles.

9.7: Review Questions

1. What is commonly the most difficult aspect of documenting a project?

   A. Writing the Statement of Work
   B. Explaining how you mitigated interference
   C. Keeping documentation up to date
   D. Getting the client to sign off your design

2. Which tool can you use to detect interference?

   A. Throughput tester
   B. Spectrum Analyzer
   C. Packet Capture tool
   D. Multimeter

3. Your least capable most important device's maximum output power is 40mW. You are using dynamic channel and power assignments. To what should you set the maximum power setting for the algorithm?

   A. 30mW
   B. 50mW
   C. 40mW
   D. 100mW

4. You have modeled a WLAN deployment using a maximum power of 25 mW but your client has elected to go with dynamic power and channel assignment. You then proceed with a validation survey which shows the power has been set on most APs to 5mW. What could explain this behavior?

   A. APs have a non-obstructed line of sight of the client
   B. APs have directional antennas
   C. APs have a non-obstructed line of sight of each other
   D. APs use the natural attenuation of the space

5. To allow wireless clients to be configured automatically for 802.1X, what can you use?

A. A webpage procedure
B. Profiling and posturing based on the RADIUS server
C. A downloadable VPN client
D. A downloadable profile which can be accessed from an open SSID set for this purpose only

6. Which of these options is a possible way to mitigate CCI in 2.4 for a dual-band deployment?

A. Only use non-adjacent channels in UNII bands
B. Turn the power up on APs
C. Turn off the band on select APs
D. Use only 5 GHz

7. What should not be included in your final report?

A. The source of all devices, parts, APs and antennas
B. The bill of materials including all parts, cables and network devices
C. The coverage maps (including signal strength, SNR, etc.)
D. The QoS profiles for voice

8. Which of these tools is the most appropriate tool to validate available airtime?

A. Site survey software
B. Spectrum Analyzer
C. Packet Capture software
D. Throughput tester

9. Which of the following items should be included in an executive summary?

A. Information on available services
B. Maps of the site
C. Site Survey report
D. Goal and purpose of the infrastructure

10. Why should you consider picking non-adjacent channels in a 5 GHz channel plan, especially if using 40MHz channels?

A. Adjacent channels may cause CCI
B. Spectral regrowth using DSSS
C. Adjacent channels may cause ACI
D. Spectral regrowth may cause CCI

9.8: Review Answers

1. The correct answer is C: Keeping documentation up to date is often much harder than writing it for a project.

2. The correct answer is B. A spectrum analyzer will allow you to detect interference.

3. The correct answer is C. At most your transmit power on APs should be at 40 mW for typical indoor deployments.

4. The correct answer is C. APs have a non-obstructed line of sight of each other and since they were detecting each other with a strong signal, the algorithm has lowered the power too aggressively.

5. The correct answer is D. A downloadable profile accessible through an SSID set up for this purpose can help in deploying the 802.1X configuration.

6. The correct answer is C. Turn off the band on select APs in 2.4 may sound like a drastic measure but it is the best choice.

7. The correct answer is A. The source of parts is not relevant. Serial number and model number are an absolute must.

8. The correct answer is B. Spectrum analyzers display airtime utilization in real time.

9. The correct answer is D. Goal and purpose of the infrastructure: if you can tie in the mission or strategic plan of the business it could make the project even more relevant.

10. The correct answer is C. Adjacent channels may cause ACI due to the major sidebands created by OFDM

Glossary: A Universal Wi-Fi Glossary

The following glossary is based upon unique CWNP definitions and explanations and is also enhanced through added definitions from the following cited resources:

211-2018. (2018). IEEE 211-2018 Standard Definitions of Terms for Radio Waves Propagation.

802.11-2020. (2020). IEEE 802.11-2020 Standard for Information Technology, Part 11: Wireless LAN Medium Access Control (MAC) and Physical Layer (PHY) Specification.

802.11ax. (2021). IEEE 802.11ax-2021 Part 11: Wireless LAN Medium Access Control (MAC) and Physical Layer (PHY) Specification, Amendment 1: Enhancements for High Efficiency WLAN.

802.1X. (2020). IEEE 802.1X-2020 Standard for Local and Metropolitan Area Networks - Port-Based Network Access Control.

Boyd, C., Mathuria, A., & Stebila, D. (2020). Protocols for Authentication and Key Establishment, Second Edition.

Butterfield, A. (2016). Oxford Dictionary of Computer Science, Seventh Edition.

Dennison, M., & Fielding, J. (2009). Radio Communication Handbook, Tenth Edition.

Gibilisco, S. (2001). The Illustrated Dictionary of Electronics, Eighth Edition.

Peterson, L. L., & Davie, B. S. (2021). Computer Networks, A Systems Approach, Sixth Edition.

20 MHz Channel: A channel using ± 10 MHz above and below the center frequency.

40 MHz Channel: A channel, with a 20 MHz primary channel and another 20 MHz above or below the primary channel.

40 MHz Intolerant: A bit potentially set in the 802.11 frame allowing STAs to indicate that 40 MHz channels should not be used in their BSS or in surrounding networks. The bit is processed only in the 2.4 GHz band.

4-Way Handshake: The process used to generate encryption keys for unicast frames (Pairwise Transient Key (PTK)) and transmit encryption keys for group (broadcast, multicast) (Group Temporal Key (GTK)) frames using material (PMK) from the 802.1X/EAP authentication, the pre-shared key (PSK), or an SAE authentication exchange. The PTK and GTK are derived from the Pairwise Master Key (PMK) and Group Master Key (GMK) respectively.

80 MHz Channel: A channel comprised of four 20 MHz channels.

160 MHz Channel: A channel comprised of eight 20 MHz channels.

802.11: A standard maintained by the IEEE for implementing and communicating with wireless local area networks (WLANs). Regularly amended, the standard continues to evolve to meet new demands. Several Physical Layer (PHY) methods are specified and the Medium Access Control (MAC) sublayer is also specified.

802.11a: An 802.11 amendment that operates in the 5GHz band. It uses OFDM modulation and is called the OFDM PHY. It can support data rates of up to 54 Mbps.

802.11aa: An 802.11 amendment that added support for robust audio and video streaming through MAC enhancements. It specifies a new category of station called a Stream Classification Service (SCS) station. The SCS implementation is optional for a WMM QoS station.

802.11ac: An 802.11 amendment that operates in the 5GHz band. It uses MU-MIMO, beamforming, and 256 QAM technology, up to 8 spatial streams and OFDM modulation. Support is included for data rates up to 6933.3 Mbps.

802.11ae: An 802.11 amendment that provides prioritization of management frames. It defines a new Quality of Service Management Frame (QMF). When the

QMF service is used, some management frames may be transmitted using an access category other than the one used for voice (AC_VO). When communicating with stations that do not support the QMF service, the station uses access category AC_VO to transmit management frames. When QMF is supported, the beacon frame includes a QMF Policy element.

802.11ah: An 802.11 amendment that specifies operations in the sub-1 GHz range. Frequencies used vary by regulatory domain. It supports 1, 2, 4, 8 and 16 MHz channels with OFDM modulation.

802.11ax: An 802.11 amendment that supports bi-directional MU-MIMO, higher modulation rates and sub-channelization. It supports operations in the 2.4 GHz, 5 GHz, and 6 GHz bands and it implements OFDMA as a modulation scheme with full backward compatibility in required bands.

802.11b: An IEEE 802.11 amendment that operates in the 2.4GHz ISM band. It uses HR/DSSS and earlier technology. It can support data rates of up to 11Mbps.

802.11e: An 802.11 amendment, now incorporated into the most recent rollup, that provided quality of service extensions to the wireless link through probabilistic prioritization based on the contention window. The Wi-Fi Multimedia (WMM) certification is based on this amendment.

802.11g: An IEEE 802.11 amendment that operates in the 2.4GHz ISM band. It uses ERP-OFDM and earlier technology. It can support data rates of up to 54Mbps.

802.11i: An 802.11 amendment, now incorporated into the most recent rollup, which provided security enhancements to the standard and resolved weaknesses in the original WEP encryption solution. It provided for TKIP/RC4 (now deprecated) and CCMP/AES cipher suites and encryption algorithms.

802.11n: An IEEE 802.11 amendment that operates in the 2.4 ISM and 5GHz UNII/ISM bands. It uses MIMO, HT-OFDM and earlier technology. It can support data rates of up to 600Mbps.

802.11k: An IEEE 802.11 amendment that specifies and defines WLAN characteristics and mechanisms.

802.11r: An IEEE 802.11 amendment that enables roaming between access points.

802.11s: An IEEE 802.11 amendment that adds mesh networking capabilities to the standard and introduced the SAE authentication method.

802.11u: An IEEE 802.11 amendment that adds features for mobile communication devices such as phones and tablets.

802.11w: An IEEE 802.11 amendment to increase security for the management frames.

802.11y: An IEEE 802.11 amendment that allows registered stations to operate at a higher power output in the 3650-3700 MHz band.

802.1X: 802.1X is an IEEE standard that uses the Extensible Authentication Protocol (EAP) framework to authenticate devices attempting to connect to the LAN or WLAN. The process involves the use of a supplicant to be authenticated, authenticator, and authentication server.

802.11 State Machine: The 802.11 state machine defines the condition of the connection of a client STA to another STA and can be in one of three states: Unauthenticated/Unassociated, Authenticated/Unassociated, or Authenticated/Associated.

802.3: A set of standards maintained by the IEEE for implementing and communicating with wired Ethernet networks and including Power over Ethernet (PoE) specifications.

AAA Framework: Authentication, Authorization, and Accounting is a framework for monitoring usage, enforcing policies, controlling access to computer resources, and providing the correct billing amount for services.

AAA Server Credential: The AAA server credential is the validation materials used for the server. When mutual authentication is required, a server certificate is typically used as the AAA server credential.

Absorption: The process of converting electromagnetic energy to heat (211-2018, 2018). Occurs when an obstacle absorbs some or all of a radio wave's energy (CWNP Explanation).

Access Category (AC): An access category is a priority class. 802.11 specifies four different priority classes – voice (AC_VO), video (AC_VI), best effort (AC_BE), and background (AC_BK).

Access Layer Forwarding: Data forwarding that occurs at the access layer, also called *distributed data forwarding*. The data is distributed from the access layer directly to the destination without passing through a centralized controller.

Access Point (AP): An entity that contains on station (STA) and provides access to the distribution system services, via the wireless medium (WM) for associated STAs. An AP Comprises a STA and a distribution system access function (DSAF) (802.11ax, 2021). At the physical device level, an AP may contain multiple such logical entities as defined in the 802.11 standard and, when doing so, the APs are considered co-located.

Access Port: An AP used for mesh networks and that connects to the wired or wireless network at the edge of the mesh, also called a mesh gate.

Acknowledgement (Ack) Frame: A frame sent by the receiving 802.11 station confirming the reception of a frame.

Access Control List (ACL): ACLs are lists that inform a STA or user what permissions are available to access files and other resources. ACLs are also used in routers and switches to control packets allowed through to other networks.

Active Mode: A power-save mode in which the station never turns the radio off.

Active Scanning: A scanning (network location) method in which the client broadcasts probe requests and records the probe responses to determine the network with which it will establish an association.

Active Survey: A wireless survey conducted on location that involves measuring throughput rates, round trip time, and/or packet loss by connecting devices to an AP and transmitting data during the survey.

Ad-Hoc Mode: The colloquial name for an Independent Basic Service Set (IBSS). STAs connect directly with each other, and an AP is not used.

Adjacent Channel: Also called adjacent non-overlapping channel, is a channel separated by at least 30 MHz for DSSS and HR/DSSS or by at least 25 MHz for later PHYs in 2.4 GHz in relation to 22 MHz and 20 MHz channels. All 5 GHz 20 MHz channels are considered adjacent as long as they are one channel above or below the comparison channel. Non-overlapping channels are channels whose bands do not interfere with neighboring channels on the primary carrier frequencies.

Adjacent Channel Interference (ACI): ACI occurs when channels near each other (in the frequency domain) interfere with one another due to either partial frequency overlap on primary carrier frequencies or excessive output power.

Adjacent Overlapping Channels: Adjacent overlapping channels are channels whose bands interfere with their neighboring channels on the primary carrier frequencies. For example, in 2.4 GHz channels 1 and 3 are adjacent overlapping channels.

AES (Advanced Encryption Standard): The encryption cipher typically used with CCMP and WPA2 and with SAE and WPA3 providing improved security over WEP/RC4 or TKIP/RC4.

AID: Association ID (AID) is an identification assigned by a wireless STA (AP) to another STA (client) in order to transmit the correct data to that device in an Infrastructure Basic Service Set.

AirTime Fairness: Transmits more frames to client STAs with higher data rates than those with lower data rates so that the STAs get fair access to the air (medium) instead of having to wait for slower data rate STAs.

Aggregate MAC Protocol Data Unit (A-MPDU): A structure that contains one or more MPDUs and is transported by a physical layer (PHY) as a single PHY service data unit (PPDU) (802.11-2020, 2020).

Aggregate MAC Protocol Data Unit (A-MPDU) Subframe: A portion of an A-MPDU that contains a delimiter and optionally contains an MPDU plus any necessary padding (802.11-2020, 2020).

Aggregate MAC Service Data Unit (A-MSDU): A structure that contains one or more MSDUs and is transmitted in one or more QoS Data frames with the same sequence number (802.11ax, 2021).

Aggregate MAC Service Data Unit (A-MSDU) Subframe: A portion of an A-MSDU that contains a header and associated MSDU (802.11-2020, 2020).

Amplification: The process of increasing a signal's power level.

Amplifier: A device intended to increase the power level of a signal.

Amplitude: The maximum peak value of a waveform (211-2018, 2018). The power level of a signal (CWNP Explanation).

Antenna: A device through which electric power (alternating current (AC)) is converted into radio waves and radio waves into electric power (CWNP Definition). In a communications system, a specialized transducer that converts incoming electromagnetic fields into alternating currents having the same frequencies (receiving antenna) or converts an alternating current at a specific frequency into an outgoing electromagnetic field at the same frequency (transmitting antenna). An antenna can be a simple wire or rod, or a complicated structure (Gibilisco, 2001).

Announcement Traffic Indication Message (ATIM): A traffic indication map (sent in a management frame) in an Ad-Hoc (IBSS) network to notify other clients of pending data transfers for power saving purposes.

Association: The condition wherein a client station is linked with an AP for frame transmission through the AP to the network and from the network through the AP to the client station (CWNP Definition). The service used to establish an access point (AP) or personal basic service set (PBSS) control point (PCP), and a station (STA) and enable STA invocation of the distribution system services (DSSs) (802.11-2020, 2020).

Attenuate: To reduce in amplitude (Gibilisco, 2001).

Attenuation: The reduction in amplitude of a signal when it passes through a medium that dissipates its energy (Butterfield, 2016). The loss of signal strength as an RF wave passes through a medium (CWNP Definition). The decrease in magnitude of a field with distance or with changes in the path in excess of the decrease due to a geometrical spreading factor (211-2018, 2018).

Attenuator: An electric component that intentionally reduces the strength of an RF signal in the path between the radio and the antenna.

Authentication: A process by which subjects, normally users, establish their identity to a system (Butterfield, 2016). The service used to establish identity of one station (STA) as a member of the set of STAs authorized to associate with another STA (802.11-2020, 2020). Security protocol by which two suspicious parties prove to each other that they are who they claim to be (Peterson & Davie, 2021).

Authentication and Key Management (AKM): The protocols used to authenticate a client STA on a WLAN and generate encryption key for use in frame encryption (CWNP Definition). Authentication is concerned with knowing that the correct parties are communicating; key establishment is concerned with obtaining good cryptographic keys to protect the communications, particularly to provide confidentiality and integrity of the data communicated (Boyd, Mathuria, & Stebila, 2020).

Authentication Server: The authentication server validates the client before allowing access to the network. In an 802.1X/EAP implementation for WLANs, the authentication server is often a RADIUS server (CWNP Definition). An entity that provides an authentication service to an Authenticator. This service determines, from the credentials provided by the Supplicant, whether the Supplicant is authorized to access the services provided by the system in which the Authenticator resides (802.1X, 2020).

Authenticator: The device that provides access to authentication services in order to allow connected devices to access network resources. In an 802.1X/EAP implementation for WLANs, the authenticator is typically the AP or controller (CWNP Definition). An entity that facilitates authentication of other entities attached to the same LAN (802.1X, 2020).

Authorization: The act of determining whether a particular right, such as access to a resource, is granted to an entity (802.11-2020, 2020).

Automatic Power Save Delivery (APSD): APSD is a power saving method which uses both scheduled (S-APSD) and unscheduled (U-APSD) frame delivery methods. S-APSD sends frames to a power save STA from the AP at a planned time. U-APSD sends frames to a power save STA from the AP when the STA sends a frame to the AP. The frame from the STA is considered a trigger frame.

Autonomous AP: An AP that can perform security functions, RF management, and configuration without the need for a centralized WLAN controller or any other control platform.

Azimuth: The direction of arrival measured along the horizon with respect to some reference direction (e.g., East) (211-2018, 2018).

Azimuth Chart: A chart showing the radiation pattern of an antenna as viewed from the top of the antenna. Also called an H-Plane Chart or H-Chart.

Backoff timer: The timer used during CSMA/CA to wait for access to the medium, which is selected from the contention window.

Band Steering: A method used by vendors to encourage STAs to connect to the 5 GHz band instead of the 2.4 GHz band, which is more congested. Typically implemented by ignoring probe requests for some period of time before allowing connection to the 2.4 GHz radio by clients known to have a 5 GHz radio based on previous connections to the AP or controller.

Bandwidth: The frequencies used for transmission of data. For example, a 20 MHz wide channel has 20 MHz of bandwidth. Bandwidth can also apply to subcarriers with OFDM and OFDMA.

Basic Service Area (BSA): The coverage area provided by an AP wherein client STAs may connect to the AP to transmit data on the WLAN or through the AP to the network (CWNP Definition). The area containing the members of a basic service set (BSS). It might contain members of other BSSs (802.11-2020, 2020).

Basic Service Set (BSS): An AP and its associated STAs. Identified by the BSSID (CWNP Definition). A set of stations (STAs) that have successfully synchronized using the JOIN service primitives and one STA that has used the START primitive. Alternatively, a set of STAs that have used the START primitive specifying matching mesh profiles where the match of the mesh profiles has been verified via the scanning procedure. Membership in a BSS does not imply that wireless communication with all other members of the BSS is possible (802.11-2020, 2020).

Basic Service Set Identification (BSSID): The ID for the BSS. Often the MAC address of the AP STA. When multiple SSIDs are used, another MAC address-like BSSID is generated.

Beacon Frame: A frame transmitted periodically from an AP that indicates the presence of a BSS network and contains capabilities and requirements of the BSS. Also colloquially called a beacon instead of the full phrase, beacon frame.

Beamformee: A station (STA) that receives a PPDU that was transmitted using a beamforming steering matrix (802.11-2020, 2020).

Beamformer: A station (STA) that transmits a PPDU using a beamforming steering matrix (802.11-2020, 2020).

Beamforming: Directing radio waves to a specific area or device by manipulating the RF waveforms within the different radio chains (CWNP Definition). A spatial filtering mechanism used at a transmitter to improve the received signal power or signal-to-noise ratio (SNR) at an intended receiver. Synonym: beam steering (802.11-2020, 2020).

Beamforming Steering Matrix: A matrix determined using knowledge of the channel between a transmitter and an intended receiver that maps from space-time streams to transmit antennas with the goal of improving the signal power or signal-to-noise ratio (SNR) at the intended receiver (802.11-2020, 2020).

Beamwidth: The width of the radiated signal lobe from the antenna in the intended direction of propagation. It is usually measured at the point where 3 dB of loss is experienced.

Bill of materials (BOM): A list of the materials and licenses required to assemble a system, in the case of WLANs, including APs, controllers, PoE injectors, licenses, etc.

Bit: A basic unit of information for computer systems. A bit can have a value of 1 or 0. Used in binary math.

Block Acknowledgement (BlockAck): An acknowledgement frame that groups together multiple ACKs instead of transmitting each individual ACK when a block transmission has been received.

Bridge: A device used to connect two networks. Wireless bridges create the connection across the wireless medium.

Broadcast Address: A unique group address that specifies all stations (STAs) (802.11-2020, 2020).

BSS Transition: Roaming that occurs between two BSSs that are part of the same ESS.

Byte: A basic unit of information that typically consists of 8 bits. Also called an octet.

Capacity: The number of clients a network or AP can accommodate including consideration of the applications used.

Captive Portal: Authentication technique that re-routes a user to a special webpage to verify their credentials before allowing access to the network. Commonly used in hotel and guest networks.

Carrier Frequencies: The frequency of a carrier signal or the frequencies used to modulate information.

Carrier Sense Multiple Access (CSMA): CSMA is a protocol that allows a node to detect the presence of traffic before sending data on a shared network. Used in CSMA/CA.

Carrier Sense Multiple Access with Collision Avoidance (CSMA/CA): CSMA/CA is the method in 802.11 networks in which a node only sends data if the shared network is idle in order to avoid collisions.

CCMP: Counter Cipher Mode with Block Chaining Message Authentication Code Protocol (CCMP) is a key management solution that provides for improved security over WEP.

CCMP/AES: CCMP used with AES, as it is in 802.11 networks, is a key management and encryption protocol that provides more security than WEP. It is based on the AES standard and uses a 128 bit key and 128 bit block size.

Centralized Forwarding: Every forwarding decision is made by a centralized forwarding engine, such as the WLAN controller.

Certificate Authority (CA): A server that validates the authenticity of a certificate used in authentication and encryption systems. The CA may issue certificates, or it may authorize other servers to do the same (CWNP Definition). An entity that vouches for the binding between a device's identity, its public key, and associated keying materials (such as key validity period and key usage) (802.11-2020, 2020).

Channel: A specified range of frequencies used in the 802.11 standard used by devices to communicate on the network. Channels are commonly 20, 40, 80 and 160 MHz in width in WLANs. Newer standards will support 1, 2, 4, 8 and 16 MHz channels in sub-1 GHz networks and channels wider than 1 GHz in the DMG PHY (CWNP Definition). An instance of the wireless medium (WM) use for the purpose of passing PPDUs between two or more stations (STAs) (802.11-2020, 2020). A means to transfer a signal (211-2018, 2018).

Channel Spacing: The difference between the center frequencies of two nonoverlapping and adjacent channels of the radio transmitter (802.11-2020, 2020).

Channel Width: The range of frequencies a single channel encompasses.

Cipher Suite: A set of one or more algorithms, designed to provide data confidentiality, data authenticity or integrity, and/or replay protection (802.11-2020, 2020).

Clear Channel Assessment (CCA): CCA is a feature defined in the IEEE 802.11 standard that allows a client to determine idle or busy state of the medium based on energy levels of a frame or raw energy levels as specified in each PHY.

Client Utilities: Software installed on devices that allows the device to connect to, authenticate with, and participate in a WLAN.

Co-Channel Interference (CCI): Congestion cause by the normal operations of CSMA/CA when multiple BSSs exist on the same channel. Commonly called co-channel congestion (CCC) today as well.

Collision Avoidance (CA): A method in which devices attempt to avoid simultaneous data transmissions in order to prevent frame collisions. Used in CSMA/CA.

Coding: A process used to encode bits to be transmitted on the wireless medium such that error recovery can be achieved. Part of forward error correction (FEC) and defined in the modulation and coding schemes (MCSs) from 802.11n forward.

Containment: A process used against a detected rogue AP to prevent any connected clients from accessing the network through it.

Contention Window: A number range defined in the 802.11 standard and varying by QoS category from which a number is selected at random for the backoff timer in the CSMA/CA process.

Control Frame: An 802.11 frame that is used to control the communications process on the wireless medium. Control frames include, RTS frames, CTS frames, PS-Poll frames, and ACK frames.

Controlled Port: In an 802.1X authentication system, the virtual port that allows all frames through to the network, but only after authentication is completed (CWNP Definition). The access point used to provide the secure MAC Service to a client of a MAC Security Entity (802.1X, 2020).

Controller-Based AP: An AP managed by a centralized controller device. Also called a lightweight AP or thin AP.

Coverage: 1) The colloquial term used for the BSA of an AP. 2) The requirement of available WLAN connectivity throughout a facility, campus or area. Often specified in minimum signal strength as dBm; for example, -67 dBm.

Clear-to-Send (CTS) Frame: A CTS frame sent from one STA to another to indicate that the other STA can transmit on the medium. The duration value in the CTS frame is used to silence all other STAs by setting their NAV timers.

Data Frame: An 802.11 frame specified for use in carrying data based on the general frame format. Also used for some signaling purposes as null data frames.

Data Rate: The rate at which data is sent across the wireless medium. Typically represented as megabits per second (Mbps) or gigabits per second (Gbps). The data rate should not be confused with throughput rate, which is a measurement of Layer 4 throughput or useful user data.

dBd (decibel to dipole): A relative measurement of antenna gain compared to a dipole antenna. Calculated as 2.14 dB greater than dBi as a dipole antenna already has 2.14 dBi gain.

dBi (decibel to isotropic): A relative measurement of antenna gain compared to a theoretical isotropic radiator. When necessary, calculated as 2.14 dB less than dBd.

dBm (decibel to milliwatt): An absolute measurement of the power of an RF signal based on the definition of 0 dBm = 1 milliwatt (mW).

Distributed Coordination Function (DCF): A protocol defined in 802.11 that uses carrier sensing, backoff timers, interframe spaces and frame duration values to diminish collisions on the wireless medium (CWNP Definition). A class of coordination function where the same coordination function logic is active in every station (STA) in the basic service set (BSS) whenever the network is in operation (802.11-2020, 2020).

Deauthentication Frame: A notification frame sent from an 802.11 STA to another STA in order to terminate a connection between them.

Decibel (dB): A logarithmic, relative unit used when measuring antenna gain, signal attenuation, and signal-to-noise ratios. Strictly defined as 1/10 of a bel.

Delay: The time it takes for a bit of data to travel from one node to another.

Delivery Traffic Indication Message (DTIM): A message sent from an AP to clients in the Beacon frame indicating that it has data to transmit to the clients specified by the AIDs.

Differentiated Services Code Point (DSCP): A Layer 3 QoS marking system. IP packets can include DSCP markings in the headers. Eight precedence levels, 0-7, are defined.

Diffraction: The deviation of the direction of energy flow of a wave, not attributable to reflection or refraction, when it passes an obstacle, a restricted aperture, or other inhomogeneities in a medium (211-2018, 2018). The bending of waves around a very large object in relation to the wave (CWNP Explanation).

Dipole Antenna: An antenna that propagates in all directions horizontally. Creates a coverage area similar to a donut (toroidal) shape. Also known as a omni-directional antenna (CWNP Definition). Also called dipole and doublet. A half-wavelength radiator fed at the center with a two-wire or coaxial transmission line. Each "leg" of the antenna is one-quarter wavelength long (Gibilisco, 2001).

Direct-Sequence Spread Spectrum (DSSS): A modulation technique where data is coupled with coding that spreads the data across a wide frequency range. Also a PHY name that provides 1 or 2 Mbps data rates in 802.11 networks.

Disassociation Frame: A frame sent from one STA to another in order to terminate the association.

Distributed Forwarding: See *Access Layer Forwarding*. Also called, *distributed data forwarding.*

Distribution System (DS): A system used to interconnect a set of basic service sets (BSSs) and integrated local area networks (LANs) to create an extended service set (ESS) (802.11-2020, 2020). The system that connects a set of BSSs and LANs such that an ESS is possible (CWNP Definition).

Distribution System Medium (DSM): The medium used to interconnect APs through the DS such that they can communicate with each other for ESS operations using either wired or wireless for the DS connection (CWNP Definition). The medium or set of media used by a distribution system (DS) for communications between access points (APs), mesh gates, and the portal of an extended service set (ESS) (802.11-2020, 2020).

Domain Name System (DNS): A protocol and service that provides host name resolution (looking up the IP address of a given host name) and recursive IP address lookups (finding the host name of a known IP address). Also, colloquially used to reference the server that provides DNS lookups.

Driver: Software that allows a computer to interact with a hardware device such as a WLAN adapter.

Duty Cycle: A measure of the time a radio is transmitting. The aggregate of all duty cycles of all transmitters seen in the channel is the channel utilization.

Dynamic Frequency Selection (DFS): A setting on radios that dynamically changes the channel selection based on detected interference from radar systems. Many 5 GHz channels require DFS operations (CWNP Definition). Facilities mandated to satisfy requirements in some regulatory domains for radar detection and uniform channel spreading in the 5 GHz band. These facilities might also be used for other purposes, such as automatic frequency planning (802.11-2020, 2020).

Dynamic Rate Switching (DRS): The process of reducing a client's data rate as frame transmission failures occur or signal strength decreases. DRS results in lower data rates but fewer transmissions required to successfully transmit a frame (CWNP Definition). Some PHYs have multiple data transfer rate capabilities that allow implementations to perform dynamic rate switching with the objective of improving performance (802.11-2020, 2020).

Eavesdropping: In a network, capturing and viewing network communications not intended for the actor (CWNP Definition). The adversary captures the information sent in the protocol. Eavesdropping is sometimes distringuished as being a passive attack since it does not require the adversary to disturb the communications of legitimate principals (Boyd, Mathuria, & Stebila, 2020).

Effective Isotropic Radiated Power (EIRP): The equivalent power of a transmitted signal in terms of an isotropic (omnidirectional) radiator (802.11-2020, 2020).

Electromagnetic Field: A time-varying field, associated with the electric or magnetic forces and described by Maxwell's equations (211-2018, 2018). A combination of alternating electric and magnetic fields (Gibilisco, 2001).

Electromagnetic Waves: Waves produced in a conductor or in space by the acceleration or oscillation of electric charge carriers. Such waves have an electric and a magnetic component acting at right angles to each other. The waves propagate at right angles to both the electric and magnetic flux lines (Gibilisco, 2001).

Elevation Chart: A chart showing the radiation pattern of an antenna as viewed from the side antenna. Also called an E-Plane Chart or E-Chart.

Encapsulate: To construct a data unit by adding relevant information before (prepending) or after (appending) the data unit applicable to the processing requirements of an acting layer of the OSI Reference Model.

Encryption: The process of converting data into a form that unauthorized users cannot understand by encoding the data with an algorithm and a key or keys (CWNP Definition). An encryption scheme defines four sets: a set of encryption keys K_E, a set of decryption keys K_D, a message set M, and a ciphertext set C, together with three algorithms... a key generation algorithm... an encryption algorithm… a decryption function (Boyd, Mathuria, & Stebila, 2020).

Enhanced Distributed Channel Access (EDCA): An enhancement to DCF introduced in 802.11e that implements priority based queuing for transmissions in 802.11 networks based on access categories.

Equivalent Isotropically Radiated Power (EIRP): The output power required of an isotropic radiator to equal the measured power output from an antenna in the intended direction of propagation.

Extended Rate Physical (ERP): A physical layer technology introduced in 802.11g that uses OFDM (from 802.11a) in the 2.4 GHz band and offers data rates up to 54 Mbps.

Extended Service Set (ESS): A group of one or more basic service sets (BSSs) that are interconnected by a distribution system (DS) (CWNP Definition). A set of one or more basic service sets (BSSs) that appears as a single BSS to the logical link control (LLC) layer at any station (STA) associated with one of those BSSs (802.11-2020, 2020).

Extensible Authentication Protocol (EAP): An authentication framework that defines message formats for authentication exchanges used by 802.1X WLAN authentication solutions (CWNP Definition). An authentication framework which supports multiple authentication methods (RFC 3748).

Fade Margin: An amount of signal strength, in dB, added to a link budget to ensure proper operations.

Fast Fourier Transform (FFT): A mathematical algorithm that takes in a waveform as represented in the time or space domain and shows it in the frequency domain. Used in spectrum analyzers to show real-time views in the frequency domain (Real-time FFT).

Flux Lines: Theoretical lines of force that extend in all directions from an electric charge (electric flux) or from a magnetic pole (magnetic flux) (Gibilisco, 2001).

Fragmentation: The process of fragmenting 802.11 frames based on the fragmentation threshold configured. Fragmented frames have a greater likelihood of successful delivery in the presence of sporadic interference.

Frame Aggregation: A feature in the IEEE 802.11n PHY and later PHYs that increases throughput by sending more than one frame in a single transmission. Aggregated MSDUs or aggregated MPDUs may be supported.

Frame: A well-defined, meaningful set of bits used to communicate management and control information on a network or transfer payloads from higher layers. Frames are defined at the MAC and PHY layer (CWNP Definition). A unit of data exchanged between peer protocol entities (802.11-2020, 2020).

Free Space: Space that is free of obstruction and characterized by the constitutive parameters of a vacuum (211-2018, 2018). Open air space without obstructions such as walls, trees, buildings, mountains, etc. (CWNP Explanation).

Free Space Path Loss: The natural loss of amplitude that occurs in an RF signal as it propagates through space and the wave front spreads.

Fresnel Zones: Ellipsoid shaped zones around the visual LoS in a wireless link. The first Freznel zone should be 60% clear and would preferably be 80% clear to allow for environmental changes.

Frequency: The speed at which a waveform cycles in a second (CWNP Definition). The number of identical cycles per second, measured in Hertz (211-2018, 2018).

Four-Way Handshake: *See 4-Way Handshake.*

Full Duplex: A communication system that allows an endpoint to send data to the network at the same time as it receives data from the network.

Gain: The increase in signal strength. Can be accomplished passively by directing energy into a smaller area (antenna gain) or actively by increasing the strength of the broadcasted signal before it is sent to the antenna (increase radio transmit power or amplifier).

Group Key Handshake: Used to transfer the group temporal key (GTK) among STAs in an 802.11 network if the GTK requires updating. Initiated by the AP/controller in a BSS.

Group Master Key (GMK): An auziliary key that might be used to derive a group temporal key (GTK) (802.11-2020, 2020). Used to generate the GTK for encryption of broadcast and multicast frames and is unique to each BSS (CWNP Explanation).

Group Temporal Key (GTK): A random value, assigned by the group source, which is used to protect group addressed MPDUs from that source. The GTK might be derived from a group master key (GMK) (802.11-2020, 2020). Used to encryption broadcast and multicast frames and is unique to each BSS (CWNP Explanation).

Guard Interval (GI): A period of time between symbols within a frame used to avoid intersymbol interference.

Guest Network: A segregated network that is designed for use by temporary visitors.

Half Duplex: A communication system that allows only sending or receiving data by an endpoint at any given time.

Hidden Node: A node that cannot be seen by all other nodes in the wireless channel. The hidden node problem arises when nodes cannot receive each other's frames though they are participating in the same network, which can lead to packet collisions and retransmissions (CWNP Definition). Hidden station in IEEE 802.11-2020.

High Density: A phrase referencing a WLAN network type that is characterized by large numbers of devices requiring access. Examples of high density networks include stadiums and conference auditoriums as well as many IoT networks.

Highly-Directional Antenna: An antenna, such as a parabolic dish or grid antenna, that has a high gain in a specified direction and a low beamwidth measurement as compared to semi-directional and omnidirectional antennas.

High Efficiency (HE): An amendment-based MAC and PHY (802.11ax) that increased data rates, added OFDMA modulation schemes and, more importantly, added support for 6 GHz operations.

High Rate Direct Sequence Spread Spectrum (HR/DSSS): An amendment-based PHY (802.11b) that increased the data rate in 2.4 GHz from the original 1 or 2 Mbps (DSSS) to 5.5 and 11 Mbps while maintaining backward compatibility with 1 and 2 Mbps.

High Throughput (HT): An amendment-based PHY (802.11n) that increased the data rate up to 600 Mbps and added support for transmit beamforming and MIMO.

Hotspot: A term referencing a wireless network connection point that is typically open to the public or to paid subscribers. Hotspot 2.0 is a Wi-Fi Alliance Certification/Standard for implementing hotspots with service provider authentication.

Independent Basic Service Set (IBSS): A set of 802.11 devices operating in ad-hoc (peer-to-peer) mode without the use of an AP (CWNP Definition). A basic service set (BSS) that forms a self-contained network, and in which no access to a distribution system (DS) is available (802.11-2020, 2020).

Institute of Electrical and Electronics Engineers (IEEE): A standardization organization that develops standard for multiple industries including the networking industry with standard such as 802.3, 802.11 and 802.16.

Intentional Radiator: Any device that is purposefully sending radio waves. Signal strength of the intentional radiator is measured at the point where energy enters the radiating antennas.

Interference: In wireless networks, an RF signal or incidental RF energy that is radiated in the same frequencies as used by the wireless network and that has sufficient amplitude and duty cycle to prevent valid network frames from successful delivery (CWNP Definition). In communications, degradation of reception caused by noise or undesired signals (Gibilisco, 2001).

Interframe Space (IFS): A time interval that must exist between frames. Varying lengths are used in 802.11 and are referenced as DIFS, SIFS, RIFS, PIFS, EIFS and AIFS in common use.

Internet Engineering Task Force (IETF): An open group of volunteers develops Internetworking standards through request for comments (RFC) documents. Examples include RADIUS, EAP and DNS.

Isotropic Radiator: A theoretical antenna that spreads the radiation equally in every direction as a sphere. None exist in reality, but the concept is used to measure relative antenna gain in dBi.

Jitter: The variance in delay between packets sent on a network. Excessive jitter can result in poor quality for real-time applications such as voice and video.

Jumbo Frame: An Ethernet frame that contains more than 1500 bytes of payload and up to 9000 to 9216 bytes.

Latency: While sometimes used synonymously with delay, it is more accurate to use latency for the round-trip time in network communications.

Layer 1: The physical layer (PHY) that is responsible for framing and transmitting bits on the medium. In 802.3 and 802.11 the entirety of Layer 1 is defined. The PHY transmits and received PPDUs.

Layer 2: The data-link layer that deals with data frames moving within a local area network (LAN). In 802.3 and 802.11, the MAC sublayer of Layer 2 is defined. The MAC transmits and received MPDUs

Layer 3: The network layer where packets of data are routed between sender and receiver. Most modern networks use Internet Protocol (IP) at Layer 3.

Layer 4: The transport layer where segmentation occurs for upper layer data and TCP (connection oriented) and UDP (connectionless) are the most used protocols.

Lightning Arrestor: A device that can redirect ambient energy from a lightning strike away from attached equipment.

Line of sight (LoS): When existing, the visual path between to ends. RF LoS is different from visual LoS. RF LoS does not require the same clear path for the remote receiver to hear the signal. When creating bridge links, visual LoS is often the starting point (CWNP Definition). The direct path from the source to the receiver (211-2018, 2018).

Link Budget: The measurement of gains and losses from the transmitting radio to the receiving radio including free space path loss, antenna gains, and cabling or adapter losses.

Loss: The reduction in the amplitude of a signal.

MAC filtering: A common setting that only allows specific MAC addresses onto a network. Ineffective against knowledgeable attackers because the MAC address can be spoofed to impersonate authorized devices.

MAC Frame: A synonym for MPDU.

Management Frame: A frame type defined in the 802.11 standard that encompasses frames used to manage access to the network including beacon, probe request, prober response, authentication, association, reassociation, deauthentication and disassociation frames.

Master Session Key (MSK): A key derived between an EAP client and EAP server and exported by the EAP method. Used to derive the PMK, which is used to derive the PTK. The MSK is used in 802.1X/EAP authentication implementations. In personal authentication implementations, the PMK is derived from the pre-shared key (CWNP Definition). Keying material that is derived between the Extensible Authentication Protocol (EAP) peer and exported by the EAP method to the Authentication Server (AS) (802.11-2020, 2020).

Maximal Ratio Combining (MRC): A method of increasing the signal-to-noise ratio (SNR) by combining signals received on multiple radio chains (multiple antennas and radios).

Mesh: A network that uses interconnecting devices to form a redundant set of connections offering multiple paths through the network. 802.11s defined mesh for 802.11 networks.

Mesh BSS (MBSS): A basic service set (BSS) that forms a self-contained network of mesh stations (STAs) that use the same mesh profile. An MBSS contains zero or more mesh gates, and can be formed from mesh STAs that are not in direct communication (802.11-2020, 2020).

Mesh Gate: A mesh STA that has direct access to a distribution system. Other mesh STAs in a Mesh BSS (MBSS) access the distribution system via the mesh gate (802.11-2020, 2020).

Mesh Link: A link from one mesh station (STA) to a neighbor mesh STA that have a mesh peering with each other (802.11-2020, 2020).

Mesh Neighborhood: The set of all neighbor mesh stations (STAs) of a particular mesh STA (802.11-2020, 2020).

Milliwatt (mW): A unit of electrical energy used in measuring output power of RF signals in WLANs. A mW is equal to 1/1000 of a watt (W).

Mobile User: A user that physically moves while connected to the network. The opposite of a stationary user.

Modulation: The process of changing a wave by changing its amplitude, frequency, and/or phase such that the changes represent data bits (CWNP

Definition). Combining two signals with the result that some aspect of one signal (carrier) is varied by and in sympathy with the other (the modulating signal). Usually, the carrier has a frequency considerably higher than that of the modulating signal (Gibilisco, 2001). A simple sine wave radio signal conveys no information except that it is there and perhaps the direction of its origin. In order to convey information we must change the signal in some agreed way, a process called modulation (Dennison & Fielding, 2009).

Modulation and Coding Scheme (MCS): Term used to describe the combination of the radio modulation scheme and the coding scheme used when transmitting data, first introduced in 802.11n.

MPDU: A MAC protocol data unit (MPDU) is a portion of data to be delivered to a MAC layer peer on a network and it is data prepared for the PHY layer by the MAC sublayer. The MAC sublayer receives the MSDU from upper layers on transmission and creates the MPDU. It receives the MPDU from the lower layer on receiving instantiation and removes the MAC header and footer to create the MSDU for the upper layers (CWNP Definition). The unity of data exchanged between two peer MAC entities using the services of the physical layer (PHY) (802.11-2020, 2020).

MSDU: A MAC service data unit is a portion of transmitted data to be handled by the MAC sublayer that has yet to be encapsulated into a MAC Layer frame (CWNP Definition). Information that is delivered as a unity between MAC service access points (SAPs) (802.11-2020, 2020).

Maximum Transmission Unit (MTU): The largest amount of data that can be sent at a particular layer of the OSI model. Typically set at layer 4 for TCP.

Multi-User MIMO (MU-MIMO): A technique by which multiple stations (STAs), each with one or more antennas, either simultaneously transmit to a single STA or simultaneously receive from a single STA independent data streams over the same subcarriers (802.11ax, 2021).

Multipath: The phenomenon that occurs when multiple copies of the same signal reach a receiver based on RF behaviors in the environment (CWNP Definition). The propagation phenomenon that results in signals reaching the receiving antenna by two or more paths (211-2018, 2018).

Multiple Channel Architecture (MCA): A wireless network design using multiple channels strategically designed so that the implemented BSSs have minimal interference with one another.

Multiple Input/Multiple Output (MIMO): A technology used to spread a stream of data bits across multiple radio chains using spatial multiplexing at the transmitter and to recombine these streams at the receiver (CWNP Definition). A communications system that has multiple points of transmission and multiple points of reception (211-2018, 2018).

Narrowband Interference: Interference that covers a very narrow band of frequencies and typically not the full with of an 802.11 channel when used in reference to WLAN interferers.

Near-Far: A problem that occurs when a high powered device is closer to the AP in a BSS and a low powered device is farther from the AP. Most near-far problems are addressed with standard CSMA/CA operations in 802.11 networks.

Network Allocation Vector (NAV): The NAV is a virtual carrier sense mechanism used in CSMA/CA to avoid collisions and is a timer set based on the duration values in frames transmitted on the medium (CWNP Definition). An indicator, maintained by each station (STA), of time periods when transmission onto the wireless medium (WM) is not initiated by the STA regardless of whether the STA's clear channel assessment (CCA) function senses that the WM is busy (802.11-2020, 2020).

Network Segmentation: The process used to separate a larger network into smaller networks often utilizing Layer 3 routers or multi-layer switches.

Noise: RF energy in the environment that is not part of the intentional signal of your intentional wireless transmissions. Noise can also be generated within the receiving circuitry of an RF receiver. This noise is measured as the Noise Figure (dB) or Noise Factor (ratio). (CWNP Definition). Higher level modulation schemes can combine even more binary digits into a symbol, allowing even higher data rates… the difference between each symbol gets progressively less and less as each symbol is used to contain more and more binary digits… in isolation this is all very well but the effects of noise and interference get progressively worse as the complexity of the system increases (Dennison & Fielding, 2009).

Noise Factor: Defines the maximum sensitivity of a receiver without regard to its pass-band (bandwidth allowed into the circuitry by filters) or its input impedance and will be determined in the front-end of the receiver. *It is* expressed logarithmically: Noise Figure NF (dB) = $10\log_{10}$(SNRin/SNRout) (Dennison & Fielding, 2009).

Noise Figure: Defines the maximum sensitivity of a receiver without regard to its pass-band (bandwidth allowed into the circuitry by filters) or its input impedance and will be determined in the front-end of the receiver. It is defined as the ration: Noise factor F = SNRin/SNRout (Dennison & Fielding, 2009). The noise factor of a circuit, expressed in decibels (Gibilisco, 2001).

Noise Floor: The amount of noise that is consistently present in the environment, which is typically measured in dBm (CWNP Definition). 1) In a receiver, the level of noise in microvolts that determines the weakest signal that can be heard or accurately received. 2) In a spectrum analyzer, the level of noise that determines the weakest signal that can be visibly displayed (Gibilisco, 2001).

Network Time Protocol (NTP): A protocol used to synchronize clocks in devices using centralized time servers.

Non-Adjacent Channel: A channel separated from another channel by more frequency space than the separation required to be considered an adjacent non-overlapping channel.

Non-High-Throughput (Non-HT): A modifier meaning not high throughput (HT), not very high throughput (VHT) and not high efficiency (HE) (802.11ax, 2021).

Octet: A group of eight ones and zeros. An 8-bit byte. Sometimes simply called a byte.

OFDM: *See Orthogonal Frequency Division Multiplexing (OFDM).*

OFDMA: *See Orghogonal Frequency Division Multiple Access (OFDMA).*

Orthogonal Frequency Division Multiplexing (OFDM): A modulation technique and a named physical layer in 802.11 that provides data rates up to 54 Mbps and operates in the 5 GHz band. The modulation is used in all bands, but the named PHY operates only in the 5 GHz band.

Orthogonal Frequency Division Multiple Access (OFDMA): An orthogonal frequency division multiple (OFDM)-based multiple access technique by which multiple stations (STAs) either simultaneously transmit to a single STA or simultaneously receive from a single STA independent data streams over different groups of subcarriers (802.11ax, 2021).

Omni-Directional Antenna: An antenna that propagates in all directions horizontally. Creates a coverage area similar to a donut shape (toroidal). Also known as a dipole antenna.

Open System Authentication: A simple frame exchange, providing no real authentication, used to move through the state machine in relation to the connection between two 802.11 STAs.

Opportunistic Key Caching (OKC): A roaming solution for WLANs wherein the keys derived from the 802.1X/EAP authentication are cached on the AP or controller such that only the 4-way handshake is required at the time of roaming.

OSI (Open Systems Interconnection) Model: A theoretical model for communication systems that works by separating the communications process into seven, well-defined layers. The seven layers are Application, Presentation, Session, Transport, Network, Data Link and Physical. In early development, many thought it would become an actual protocol of use; however, TCP/IP, in large part thanks to the Internet, has become the true operational model for most networks.

Packet: Data as represented at the network layer (Layer 3) for IP communications.

Pairwise: Referring to, or an attribute of, two entities that are associated with each other, e.g., an access point (AP) and an associated station (STA), or two STAs in an independent basic service set (IBSS) network. The term is used to refer to a type of encryption key hierarchy pertaining to keys shared by only two entities (802.11-2020, 2020).

Passive Gain: An increase in strength of a signal by focusing the signal's energy using an antenna rather than increasing the actual energy available, such as with an amplifier (CWNP Definition). Antennas achieve gain by focusing the radiated energy in a particular direction. This leads to the concept of Effective Radiated Power (ERP) which is the product of the power fed to the antenna multiplied by the antenna gain. It should be noted that this gain also applies on receive because antennas are passive, reciprocal devices that work identically on transmit or receive (Dennison & Fielding, 2009).

Passive scanning: A scanning (network location) method wherein a STA waits to receive beacon frames from an AP which contain information about the WLAN.

Passive survey: A survey conducted on location that gathers information about RF interference, signal strength and coverage areas by monitoring RF activity without active communications.

Passphrase/Pass-Phrase: A secret text string employed to corroborate the user's identity (802.11-2020, 2020). May also be used for device identity corroboration (CWNP Explanation).

Passphrase Authentication: A type of access control that uses a phrase as the pass key. Also called personal in WPA and WPA2.

Password: A shared, secret, and potentially low-entropy word, phrase, code, or key used as a credential for authentication purposes (802.11-2020, 2020).

Path Loss: The loss in power density experienced by a wave as it traverses the path between the transmitter and receiver (211-2018, 2018).

Phase: A measurement of the variance in arrival or departure state between to copies of a wave form. Waves are said to be in phase or out of phase by some degree. The phase can be manipulated for modulation (CWNP Definition). Two waveforms that are of the same frequency are in phase if they both start at the same point in time. If one waveform it delayed with respect to the other then that are not in phase and the phase difference is usually expressed as a proportion of a complete cycle of 360° (Dennison & Fielding, 2009).

PHY: A shorthand notation for physical layer which is the physical means of communication on a network to transmit bits.

Physical (PHY) Layer: The physical (PHY) layer refers to the physical means by which a message is communicated. Layer one of the OSI model.

Physical Layer (PHY) Protocol Data Unit (PPDU): *See PPDU.*

Physical Layer (PHY) Frame: Synonym for PPDU. *See PPDU.*

PMK Caching: Stores the PMKSA information, including the PMK, so a device has only to perform the 4-way handshake when connecting to an AP to which it has already connected (CWNP Definition). A STA can supply a list of PMK identifiers in the (Re)Association Request frame… each key identifier names a PMKSA… the Authenticator can specify the selected PMK identifier in message 1 of the 4-way handshake (802.11-2020, 2020).

Pairwise master Key (PMK): The key derived from the MSK, which is generated during 802.1X/EAP authentication. Used to derive the PTK. Used in unidirectional communications with a single peer.

PoE Injector: Any device that adds Power over Ethernet (PoE) to ethernet cables. Come in two variants, endpoint (such as switches) and midspan (such as inline injectors).

Point-to-Multipoint (PtMP): A connection between a single point and multiple other points for wireless bridging or WLAN access.

Point-to-Point (PtP): A connection between two points often used to connect two networks via wireless bridging.

Polarization: The locus of the top of the electric field vector observed over time at a fixed point in space in a plane orthogonal to the wave normal (211-2018, 2018). The technical term used to reference the orientation of antennas related to the electric field in the electromagnetic wave (CWNP Explanation).

Power over Ethernet (PoE): A method of providing power to certain hardware devices that can be powered across the Ethernet cables. Specified in IEEE 802.3 as a standard. Various classes are defined based on power requirements.

PPDU: The unit of data exchanged between PHY entities to provide the PHY data service (802.11ax, 2021).

Preauthentication: An authentication process where the authentication service is invoked independently from the association service. Typically performed by a station that is already associated with an AP.

Pre-shared Key (PSK): Refers to any security protocol that uses a password or passphrase or string as the key from which encryption materials are derived (CWNP Definition). A static key that is distributed to the units in the system by some out-of-band means (802.11-2020, 2020).

Primary Channel: When implementing channels wider than 20 MHz in 802.11n, 802.11ac and 802.11ax, the 20 MHz channel on which management and control frames are sent and the channel used by STAs not supporting the wider channel (CWNP Definition). The common channel of operation for all stations (STAs) that are members of the basic service set (BSS). For example, in a 20 MHz, 40 MHz, 80 MHz, 160 MHz, or 80+80 MHz BSS the primary channel is a primary 20 MHz channel (802.11-2020, 2020).

Probe Request: A type of frame sent when a client device wants information about APs in the area or is seeking a specific SSID to which it desires to connect.

Probe Response: A type frame sent in response to a probe request that contains information about the AP and the requirements of BSSs it provides.

Protected Management Frame (PMF): Frames used for managing a wireless network that are protected from spoofing using encryption. Protocol defined in the 802.11w amendment.

Protocol Analyzer: Hardware or software used to capture and analyze networking communications. WLAN protocol analyzers have the ability to capture 802.11 frames from the RF medium and decode them for display and analysis.

Protocol Decodes: The way information in captured packets or frames is interpreted for display and analysis.

PSDU: PHY Service Data Unit (PSDU) is the name for the contents that are contained within the PPDU, the PHY Protocol Data Unit. It is the same as the MPDU as perceived and received by the PHY.

PTK (Pairwise Transient Key): A key derived during the 4-way handshake and used for encryption only between two specific endpoints, such as an AP and a single client.

Quality of Service (QoS): Traffic prioritization and other techniques used to improve the end-user experience. IEEE 802.11e and therefore 802.11-2020 includes QoS protocols for wireless networks based on access categories (CWNP Definition). Packet delivery guarantees provided by network architecture. Usually related to performance guarantees, such as bandwidth and delay. The Internet offers a best-effort delivery service, meaning that every effort is made to deliver a packet but delivery is not guaranteed (Peterson & Davie, 2021).

Quality of Service (QoS) Access Point (AP): An Ap that supports the QoS facility (802.11-2020, 2020).

Quality of Service (QoS) Basic Service Set (BSS): A BSS supporting 802.11 QoS features (CWNP Definition). A BSS that provides the QoS facility. An infrastructure QoS BSS contains a QoS access point (AP) (802.11-2020, 2020).

Quality of Service (QoS) Facility: The set of enhanced functions, channel access rules, frame formats, frame exchange sequences and managed objects used to provide parameterized and prioritized QoS (802.11-2020, 2020).

Quality of Service (QoS) Station (STA): A STA that implements the QoS facility (802.11-2020, 2020).

Radio Chain: A reference to the radio and antenna used together to transmit in a given frequency range. Multi-stream devices have multiple radio chains as one radio chain is required for each stream (CWNP Definition). Radio Frequency (RF) Chain: The physical entity that is able to act as a receive chain or transmit chain, or both (802.11-2020, 2020).

Radio Frequency (RF): The electromagnetic wave frequency range used in WLANs and many other wireless communication systems (CWNP Definition). A frequency in the radio spectrum (211-2018, 2018).

Radio Resource Management (RRM): Automatic management of various RF characteristics like channel selection and output power. Known by different terms among the many WLAN vendors, but referencing the same basic capabilities.

Radio Spectrum: The radio frequency portion of the electromagnetic spectrum (211-2018, 2018). The range is from 3 Hz to 3 THz according to IEEE 211-2018 (CWNP Explanation).

RADIUS: Remote Authentication Dial-In User Service (RADIUS) refers to a network protocol that handles AAA management which allows for authentication, authorization and accounting (auditing). Used in 802.11 WLANs as the authentication server in an 802.1X/EAP implementation.

RC4 (Rivest Cipher 4): An encryption cipher used in WEP and with TKIP. A stream cipher.

Real-Time Location Service (RTLS): A function provided by many WLAN infrastructure and overlay solutions allowing for device location based on triangulation and other algorithms.

Reassociation: The process used to associate with another AP in the same ESS. May also be used when a STA desires to reconnect to an AP to which it was formerly connected (CWNP Definition). The service that enables an established association to be transferred from on AP to another (or the same) AP (802.11-2020, 2020).

Received Channel Power Indicator (RCPI): Introduced in 802.11k, a power measurement calculated as INT((dBm + 110) * 2). Expected accuracy is +/- 5 dB. Ranges from 0-220 are available with 0 equaling or less than -110 dBm and 220 equaling or greater than 0 dBm. The value is calculated as an average of all received chains during the reception of the data portion of the transmission. All PHYs support RCPI and, though 802.11ac does not explicitly list its formulation, it references the 802.11n specification for calculation procedures (CWNP Definition). An indication of the total channel power (signal, noise, and interference) of a received frame measured on the channel and at the antenna connector used to receive the frame (802.11-2020, 2020).

Received Signal Strength Indicator (RSSI): A relative measure of signal strength for a wireless network. The method to measure RSSI is not standardized though it is constrained to a limited number of values in the 802.11 standard. Many use the term RSSI to reference dBm, and the 802.11 standard uses terms like DataFrameRSSI and BeaconRSSI and defines them as the signal strength in dBm of the specified frames, so the common vernacular is understandable. However, according to the standard, "absolute accuracy of the RSSI reading is not specified" (802.11-2012, Clause 14.3.3.3).

Reflection: An RF behavior that occurs when a wave meets a reflective obstacle larger than the wavelength, similar to light waves in a mirror (CWNP Definition). Reflected Wave: For two media, separated by a planar interface, that part of the incident wave that is returned to the first medium (211-2018, 2018).

Refraction: An RF behavior that occurs as an RF wave passes through material causing a bending of the wave and possible redirection of the wave front (CWNP Definition). Refracted Wave: For two media, the part of the incident wave that travels from the first medium into the second medium (211-2018, 2018).

Regulatory Domain: A reference to geographic regions management by organizations like the FCC and ETSI that determine the allowed frequencies, output power levels and systems to be used in RF communications.

Remote AP: An AP designed to be implemented at a remote location and managed across a WAN link using special protocols.

Resolution Bandwidth (RBW): The smallest frequency that can be extracted from a received signal by a spectrum analyzer or the configuration of that frequency. Many spectrum analyzers allow for the adjustment of the RBW within the supported range of the analyzer.

Retry: That which occurs when a frame fails to be delivered successfully. A bit set in the frame header to specify that it is a repeated attempt at delivery.

Return Loss: A measure of how much power is lost in delivery through a transmission line and connectors to an antenna due to reflection within the system.

RF Cables: A cable, typically coaxial, that allows for the transmission of electromagnetic waves between a transceiver and an antenna.

RF Calculator: A software application used to perform calculations related to RF signal strength values.

RF Connector: A component used to connect RF cables, antennas and transmitters. RF connectors come in many standardized forms and should match in type and resistance.

RF Coverage: Synonymous with coverage in WLAN vernacular. Reference to the BSA provided by an AP. May reference, colloquially, the availability of RF signals at a specific location.

RF Link: An established connection between two radios.

RF Line of Sight (LoS): The existence of a path, possibly including reflections, refractions and pass-through of materials, between two RF transceivers. Contrast with simple Line of Sight, which is the direct path from the source to the receiver.

RF Propagation: The process by which RF waves move throughout an area including reflection, refraction, scattering, diffraction, absorption and free space path loss.

RF Propagation Delay: The time required for energy to propagate between two specified points, determined by multiplying the group velocity of the wave by the distance between the two points projected onto the direction of propagation (211-2018, 2018).

RF Signal Splitter: An RF component that splits the RF signal with a single input and multiple outputs. Historically used with some antenna arrays, but less common today in WLAN implementations.

RF Site Survey: The process of physically measuring the RF signals within an area to determine resulting RF behavior and signal strength. Often performed as a validation procedure after implementation based on a predictive model.

Roaming: That which occurs when a wireless STA moves from one AP to another either because of end user mobility or changes in the RF coverage.

Robust Security Network (RSN): A network that supports CCMP/AES or WPA2 and optionally TKIP/RC4 or WPA. To be an RSN, the network must support only RSN Associations (RSNAs), which are only those associations that use the 4-way handshake. WEP is not supported in an RSN.

Robust Security Network Association (RSNA): An association between a client STA and an AP that was established through authentication resulting in a 4-way handshake to derive unicast keys and transfer group keys. WEP is not supported in an RSNA.

Rogue Access Point: An access point that is connected to a network without permission from a network administrator or other official.

Rogue Containment: Procedures used to prevent clients from associating with a rogue AP or to prevent the rogue AP from communicating with the wired network.

Rogue Detection: Procedures used to identify rogue devices. May include simple identification of unclassified APs or algorithmic processes that identify likely rogues.

Role-Based Access Control (RBAC): An authorization system that assigns permissions and rights based on user roles. Similar to group management of authorization policies.

RSN Information Element: A portion of the beacon frame that specifies the security used on the WLAN.

Request to Send/Clear to Send (RTS/CTS): A frame exchange used to clear the channel before transmitting a frame in order to assist in the reduction of collisions on the medium. Also used as a backward compatible protection mechanism.

RTS Threshold: The minimum size of a frame required to use RTS/CTS exchanges before transmission of the frame.

S-APSD: See *Automatic Power Save Delivery*.

Scattering: An RF behavior that occurs when an RF wave encounters reflective obstacles that are smaller than the wavelength. The result is similar to multiple reflections or scattering of the wave front (CWNP Definition). A process by which the energy of a traveling wave is dispersed in direction by means other than reflection and refraction (211-2018, 2018). Millimeter waves, which represent the present frontier of practical technology, suffer attenuation due to scattering and absorption by clouds, fog, and water droplets in the atmosphere (Dennison & Fielding, 2009).

Secondary Channel: When implementing channels wider than 20 MHz in 802.11n and 802.11ac, the second channel used to form a 40 MHz channel for data frame transmissions to and from supporting client STAs.

Semi-Directional Antenna: An antenna such as a yagi or a patch that has a propagation pattern which maximizes gain in a given direction rather than an omni-directional pattern, having a larger beamwidth than highly directional antennas.

Service Set Identifier (SSID): The BSS and ESS name used to identify WLAN. Conventionally made to be readable by humans. Maximum of 32 bytes long (CWNP Definition). A string used to identify the infrastructure basic service sets

(BSSs) that comprise an extended service set (ESS), or to identify a non-infrastructure BSS (802.11-2020, 2020).

Signal Strength: A measure of the amount of RF energy being received by a radio. Often specified as the RSSI, but referenced in dBm, which is not the proper definition of RSSI from the 802.11 standard. However, the colloquial use of RSSI as synonymous with dBm has become popular.

Single Channel Architecture (SCA): A WLAN architecture that places all APs on the same channel and uses a centralized controller to determine when each AP can transmit a frame. No control of client transmissions to the network is provided.

Single Input, Single Output (SISO): A technique by which a station (STA) transmits to or received from a single STA a single space-time stream (802.11ax, 2021).

Signal: An electrical quantity, such as a current or voltage, that can be used to convey information for communication, control, calculation, etc. (Gibilisco, 2001)

Signal to Noise Ratio (SNR): A comparison between the received signal strength and the noise floor. Typically presented in dB. For example, given a noise floor of -95 dBm and a signal strength of -70 dBm, the SNR is 25 dB (CWNP Definition). In a receiver, the ratio of the combined signal amplitude to noise amplitude. It is usually expressed in decibels (dB) (Gibilisco, 2001).

SISO: *See Single Input, Single Output (SISO).*

Space-Time Block Coding (STBC): The use of multiple streams of the same data across multiple radio chains to improve reliability of data transfer through redundancy.

Spatial Multiplexing (SM): Used with MIMO technology to send multiple spatial streams of data across the channel using multiple radio chains (radios coupled with antennas) (CWNP Definition). A transmission technique in which data streams are transmitted on multiple spatial channels that are provided through the use of multiple antennas at the transmitter and the receiver (802.11-2020, 2020).

Spatial Multiplexing Power Save (SMPS): A power saving feature from 802.11n that allows a station to use only one radio (or spatial stream).

Spatial Streams: The partitioning of a stream of data bits into multiple streams transmitted simultaneously by multiple radio chains in an AP or client STA (CWNP Definition). Spatial Stream: One of several streams of bits or modulation symbols that might be transmitted over multiple spatial dimensions that are created by the use of multiple antennas at both ends of a communications link (802.11-2020, 2020).

Spectrum Analysis: The inspection of raw RF energy to determine activity in an area on monitored frequencies. Useful in troubleshooting and design planning.

Spectrum Analyzer: A hardware and software solution that allows the inspection of raw RF energy.

Station (STA): Any entity that can use IEEE 802.11 protocol. Includes entities within both APs and clients (CWNP Definition). A logical entity that is a singly addressable instance of a medium access control (MAC) and physical layer (PHY) interface to the wireless medium (WM) (802.11-2020, 2020).

Supplicant: In 802.1X, the device attempting to be authenticated. Also the term used for the client software on a device that is capable of connecting to a WLAN (CWNP Definition). An entity at one end of a point-to-point LAN segment that seeks to be authenticated by an Authenticator attached to the other end of that link (802.1X, 2020).

Sweep Cycle: The time it takes a spectrum analyzer to sweep across the frequencies monitored. Often a factor of the number of frequencies scanned and the RBW.

System Operating Margin (SOM): The actual positive difference in the required link budget for a bridge link to operate properly and the received signal strength in the link.

Temporal Key Integrity Protocol (TKIP): The authentication and key management protocol supported by WPA systems and implemented as an interim solution between WEP and CCMP in IEEE 802.11i.

Transition Security Network (TSN): A network that allows WEP connections during the transition period over to more secure protocols and an eventual RSN. An RSN does not allow WEP connections.

Transmit Beamforming (TxBF): The use of multiple antennas to transmit a signal strategically with varying phases so that the communication arrives at the receiver such that the signal strength is increased.

Transmit Power Control (TPC): A process implemented in WLAN devices allowing for the output power to be adjusted according to local regulations or by an automated management system.

U-APSD: See *Automatic Power Save Delivery*.

Uncontrolled Port: In an 802.1X authentication system, the virtual port that allows only authentication frames/packets through to the network and, when authentication is successfully completed, provides the 802.1X service with the needed information to open the controlled port (CWNP Definition). The access point used to provide the insecure MAC Service to a client of a MAC Security Entity (802.1X, 2020).

User Priority (UP): A value (from 0-7) assigned to prioritize traffic that correspond to different access categories for WMM QoS (CWNP Definition). A value associated with an medium access control (MAC) service data unit (MSDU) that indicates how the MSDU is to be handled. The UP is assigned to an MSDU in the layers above the MAC (802.11-2020, 2020).

Virtual Carrier Sense: The 802.11 standard currently defines the Network Allocation Vector (NAV) for use in virtual carrier sensing. The NAV is set based on the duration value in perceived frames within the channel.

Voltage Standing Wave Ratio (VSWR): The Voltage Standing Wave Ratio is the ratio between the voltage at the maximum and minimum points of a standing wave (CWNP Definition). In a standing-wave system, the ratio of the maximum voltage to the minimum voltage (Gibilisco, 2001).

Watt (W): A unit of power. Strictly defined as the energy consumption rate of one joule per second such that 1 W is equal to 1 joule per 1 second.

Wavelength: The distance between two repeating points on a wave. Wavelength is a factor of the frequency and the constant of the speed of light (CWNP Definition). The displacement in one complete wave of an alternating or vibrating phenomenon, generally measured from crest to crest or from trough to trough of successive waves. For electromagnetic waves in free space, the wavelength in meters is equal to 3×10^8 divided by the frequency in Hertz (Gibilisco, 2001).

Wired Equivalent Privacy (WEP): A legacy method of security defined in the original IEEE 802.11 standard in 1997. Used the RC4 cipher like TKIP (WPA), but implemented it poorly. WEP is deprecated and should no longer be used.

Wi-Fi Alliance: An association that certifies WLAN equipment to interoperate based on selected portions of the 802.11 standard and other standards. Certifications include those based on each PHY as well as QoS and security.

Wi-Fi Multimedia (WMM): A QoS certification created and tested by the Wi-Fi Alliance using traffic prioritizing methods defined in the IEEE 802.11e.

Wi-Fi Multimedia Power Save (WMM-PS): A power saving certification designed by the Wi-Fi Alliance and optimized for mobile devices and implementing methods designated in the IEEE 802.11e amendment.

Wireless Intrusion Prevention System (WIPS): A system used to detect and prevent unwanted intrusions in a WLAN by detecting and preventing rogue APs and other WLAN threats.

Wireless Local Area Network (WLAN): A local area network that connects devices using wireless signals. In common colloquial use today, a wireless local area network based on the 802.11 protocol rather than wires and the common 802.3 protocol.

WPA-Enterprise: A security protocol designed by the Wi-Fi Alliance. Requires an 802.1X authentication server. Uses the TKIP encryption protocol with the RC4 cipher. Implements a portion of 802.11i and the older, no deprecated TKIP/RC4 solution.

WPA-Personal: A security protocol designed by the Wi-Fi Alliance. Does not require an authentication server. Uses the TKIP encryption protocol with the RC4 cipher. Also known as WPA-PSK (Pre-Shared Key).

WPA2-Enterprise: A security protocol designed by the Wi-Fi Alliance. Requires an 802.1X authentication server. Uses the CCMP key management protocol with the AES cipher. Also known as WPA2-802.1X. Implements the non-deprecated portion of 802.11i.

WPA2-Personal: A security protocol designed by the Wi-Fi Alliance. Does not require an authentication server. Uses the CCMP key management protocol with the AES cipher. Also known as WPA2-PSK (Pre-Shared Key).

Wi-Fi Protected Setup (WPS): A standard designed by the Wi-Fi Alliance to secure a network without requiring much user knowledge. Users connect either by entering a PIN associated with the device or by Push-Button which allows users to connect when a real or virtual button is pushed.

Printed in the USA
CPSIA information can be obtained
at www.ICGtesting.com
CBHW041946040424
6055CB00020B/4

9 781737 216636